T0191853

$\mathbb{Z}_2\mathbb{Z}_4$-Linear Codes

Joaquim Borges • Cristina Fernández-Córdoba
Jaume Pujol • Josep Rifà • Mercè Villanueva

$\mathbb{Z}_2\mathbb{Z}_4$-Linear Codes

Springer

Joaquim Borges
Department of Information
and Communications Engineering
Universitat Autònoma de Barcelona
Bellaterra, Spain

Jaume Pujol
Department of Information
and Communications Engineering
Universitat Autònoma de Barcelona
Bellaterra, Spain

Mercè Villanueva
Department of Information
and Communications Engineering
Universitat Autònoma de Barcelona
Bellaterra, Spain

Cristina Fernández-Córdoba
Department of Information
and Communications Engineering
Universitat Autònoma de Barcelona
Bellaterra, Spain

Josep Rifà
Department of Information
and Communications Engineering
Universitat Autònoma de Barcelona
Bellaterra, Spain

ISBN 978-3-031-05443-3   ISBN 978-3-031-05441-9   (eBook)
https://doi.org/10.1007/978-3-031-05441-9

This Springer imprint is published by the registered company Springer Nature Switzerland AG
The registered company address is: Gewerbestrasse 11, 6330 Cham, Switzerland

Foreword

I am very happy that Cristina Fernández asked me to write a preface to this book coauthored with her colleagues at UAB Borges, Pujol, Rifà and Villanueva. It seems to me it was only yesterday that Cristina was my intern PhD student in Sophia Antipolis. Since then, she has become a very active coding theorist with many publications to her name.

Philippe Delsarte created the field of Algebraic combinatorics in his monumental 1973 thesis [57]. In this work, he studied codes over abelian groups. In a seminal paper [134], Rifà and Pujol characterized so-called propelinear codes over the binary field as subgroups of certain abelian groups. The groups involved in that result can only be the direct products of multiple copies of the cyclic groups of order 2 and 4 and of the quaternionic group of order 8. This result was very timely, three years after the famous $\mathbb{Z}_4$ paper [92], and the avalanche of codes over rings papers it triggered [143]. This prompted the group at UAB to study $\mathbb{Z}_2\mathbb{Z}_4$ codes, that is to say additive subgroups of $\mathbb{Z}_2^m\mathbb{Z}_4^n$ for some integers m, n. The present book which compiles a quarter century of research, is devoted to this class of codes.

At a structural level, this book contains studies on the algebraic structure of $\mathbb{Z}_2\mathbb{Z}_4$ codes (cyclicity) and also to their arithmetic structure (orthogonality, self-duality, build up construction). From a constructive standpoint it also considers special families of such codes like perfect codes and Hadamard codes, a concept introduced in [57]. Another important family is that of Reed-Muller-type codes, which are relevant to Boolean functions and cryptography. To allow for numerical experimentation, the UAB group has written a package of the software Magma, a general resource for mathematical formal computations. Most paragraphs of that book contains short programs written in that package. This is a welcome innovation in the often too abstract literature on Coding Theory.

In a last chapter several generalizations to other mixed rings alphabets have been considered. Some applications to engineering (steganography) are also considered.

To conclude, this is a deep book written by experts of the field. Its lectorship combines mathematicians, computer scientists, and engineers. It can be the support of a short course at master level.

Patrick Solé
Directeur de Recherche au CNRS,
Institut de Mathématiques de Marseille,
Luminy, le 2 Septembre 2021

Preface

Prior to 1948, systems for the digital transmission of information already existed, such as the telegraph, where the Morse code (1830's) is used. However, it was not until 1948 that Claude Shannon developed the Information Theory that deals with the problem of the transmission of information through noisy channels. At the same time, his colleague at Bell Labs, Richard Hamming, gave the first construction of what is now known as the 1-error-correcting and 2-error-detecting binary Hamming code, and also the 1-error-correcting and 3-error-detecting extended binary Hamming code. In today's technology, the messages are transmitted in sequences of 0's and 1's and, since errors can be produced in the transmission channel, it is very necessary to use these codes that correct errors (for example, in e-mail, mobile, remote sensing, IoT, etc.). Using linear algebra, we have the remarkable Hamming codes and all linear codes constructed later, most of them binary linear and their generalizations to codes over finite fields. The most representative codes are the BCH and Reed-Solomon that can be found in many applications, from the first CDs, to Blu-ray, QR codes, WiMax, satellite communication or storage systems.

From a historical point of view, linear codes over finite fields have been the most important codes since they are easier to construct, encode, and decode. Ring theory has been the next step of coding theory. Linear codes over rings are characterized because the underlying alphabet has the structure of a finite ring. The first codes of this type are found in 1963 (E. F. Assmus and H. F. Mattson) and, later in 1979 (P. Shankar), constructions which are analogous to the BCH or Reed-Solomon codes, but over rings, are given. Also, Lee metric codes were introduced in 1968 (E. R. Berlekamp) where, instead of the usual Hamming metric, Lee's metric is considered. The first examples of codes over rings that are cyclic appeared in 1972 (I. F. Blake). In 1991 (Nechaev), it was discovered that all Kerdock codes can be considered as cyclic linear codes over $\mathbb{Z}_4$, and in 1994 (Hammons et al.) it came up the explanation that the families of the well-known codes as Preparata, Kerdock, Goethals and Goethals–Delsarte, which are non-linear, can be represented as $\mathbb{Z}_4$-linear codes. Using the Lee weight and the appropriate definition of duality, they showed the unexplained for a long time relationship of their weight enumerators, which fulfil the MacWilliams transform. Since the 1990's, there have been many papers in the literature dealing with the design of codes over $\mathbb{Z}_4$, in particular, linear and cyclic codes over $\mathbb{Z}_4$ have been studied and their structure has been analysed and researched intensively.

On the other hand, in 1987, our research group CCSG (*Combinatorics, Coding and Security Group*) started to work on the so called binary propelinear codes, that is, codes such that their group of isometries contains a regular subgroup acting transitively on the code. Later, in 1997, we realized that in the abelian case this class of codes coincides with the additive codes defined by Delsarte in 1973 in terms of association schemes for the case of the binary Hamming scheme. An additive code, in a translation association scheme, is defined as a subgroup of the underlying abelian group. In the case of a Hamming scheme, the only structures for the abelian group are those of the form $\mathbb{Z}_2^\alpha \times \mathbb{Z}_4^\beta$. We refer to this class of codes as $\mathbb{Z}_2\mathbb{Z}_4$-additive codes. After using an extended Gray map, they can be seen as binary codes (not necessarily linear) of length $n = \alpha + 2\beta$. These codes include the binary codes (when $\beta = 0$) and also the linear codes over $\mathbb{Z}_4$ (when $\alpha = 0$).

This book aims to present the basics of this class of $\mathbb{Z}_2\mathbb{Z}_4$-additive codes and it is intended for people involved in coding theory, algorithms, computer science, discrete mathematics or algebra. We have tried to make the content accessible to a wide audience, although sometimes a minimum background in coding theory or algebra is required. The first chapters are an introduction to the topic, describing the basic parameters, generator matrices, parity check matrices and studying the concept of duality, including a chapter dedicated to $\mathbb{Z}_2\mathbb{Z}_4$-additive self-dual codes. In general, after applying the Gray map, these codes are not linear over $\mathbb{Z}_2$, which makes the rank and dimension of the kernel relevant parameters to be studied. We then proceed to present families of these $\mathbb{Z}_2\mathbb{Z}_4$-additive codes which have some additional properties, such as the (extended) perfect, Hadamard, Reed-Muller, and MDS codes. In a next chapter, we study the cyclic $\mathbb{Z}_2\mathbb{Z}_4$-additive codes and deal with the values of the rank and dimension of the kernel. In the last two chapters, we present some procedures for encoding using these codes, as well as decoding via syndrome or using the permutation decoding method. We also give an application of these codes to stenography and introduce some variants and generalizations of them that have lately appeared in the literature.

Throughout the book, there are many examples to easily follow the described concepts. There are also examples by using MAGMA functions that can be found in a MAGMA package implemented by the members of the CCSG group for the $\mathbb{Z}_2\mathbb{Z}_4$-additive codes. The latest version of this package for $\mathbb{Z}_2\mathbb{Z}_4$-additive codes and the manual with the description of all functions can be downloaded from the CCSG web page (`http://ccsg.uab.cat`).

The first step that has led us to this book was an article prepared at the request of professors Victor A. Zinoviev and Thomas Ericson in a seminar that our research group gave at the Autonomous University of Barcelona (UAB). The mentioned article was a review of the research that the CCSG group had done on the subject of $\mathbb{Z}_2\mathbb{Z}_4$-additive codes. Later, at the Academy Contact Forum "Galois geometries and applications", organized by The Royal Flemish Academy of Belgium for Science and the Arts in 2012, we presented part of the content of the book that has been revised and updated to reach the current form.

Most of the results in this book come from the research of the members of the CCSG group at the UAB with the collaboration of postdocs and professors from other research groups to whom we are indebted and to whom we would like to thank for visiting our group and/or inviting us to their respective universities: postdocs Iván Bailera, Roland Barrolleta, Nasreddine Benbelkacem, Dipak K. Bhunia, José Joaquín Bernal, Muhammad Bilal, Pere Montolio, Jaume Pernas, Lorena Ronquillo, Emilio Suárez-Canedo, Roger Ten-Valls, Carlos Vela, Fanxuan Zeng; and Professors John J. Cannon, Ángel del Rio, Steven T. Dougherty, Denis S. Krotov, Kevin T. Phelps, Helena Rifà-Pous, Faina I. Soloveva. In addition, the authors are grateful to the research fellow Adrián Torres-Martín for helpful comments on an early version of the text. The work on this book has been partially funded by the Spanish Government (grant reference PID2019-104664GB-I00/AEI/ 10.13039/501100011033).

We would like to end this preface with a sentence from the excellent book of error-correcting codes by Professors F. J. MacWilliams and N. J. A. Sloane: *When reading the book, if you get stuck on a section, skip it, but keep reading! Don't hesitate to skip the proof of a theorem: we often do.*

Barcelona, 2022

J. Borges
C. Fernández-Córdoba
J. Pujol
J. Rifà
M. Villanueva

Contents

Chapter 1

Introduction

In this chapter, some preliminary concepts are introduced as well as a brief description of the context where $\mathbb{Z}_2\mathbb{Z}_4$-linear codes were first defined. In Section 1.2, we describe propelinear codes, and we focus on translation-invariant propelinear codes. Then, we see that $\mathbb{Z}_2\mathbb{Z}_4$-linear codes are precisely the subclass of abelian translation-invariant propelinear codes. Moreover, we mention several properties showing the importance of $\mathbb{Z}_2\mathbb{Z}_4$-linear codes. In Section 1.3, we study $\mathbb{Z}_{2^k}$-linear codes as propelinear codes and we prove that, if we consider that the Gray image is translation-invariant, then necessarily $k = 1$ or 2. Finally, in Section 1.4, we present a package developed mainly by the authors, within the software MAGMA, to work with $\mathbb{Z}_2\mathbb{Z}_4$-linear codes.

1.1 Preliminaries

Let $\mathbb{Z}_2$ and $\mathbb{Z}_4$ be the ring of integers modulo 2 and 4 respectively. Let $\mathbb{Z}_2^n$ denote the set of all binary vectors of length n and let $\mathbb{Z}_4^n$ be the set of all n-tuples over the ring $\mathbb{Z}_4$. In this book, the elements of $\mathbb{Z}_4^n$ will also be called quaternary vectors of length n. We denote by $\mathbf{0}^\ell$ and $\mathbf{1}^\ell$ the all-zero and the all-one vectors, respectively, of length ℓ. If the length of such vectors is clear from the context, we omit the parameter ℓ.

The *Hamming distance* $d_H(u,v)$ between two vectors $u,v \in \mathbb{Z}_2^n$ is the number of coordinates in which u and v differ, and the *Hamming weight* of a vector $u \in \mathbb{Z}_2^n$, denoted by $\mathrm{wt}_H(u)$, is the number of non-zero coordinates of u, so $d_H(u,v) = \mathrm{wt}_H(u-v)$. On the other hand, the *Lee weights* over the elements in $\mathbb{Z}_4$ are defined as $\mathrm{wt}_L(0) = 0$, $\mathrm{wt}_L(1) = \mathrm{wt}_L(3) = 1$, and $\mathrm{wt}_L(2) = 2$. Then, the *Lee weight* of a vector $u \in \mathbb{Z}_4^n$, denoted by $\mathrm{wt}_L(u)$, is the addition of the weights of its coordinates, whereas the *Lee distance* $d_L(u,v)$ between two vectors $u,v \in \mathbb{Z}_4^n$ is $d_L(u,v) = \mathrm{wt}_L(u-v)$.

© Springer Nature Switzerland AG 2022
J. Borges et al., *$\mathbb{Z}_2\mathbb{Z}_4$-Linear Codes*, https://doi.org/10.1007/978-3-031-05441-9_1

Any non-empty subset C of $\mathbb{Z}_2^n$ is a binary code and a subgroup of $\mathbb{Z}_2^n$ is called a *binary linear code* or a $\mathbb{Z}_2$-*linear code*. Equivalently, any non-empty subset $\mathcal{C}$ of $\mathbb{Z}_4^n$ is a quaternary code and a subgroup of $\mathbb{Z}_4^n$ is called a *quaternary linear code*. The elements of a code are called *codewords*. If C is a binary linear code, it is isomorphic to an additive group $\mathbb{Z}_2^k$, so C has dimension k and it has 2^k codewords. Equivalently, if $\mathcal{C}$ is a quaternary linear code, since it is a subgroup of $\mathbb{Z}_4^n$, it is isomorphic to an abelian structure $\mathbb{Z}_2^\gamma \times \mathbb{Z}_4^\delta$. Therefore, we have that $\mathcal{C}$ is of type $2^\gamma 4^\delta$ as a group, and it has $2^{\gamma+2\delta}$ codewords.

Quaternary codes can be viewed as binary codes under the usual Gray map $\phi : \mathbb{Z}_4^n \longrightarrow \mathbb{Z}_2^{2n}$ defined as $\phi(0) = (0,0)$, $\phi(1) = (0,1)$, $\phi(2) = (1,1)$, and $\phi(3) = (1,0)$ in each coordinate. If $\mathcal{C}$ is a quaternary linear code, then the binary code $C = \phi(\mathcal{C})$ is called a $\mathbb{Z}_4$-*linear code*. Note that $\mathbb{Z}_4$-linear codes are not necessarily linear.

The *minimum Hamming distance* $d_H(C)$ of a binary code C is

$$d_H(C) = \min\{d_H(u,v) : u,v \in C, u \neq v\},$$

and the *minimum Hamming weight* of C, denoted by $\mathrm{wt}_H(C)$, is

$$\mathrm{wt}_H(C) = \min\{\mathrm{wt}_H(u) : u \in C \backslash \{\mathbf{0}\}\}.$$

It is well known that if C is a binary linear code, $d_H(C) = \mathrm{wt}_H(C)$ [112]. Equivalently, the *minimum Lee distance* $d_L(\mathcal{C})$ of a quaternary code $\mathcal{C}$ and the *minimum Lee weight* of $\mathcal{C}$, denoted by $\mathrm{wt}_L(\mathcal{C})$, are the values

$$d_L(\mathcal{C}) = \min\{d_L(u,v) : u,v \in \mathcal{C}, u \neq v\},$$

$$\mathrm{wt}_L(\mathcal{C}) = \min\{\mathrm{wt}_L(u) : u \in \mathcal{C} \backslash \{\mathbf{0}\}\}.$$

Again, if $\mathcal{C}$ is a quaternary linear code, $d_L(\mathcal{C}) = \mathrm{wt}_L(\mathcal{C})$. Note that the Gray map ϕ is an isometry which transforms Lee distances over $\mathbb{Z}_4^n$ into Hamming distances over $\mathbb{Z}_2^{2n}$. Therefore, the minimum Lee weight of a quaternary code $\mathcal{C}$ coincides with the minimum Hamming weight of $C = \phi(\mathcal{C})$, that is,

$$\mathrm{wt}_L(\mathcal{C}) = \mathrm{wt}_H(\phi(\mathcal{C})).$$

The *dual code* of a binary linear code C of length n is the code

$$C^\perp = \{x \in \mathbb{Z}_2^n : x \cdot v = 0, \;\; \forall v \in C\},$$

where $\cdot$ denotes the standard inner product between binary vectors. For a binary code C of length n, let A_i be the number of codewords of weight i ($0 \leq i \leq n$). The *(Hamming) weight enumerator polynomial* of C is

$$W_C(X,Y) = \sum_{i=0}^{n} A_i X^{n-i} Y^i.$$

The *MacWilliams identity* relates the weight enumerator polynomials of dual codes in the following way:

$$W_{\mathcal{C}^\perp}(X) = \frac{1}{|C|} W_C(X + Y, X - Y). \qquad (1.1)$$

The dual of a quaternary linear code $\mathcal{C}$, denoted by $\mathcal{C}^\perp$, is called the *quaternary dual code* and is defined in the standard way [112] in terms of the usual inner product for quaternary vectors [92]. The binary code $C_\perp = \phi(\mathcal{C}^\perp)$ is called the $\mathbb{Z}_4$-*dual code* of $C = \phi(\mathcal{C})$. As can be seen in [92], the weight enumerator polynomials of C and $C_\perp$ satisfy the MacWilliams identity (1.1).

Since 1994, quaternary linear codes have been studied and became significant since, after applying the Gray map, they can be transformed into binary non-linear codes better than any known binary linear code with the same parameters. More specifically, Hammons et al. [92] show how to construct well-known binary non-linear codes like the Nordstrom-Robinson code, Kerdock codes, Preparata-like codes, and Delsarte-Goethals codes as $\mathbb{Z}_4$-linear codes, that is, as the Gray map image of quaternary linear codes. Furthermore, they solve a long-standing open problem in coding theory regarding that the Hamming weight enumerators of the non-linear Kerdock and Preparata codes satisfy the MacWilliams identity. Actually, they prove that the Kerdock codes and some Preparata-like codes are $\mathbb{Z}_4$-linear codes and, moreover, the $\mathbb{Z}_4$-dual code of a Kerdock code is a Preparata-like code. Later, several other $\mathbb{Z}_4$-linear codes with the same parameters as some well-known families of binary linear codes (for example, extended Hamming, Hadamard, and Reed-Muller codes) have been studied and classified [34, 36, 43, 44, 102, 118, 119, 122, 130, 150].

Additive codes were first defined by Delsarte in 1973 in terms of association schemes [57, 58]. According to this definition, an additive code is a subgroup of the underlying abelian group in a translation association scheme. Later, in 1997, translation-invariant propelinear codes were defined by Rifà et al. [133, 134]. They also proved that these codes are group isomorphic to subgroups of $\mathbb{Z}_2^\alpha \times \mathbb{Z}_4^\beta \times \mathbb{Q}_8^\sigma$, being $\mathbb{Q}_8$ the non-abelian quaternion group on eight elements. In the special case of a binary Hamming scheme, that is, when the underlying abelian group is of order 2^n, the additive codes coincide with the abelian translation-invariant propelinear codes. Hence, as it is pointed out in [58, 134], the only structures for the abelian group are those of the form $\mathbb{Z}_2^\alpha \times \mathbb{Z}_4^\beta$, with $\alpha + 2\beta = n$. Therefore, the codes that are subgroups of $\mathbb{Z}_2^\alpha \times \mathbb{Z}_4^\beta$ are the only additive codes in the binary Hamming scheme. In order to distinguish them from additive codes over finite fields [26], from now on, we will call them $\mathbb{Z}_2\mathbb{Z}_4$-*additive codes*. Note that one could think of other families of codes with an algebraic structure that also include the $\mathbb{Z}_2\mathbb{Z}_4$-additive codes, such as mixed group codes [93, 109].

Since $\mathbb{Z}_2\mathbb{Z}_4$-additive codes are subgroups of $\mathbb{Z}_2^\alpha \times \mathbb{Z}_4^\beta$, they can be seen as a generalization of binary (when $\beta = 0$) and quaternary (when $\alpha = 0$) linear codes. As for quaternary linear codes, after applying the Gray map to the $\mathbb{Z}_4$ coordinates of a $\mathbb{Z}_2\mathbb{Z}_4$-additive code, we obtain binary codes called $\mathbb{Z}_2\mathbb{Z}_4$-*linear codes*. As $\mathbb{Z}_4$-linear codes, $\mathbb{Z}_2\mathbb{Z}_4$-linear codes are not necessarily linear. There are $\mathbb{Z}_2\mathbb{Z}_4$-linear codes into several important classes of binary codes. For example, $\mathbb{Z}_2\mathbb{Z}_4$-linear perfect single error-correcting codes (or 1-perfect codes) are found in [134] and fully characterized in [45]. Also, in subsequent papers [43, 102, 122, 124], $\mathbb{Z}_2\mathbb{Z}_4$-linear extended perfect and Hadamard codes are studied and classified. Note that $\mathbb{Z}_2\mathbb{Z}_4$-additive codes have allowed to classify more binary non-linear codes, giving them a structure as $\mathbb{Z}_2\mathbb{Z}_4$-additive codes. Although it is not easy to determine whether a code has a $\mathbb{Z}_2\mathbb{Z}_4$-additive structure, and whether it is unique or not, it seems that there are many more $\mathbb{Z}_2\mathbb{Z}_4$-linear codes than linear. In this sense, a preliminary proposal about counting $\mathbb{Z}_2\mathbb{Z}_4$-additive codes can be found in [64]. Finally, we mention that a permutation decoding method for $\mathbb{Z}_2\mathbb{Z}_4$-linear codes is described in [22].

Part of the research developed by the Combinatorics, Coding and Security Group (CCSG) deals with quaternary linear codes, as well as $\mathbb{Z}_2\mathbb{Z}_4$-additive codes. Since there is not any symbolic software to work with $\mathbb{Z}_2\mathbb{Z}_4$-additive codes, the members of CCSG have been developing a new package [18, 29, 46] in MAGMA [46] that supports the basic facilities for these codes. Specifically, this new MAGMA package generalizes most of the functions for quaternary linear codes to $\mathbb{Z}_2\mathbb{Z}_4$-additive codes, and includes new functions specific for this kind of codes [29]. A beta version of this package and the manual with the description of all functions can be downloaded from the CCSG web page (`http://ccsg.uab.cat`). In Section 1.4, we also give a brief description of this package.

1.2   From Propelinear Codes to $\mathbb{Z}_2\mathbb{Z}_4$-Linear Codes

In Section 1.1, we mention that $\mathbb{Z}_2\mathbb{Z}_4$-additive codes can be seen as additive codes in a binary Hamming scheme. In this section, we see that these codes can be seen as abelian translation-invariant propelinear codes [134].

In 1988, in the paper [133], a link between a class of distance-regular graphs [47, 57] and a class of completely regular codes [57, 92] was established.

Such codes have a group structure and were called *propelinear*.

Let $\mathcal{S}_n$ be the symmetric group of permutations on the set $\{1, \dots, n\}$. The group operation in $\mathcal{S}_n$ is the function composition, $\sigma_1 \circ \sigma_2$, which maps any element x to $\sigma_1(\sigma_2(x))$, $\sigma_1, \sigma_2 \in \mathcal{S}_n$. A permutation $\sigma \in \mathcal{S}_n$ acts linearly on vectors of length n by permuting their coordinates as follows: $\sigma(v_1, \dots, v_n) = (v_{\sigma^{-1}(1)}, \dots, v_{\sigma^{-1}(n)})$.

A binary code C of length n is said to be *propelinear* if for any codeword $x \in C$ there is a coordinate permutation $\pi_x \in \mathcal{S}_n$ verifying the following properties:

i) $x + \pi_x(y) \in C$, if $y \in C$,

ii) $\pi_x \circ \pi_y = \pi_z$ $\forall y \in C$, where $z = x + \pi_x(y)$.

If C is a propelinear code, we can define the operation $\star : C \times \mathbb{Z}_2^n \longrightarrow \mathbb{Z}_2^n$ as

$$x \star y = x + \pi_x(y) \quad \forall x \in C \ \ \forall y \in \mathbb{Z}_2^n.$$

This operation is clearly associative and closed in C. For any codeword $x \in C$, $x \star y = x \star z$ implies that $y = z$. Therefore, we have that $x \star y \in C$ if and only if $y \in C$. Thus, there must be a codeword e such that $x \star e = x$. It follows that $e = \mathbf{0}$ is a codeword and, from property ii), we deduce that $\pi_{\mathbf{0}}$ is the identity permutation. Hence, $(C, \star)$ is a group, which is not abelian in general; $\mathbf{0}$ is the identity element in C and $x^{-1} = \pi_x^{-1}(x)$ for all $x \in C$. Note that $\Pi = \{\pi_x : x \in C\}$ is a subgroup of $\mathcal{S}_n$ with the usual composition of permutations.

Let $(C, *)$ be a code in $\mathbb{Z}_2^n$, where $*$ is a group operation which induces an action $* : C \times \mathbb{Z}_2^n \longrightarrow \mathbb{Z}_2^n$. The action $*$ is *Hamming-compatible* if $d_H(x, x * v) = \mathrm{wt}_H(v)$ for all $x \in C$ and $v \in \mathbb{Z}_2^n$. In such case, we say that $(C, *)$ is a *Hamming-compatible group code*.

Of course, given a binary code $C \subseteq \mathbb{Z}_2^n$, among all the different group structures, we are interested in those being Hamming-compatible. A very general class of such codes are the class of propelinear codes.

Lemma 1.1. *A propelinear code is a Hamming-compatible group code.*

Proof. Let $(C, \star)$ be a propelinear code. We only have to prove that the action $\star : C \times \mathbb{Z}_2^n \longrightarrow \mathbb{Z}_2^n$ is Hamming-compatible.

Let $x \in C$ and $v \in \mathbb{Z}_2^n$. We have $d_H(x, x \star v) = d_H(x \star \mathbf{0}, x \star v)$. It is easy to check that $d_H(x \star \mathbf{0}, x \star v) = d_H(\mathbf{0}, v)$ and, therefore, $d_H(x, x \star v) = d_H(\mathbf{0}, v) = \mathrm{wt}_H(v)$. $\qquad\square$

If $(C, \star)$ is a propelinear code, we have that $x \star C = C$ for any $x \in C$, that is, $x + \pi_x(C) = C$. Thus, $C + x = \pi_x(C)$ from which we deduce that the weight distribution of $C + x$ is the same as the weight distribution of $\pi_x(C)$ or C. This means that propelinear codes are *distance-invariant* [112] and hence the minimum distance and the minimum weight coincide.

The following property is straightforward.

Lemma 1.2. *If $(C, \star)$ is a propelinear code in $\mathbb{Z}_2^n$, then*

$$d_H(x \star v, x \star u) = d_H(v, u) \quad \forall u, v \in \mathbb{Z}_2^n \ \forall x \in C.$$

A propelinear code $(C, \star)$ is called *translation-invariant* [134] if

$$d_H(x, y) = d_H(x \star u, y \star u) \quad \forall x, y \in C \quad \forall u \in \mathbb{Z}_2^n.$$

In [134], it is shown that binary linear and $\mathbb{Z}_4$-linear codes are subclasses of the more general class of translation-invariant propelinear codes. More precisely, in [134], it is proved that every translation-invariant propelinear code of length n can be viewed as a group isomorphic to a subgroup of $\mathbb{Z}_2^\alpha \times \mathbb{Z}_4^\beta \times \mathbb{Q}_8^\sigma$, where $\alpha + 2\beta + 4\sigma = n$ and $\mathbb{Q}_8$ is the quaternion group of order eight. We can take each $\mathbb{Q}_8$ as a quaternion group generated by $a = (1, 0, 1, 0)$ and $b = (1, 0, 0, 1)$ with $\pi_a = (1, 2)(3, 4)$ and $\pi_b = (1, 3)(2, 4)$ in such a way that $\mathbb{Q}_8 = \{\mathbf{0}, a, a^2, a^3, b, a \star b, a^2 \star b, a^3 \star b\}$ and $a^4 = \mathbf{0}, c = a^2 = b^2, a \star b \star a = b$.

Let $Id : \mathbb{Z}_2 \longrightarrow \mathbb{Z}_2$ be the identity map, $\phi : \mathbb{Z}_4 \longrightarrow \mathbb{Z}_2^2$ the Gray map defined by $\phi(0) = (0, 0)$, $\phi(1) = (0, 1)$, $\phi(3) = (1, 0)$, $\phi(2) = (1, 1)$ in [92], and $\phi_8 : \mathbb{Q}_8 \longrightarrow \mathbb{Z}_2^4$ the map defined by $\phi_8(a) = (1, 0, 1, 0)$, $\phi_8(a^2) = (1, 1, 1, 1)$, $\phi_8(a^3) = (0, 1, 0, 1)$, $\phi_8(b) = (1, 0, 0, 1)$, $\phi_8(a \star b) = (1, 1, 0, 0)$, $\phi_8(a^2 \star b) = (0, 1, 1, 0)$, $\phi_8(a^3 \star b) = (0, 0, 1, 1)$, $\phi_8(a^4) = (0, 0, 0, 0)$. The isomorphism that links the group $\mathbb{Z}_2^\alpha \times \mathbb{Z}_4^\beta \times \mathbb{Q}_8^\sigma$ with a subgroup in $\mathbb{Z}_2^{\alpha + 2\beta + 4\sigma}$ is given by

$$(Id, \phi, \phi_8) : \mathbb{Z}_2^\alpha \times \mathbb{Z}_4^\beta \times \mathbb{Q}_8^\sigma \longrightarrow \mathbb{Z}_2^{\alpha + 2\beta + 4\sigma}, \tag{1.2}$$

where Id, ϕ and ϕ_8 also denote the coordinate-wise extended map to elements in $\mathbb{Z}_2^\alpha, \mathbb{Z}_4^\beta$ and $\mathbb{Q}_8^\sigma$, respectively.

There are several reasons to pay special attention to translation-invariant propelinear codes which are group isomorphic to subgroups of $\mathbb{Z}_2^\alpha \times \mathbb{Z}_4^\beta$; that is, to $\mathbb{Z}_2\mathbb{Z}_4$-linear codes:

i) Since $\mathbb{Q}_8$ is non-abelian, we conclude that any abelian translation-invariant propelinear code is a $\mathbb{Z}_2\mathbb{Z}_4$-linear code.

ii) If $(C, \star)$ is a $\mathbb{Z}_2\mathbb{Z}_4$-linear code of length n, then the operation $\star$ can be extended to the whole space $\mathbb{Z}_2^n$. In other words, we can consider $\mathbb{Z}_2^n$ as the binary Gray map image of $\mathbb{Z}_2^\alpha \times \mathbb{Z}_4^\beta$, where $\alpha + 2\beta = n$. Then, we can see the operation $\star$ defined over $\mathbb{Z}_2^n \times \mathbb{Z}_2^n$. As a result, $(C, \star)$ is a subgroup of $(\mathbb{Z}_2^n, \star)$. Note that this is not true for general translation-invariant propelinear codes.

iii) For $\mathbb{Z}_2\mathbb{Z}_4$-linear codes we can define duality. This is not possible, in the usual sense, for general translation-invariant propelinear codes.

iv) In [38], it was also shown that if $\mathcal{C} \leq G$, where $G = \mathbb{Z}_{2^{i_1}}^{k_1} \times \cdots \times \mathbb{Z}_{2^{i_r}}^{k_r}$, and $C = \Phi(\mathcal{C})$, where Φ is the coordinate-wise extended Gray map defined in (1.4), is translation-invariant or it is a perfect single error-correcting code, then $i_j \leq 2$, for all $j = 1, \ldots, r$. That is, $\mathcal{C} \leq \mathbb{Z}_2^\alpha \times \mathbb{Z}_4^\beta$.

v) Several classes of important codes fall into the category of $\mathbb{Z}_2\mathbb{Z}_4$-linear codes such as binary linear codes, $\mathbb{Z}_4$-linear codes, some perfect single error-correcting codes (briefly, 1-perfect codes), some extended 1-perfect codes, some Hadamard codes, some self-dual codes, etc., as we see in Chapter 6.

(vii) Finally, $\mathbb{Z}_2\mathbb{Z}_4$-linear codes can be used in steganography [135, 140] as also shown in Section 9.3.

From the birth of $\mathbb{Z}_2\mathbb{Z}_4$-linear codes, a lot of papers have been published about the topic. For general theory and duality, the main paper is [37]. In [43, 45, 124, 137], extended and non-extended 1-perfect $\mathbb{Z}_2\mathbb{Z}_4$-linear codes are studied. Maximum distance separable $\mathbb{Z}_2\mathbb{Z}_4$-linear codes are classified in [28]. Self-dual and formally self-dual $\mathbb{Z}_2\mathbb{Z}_4$-linear codes are studied in [32, 71]. Bounds for the rank and the dimension of the kernel of $\mathbb{Z}_2\mathbb{Z}_4$-linear codes are given in [33]. In [105], a classification of Hadamard $\mathbb{Z}_2\mathbb{Z}_4$-linear codes is given. The permutation decoding method can be used for $\mathbb{Z}_2\mathbb{Z}_4$-linear codes, as it is proved in [22]. More recently, a number of papers about $\mathbb{Z}_2\mathbb{Z}_4$-additive cyclic codes have been published [1, 31, 33, 39]. Also, applications to steganography are presented in [135, 140].

1.3 $\mathbb{Z}_{2^k}$-Linear Codes as Propelinear Codes

In [134], it is shown that $\mathbb{Z}_4$-linear codes are translation-invariant propelinear codes. In this section, we prove that any $\mathbb{Z}_{2^k}$-linear code is a propelinear code, but not translation-invariant for $k > 2$. This result was published in [38] without including the proofs, which can be found in this section.

There are different ways of giving a generalization of the Gray may ϕ from $\mathbb{Z}_4$ to $\mathbb{Z}_2^2$ defined in [92] and given also in Section 1.1. For instance, different generalizations of this Gray map are shown in Section 9.1.3. A basic property of the Gray map ϕ is that the Hamming distance between the images of two consecutive elements in $\mathbb{Z}_4$ is exactly one [92]. In this section, we consider generalizations of the Gray map preserving this property. Then, we define a *Gray map* as an application $\varphi : \mathbb{Z}_r \longrightarrow \mathbb{Z}_2^m$ such that

i) φ is one-to-one,

ii) $d_H(\varphi(i), \varphi(i+1)) = 1, \forall i \in \mathbb{Z}_r$.

Lemma 1.3. *Let $\varphi : \mathbb{Z}_r \longrightarrow \mathbb{Z}_2^m$ be a Gray map. Then, r is even.*

Proof. Let $\psi : \mathbb{Z}_r \longrightarrow \mathbb{Z}_2$ defined as $\psi(i) = \text{wt}_H(\varphi(i)) \bmod 2$. Clearly, we can write $\psi(i) = \psi(0) + i \bmod 2$. By definition of Gray map, we have $d_H(\varphi(r-1), \varphi(0)) = 1$. If r is odd, then $\psi(r-1) = \psi(0) + r - 1 = \psi(0) \bmod 2$, which is a contradiction. □

The Lee weight of an element $i \in \mathbb{Z}_{2k}$ is $\text{wt}_L(i) = \min\{i, 2k - i\}$ and the Lee distance between $i, j \in \mathbb{Z}_{2k}$ is $d_L(i, j) = \text{wt}_L(j - i)$. Let $\varphi : \mathbb{Z}_{2k} \longrightarrow \mathbb{Z}_2^m$ be a Gray map. The map φ is *distance-preserving* if $d_H(\varphi(i), \varphi(j)) = d_L(i, j)$, and it is *weight-preserving* if $\text{wt}_H(\varphi(i)) = \text{wt}_L(i)$. By definition of wt_H and wt_L, if φ is distance-preserving, then it is weight-preserving.

We define the operation $\cdot$ in $\varphi(\mathbb{Z}_{2k})$ as follows:

$$\varphi(i) \cdot \varphi(j) = \varphi(i + j) \qquad (1.3)$$

for all $i, j \in \mathbb{Z}_{2k}$. Note that $\varphi(i) = \varphi(1)^i$. We are interested in those Gray maps such that $(\varphi(\mathbb{Z}_{2k}), \cdot)$ is a Hamming-compatible group code. In these cases, if φ is weight-preserving, then

$$d_H(\varphi(i), \varphi(j)) = d_H(\varphi(i), \varphi(i)\varphi(j-i)) = \text{wt}_H(\varphi(j-i)) = \text{wt}_L(j-i) = d_L(i, j)$$

for $i, j \in \mathbb{Z}_{2k}$ and, therefore, φ is distance-preserving. Moreover, $\varphi(0) = \mathbf{0}$ since, for $i \in \mathbb{Z}_{2k}$, $0 = d_H(\varphi(i), \varphi(i+0)) = d_H(\varphi(i), \varphi(i) \cdot \varphi(0)) = \text{wt}_H(\varphi(0))$.

Theorem 1.4. *Let $\varphi : \mathbb{Z}_{2k} \longrightarrow \mathbb{Z}_2^m$ be a Gray map. If $(\varphi(\mathbb{Z}_{2k}), \cdot)$ is a Hamming-compatible group code where $\cdot$ is the operation defined in (1.3), then φ is distance-preserving.*

Proof. For $j \in \{1, \ldots, m\}$, let $e_j \in \mathbb{Z}_2^m$ be the vector with one in the coordinate j and zero elsewhere. By definition of φ and due to the fact that $\varphi(0) = \mathbf{0}$, it is easy to check that $\varphi(1) = e_{j_1}$ and $\varphi(2) = e_{j_1} + e_{j_2}$ ($j_1 \neq j_2$). Since $\mathrm{wt}_H(\varphi(2k-1)) = 1$, there exists r, $2 \leq r \leq 2k - 2$ such that $\varphi(i) = e_{j_1} + \cdots + e_{j_i}$, where $j_1, \ldots, j_i$ are all distinct, for all $i \leq r$ and $\varphi(r+1) = e_{j_1} + \cdots + e_{j_r} + e_s$, where $s \in \{j_1, \ldots, j_r\}$. As $d_H(\varphi(1), \varphi(r+1)) = \mathrm{wt}_H(\varphi(r)) = r$, then $s = j_1$ and $\varphi(r+1) = e_{j_2} + \cdots + e_{j_r}$. We will proof by induction that for all $i \in \{2, \ldots, r\}$, $\varphi(r+i) = e_{j_{i+1}} + \cdots + e_{j_r}$.

Let $i \in \{2, \ldots, r\}$ such that $\varphi(r+i-1) = e_{j_i} + \cdots + e_{j_r}$. As $d_H(\varphi(i-1), \varphi(r+i-1)) = \mathrm{wt}_H(\varphi(r)) = r$ and $d_H(\varphi(i-1), \varphi(r+i)) = \mathrm{wt}_H(\varphi(r+1)) = r - 1$, we obtain $\varphi(r+i) = e_{j_i} + \cdots + e_{j_r} + e_s$, where $s \in \{j_i, \ldots, j_r\}$. Moreover, as $d_H(\varphi(i), \varphi(r+i)) = \mathrm{wt}_H(\varphi(r)) = r$, then $s = j_i$ and $\varphi(r+i) = e_{j_{i+1}} + \cdots + e_{j_r}$.

Note that $\varphi(2r) = \mathbf{0}$ and therefore $r = k$. Then, by construction, $\mathrm{wt}_H(\varphi(i)) = i = \mathrm{wt}_L(i)$ and $\mathrm{wt}_H(\varphi(k+i)) = k - i = \mathrm{wt}_L(k+i)$ for all $i \leq k$. That way φ is weight-preserving and hence it is distance-preserving. $\square$

Example 1.5. Let $\mathbf{0}^i$ be the all-zero vector of length i and let $\mathbf{1}^j$ be the all-one vector of length j. Define the Gray map $\phi : \mathbb{Z}_{2k} \longrightarrow \mathbb{Z}_2^k$ such that for all $i = 0, \ldots, k-1$

$$
\begin{aligned}
&(i) \quad \phi(i) = (\mathbf{0}^{k-i}, \mathbf{1}^i), \text{ and} \\
&(ii) \quad \phi(i+k) = \phi(i) + \mathbf{1}^k.
\end{aligned}
\tag{1.4}
$$

Note that this Gray map ϕ is distance-preserving and weight-preserving. $\triangle$

Theorem 1.6. Let $\varphi : \mathbb{Z}_{2k} \longrightarrow \mathbb{Z}_2^m$ be a Gray map. If $(\varphi(\mathbb{Z}_{2k}), \cdot)$ is a Hamming-compatible group code where $\cdot$ is the operation defined in (1.3), then φ is unique up to a permutation of coordinates.

Proof. Let $e_j \in \mathbb{Z}_2^m$ be the vector with one in the coordinate j and zero elsewhere, $j \in \{1, \ldots, m\}$. By the proof of Theorem 1.4, for $i = 1, \ldots, k$, $\varphi(i) = e_{j_1} + \cdots + e_{j_i}$ and $\varphi(i+k) = e_{j_1} + \cdots + e_{j_k} + e_{j_1} + \cdots + e_{j_i} = e_{j_{i+1}} + \cdots + e_{j_k}$. For $i = 1, \ldots, k$, let μ_i be the transposition such that $\mu_i(e_{j_i}) = e_{k-i+1}$ and $\mu = \mu_1 \circ \cdots \circ \mu_k$. Now, it is easy to check that $\phi = \mu \circ \varphi$, where ϕ is the map defined in (1.4). $\square$

From last theorem, if $\varphi : \mathbb{Z}_{2k} \longrightarrow \mathbb{Z}_2^m$ is a Gray map, we have $m \geq k$ and, if $m > k$, there are useless coordinates. Thus we can assume $m = k$. Two Gray maps, φ and φ', are permutation equivalent if there is a coordinate permutation τ such that $\varphi(i) = \tau(\varphi'(i))$ for any $i \in \mathbb{Z}_{2k}$. Then, any Gray map is permutation equivalent to the Gray map defined in (1.4). Therefore, from now on, we will refer to ϕ as the Gray map, $\phi : \mathbb{Z}_{2k} \longrightarrow \mathbb{Z}_2^k$. Note that

$\phi : \mathbb{Z}_2 \longrightarrow \mathbb{Z}_2$ is the identity and $\phi : \mathbb{Z}_4 \longrightarrow \mathbb{Z}_2^2$ coincide with the usual one defined in [92].

Let $\mathcal{C}$ be a subgroup of $(\mathbb{Z}_{2k}^n, +)$ for some $k, n \geq 1$. We say that $\mathcal{C}$ is a $\mathbb{Z}_{2k}$-additive code. The binary code $C = \phi(\mathcal{C})$ is called $\mathbb{Z}_{2k}$-linear code. We see that any $\mathbb{Z}_{2k}$-linear code has a representation as a propelinear code.

Define the permutation on k coordinates

$$\sigma_i = (1, k, k-1, \ldots, 2)^i \tag{1.5}$$

(i.e., i left shifts) for all $i = 0, \ldots, 2k-1$.

Proposition 1.7. *Let ϕ be the Gray map defined in (1.4), and σ_i the permutation defined in (1.5). Then,*

$$\phi(i) \cdot \phi(j) = \phi(i) + \sigma_i(\phi(j)), \tag{1.6}$$

where $\cdot$ is the operation defined in (1.3).

Proof. Let $\star$ be the operation defined as $\phi(i) \star \phi(j) = \phi(i) + \sigma_i(\phi(j))$. It is easy to verify that $\phi(i) = \phi(i-1) \star \phi(1) = \phi(1) \star \phi(i-1)$ and, therefore, $\phi(i) = \phi(1) \star \cdots \star \phi(1)$ (i times). Then, $\phi(i) \star \phi(j) = \phi(i+j)$, which is the definition of $\phi(i) \cdot \phi(j)$. $\square$

We define the extended map $\Phi : \mathbb{Z}_{2k}^n \longrightarrow \mathbb{Z}_2^{kn}$ such that $\Phi(j_1, \ldots, j_n) = (\phi(j_1), \ldots, \phi(j_n))$, where ϕ is the map defined in (1.4). Finally, we define the permutations $\pi_x = (\sigma_{j_1} \mid \cdots \mid \sigma_{j_n})$, for $x = \Phi(j_1, \ldots, j_n)$, where σ_i is defined in (1.5).

The next theorem proves that given a $\mathbb{Z}_{2k}$-additive code of length n, there exists a propelinear code of length kn such that both codes are isomorphic. The group isomorphism between them extends the usual structure in $(\mathbb{Z}_{2k}, +)$ to the propelinear structure in $(\mathbb{Z}_2^k, \star)$.

Theorem 1.8. *If $\mathcal{C}$ is a $\mathbb{Z}_{2k}$-additive code, then $\Phi(\mathcal{C})$ is a propelinear code with associated permutation π_x for all codeword $x \in \Phi(\mathcal{C})$.*

Proof. Let $x = \Phi(j_1, \ldots, j_n)$ and $y = \Phi(i_1, \ldots, i_n)$ be two codewords, which can be written as $(\phi(j_1), \ldots, \phi(j_n))$ and $(\phi(i_1), \ldots, \phi(i_n))$, respectively. Then,

$$x + \pi_x(y) = (\phi(j_1) + \sigma_{j_1}(\phi(i_1)), \ldots, \phi(j_n) + \sigma_{j_n}(\phi(i_n))).$$

For any coordinate, say r, we have that

$$\phi(j_r) + \sigma_{j_r}(\phi(i_r)) = \phi(1)^{j_r} \phi(1)^{i_r} = \phi(1)^{j_r + i_r} = \phi(j_r + i_r).$$

Thus,

$$x + \pi_x(y) = (\phi(j_1 + i_1), \dots, \phi(j_n + i_n)) = \Phi((j_1, \dots, j_n) + (i_1, \dots, i_n)).$$

Therefore, it is clear that $x + \pi_x(y) \in \Phi(\mathcal{C})$.

On the other hand, the permutation associated to $\phi(j_r + i_r)$ is

$$\sigma_{j_r + i_r} = (1, k, k - 1, \dots, 2)^{j_r + i_r} = \sigma_{j_r} \circ \sigma_{i_r},$$

hence, if $z = x + \pi_x(y)$, then $\pi_z = \pi_x \circ \pi_y$.   □

Corollary 1.9. *The map* $\Phi : (\mathcal{C}, +) \longrightarrow (\Phi(\mathcal{C}), \star)$ *is a group isomorphism, where* $x \star y = x + \pi_x(y)$ *for all* $x, y \in \Phi(\mathcal{C})$.

Proof. As we have seen in the previous proof, $x \star y = \Phi(\Phi^{-1}(x) + \Phi^{-1}(y))$ and, clearly, Φ is bijective.   □

Finally, we show that for $k > 2$ any $\mathbb{Z}_{2k}$-linear code, viewed as a binary propelinear code, is not translation-invariant.

Proposition 1.10. *If* $k > 2$ *and* $\mathcal{C} \in \mathbb{Z}_{2k}^n$, *then* $\Phi(\mathcal{C})$ *is a propelinear but not translation-invariant code.*

Proof. Consider the vector $z = (1, 0, \dots, 0, 1) \in \mathbb{Z}_2^{kn}$. Then it is easy to check that $d_H(\mathbf{0}^{kn} \star z, \phi(1) \star z) = 3 \neq d_H(\mathbf{0}^{kn}, \phi(1)) = 1$.   □

1.4 MAGMA Package

Along this book, we include examples by using the functions developed in a new MAGMA package to deal with $\mathbb{Z}_2\mathbb{Z}_4$-additive codes and the corresponding $\mathbb{Z}_2\mathbb{Z}_4$-linear codes [29]. This new package has been developed by the authors of this book with the collaboration of some students. In this section, we give a brief description of this package. For more information, the manual including the description of all functions can be downloaded from the CCSG web page http://ccsg.uab.cat.

MAGMA currently supports the basic facilities for linear codes over integer residue rings and Galois rings [46, Chapter 162], including cyclic codes and the complete weight enumerator computation. Moreover, some functions are available for the special case of linear codes over $\mathbb{Z}_4$ (also called $\mathbb{Z}_4$-codes, quaternary linear codes, or $\mathbb{Z}_4$-additive codes), which are subgroups of $\mathbb{Z}_4^n$ [46, Chapter 163].

MAGMA also supports functions for additive codes over a finite field, which are a generalization of the linear codes over a finite field [46, Chapter

164] in a Hamming scheme. However, for $\mathbb{Z}_2\mathbb{Z}_4$-additive codes which are subgroups of $\mathbb{Z}_2^\alpha \times \mathbb{Z}_4^\beta$ with $\alpha, \beta > 0$, MAGMA does not provide any specific function in the default distribution. A new package to deal with these codes has been developed [29], by the members of CCSG. Specifically, this MAGMA package generalizes most of the known functions for codes over the ring $\mathbb{Z}_4$ to $\mathbb{Z}_2\mathbb{Z}_4$-additive codes, maintaining all the functionality for codes over $\mathbb{Z}_4$ and adding new functions which, not only generalize the previous ones, but introduce new variants when it is needed.

More specifically, the package includes constructions for some families of $\mathbb{Z}_2\mathbb{Z}_4$-additive codes and standard constructions of new $\mathbb{Z}_2\mathbb{Z}_4$-additive codes from other ones; functions to find the minimum number of generators of order two and four in a $\mathbb{Z}_2\mathbb{Z}_4$-additive code, as well as a permutation equivalent code in standard form; and functions to compute its dual and a parity check matrix. In addition, there are functions to compute related binary and quaternary codes such as the corresponding $\mathbb{Z}_2\mathbb{Z}_4$-linear code from the Gray map image of a $\mathbb{Z}_2\mathbb{Z}_4$-additive code; efficient functions for computing the rank, dimension of the kernel and coset representatives of a $\mathbb{Z}_2\mathbb{Z}_4$-linear code; functions related to the minimum Hamming and Lee distances, and the weight distribution of such codes. There are also functions to construct and deal with cyclic $\mathbb{Z}_2\mathbb{Z}_4$-additive codes. Finally, functions for encoding and decoding $\mathbb{Z}_2\mathbb{Z}_4$-additive codes by using two different methods are also provided.

Since $\mathbb{Z}_2\mathbb{Z}_4$-additive codes are subgroups of $\mathbb{Z}_2^\alpha \times \mathbb{Z}_4^\beta$, they can be seen as a generalization of binary and quaternary linear codes. Moreover, note that $\mathbb{Z}_2\mathbb{Z}_4$-additive codes can be represented as quaternary linear codes changing the ones by twos in the first α binary coordinates. Thus, in this way, they would be subgroups of $\mathbb{Z}_4^{\alpha+\beta}$. However, these codes are not equivalent to quaternary linear codes, since the inner product defined in $\mathbb{Z}_2^\alpha \times \mathbb{Z}_4^\beta$ gives us that the dual code of a $\mathbb{Z}_2\mathbb{Z}_4$-additive code is not equivalent to the dual code of the corresponding quaternary linear code as it is shown in Chapter 3. After the Gray map, the binary codes are not equivalent either, since they have different length if $\alpha \neq 0$. The length of a $\mathbb{Z}_2\mathbb{Z}_4$-linear code is $\alpha + 2\beta$, and on the other hand, for the corresponding $\mathbb{Z}_4$-linear code, it would be $2(\alpha + \beta)$.

Every object created in MAGMA is associated with a unique parent algebraic structure. The type of an object is then simply its parent structure. Each type contains some relevant attributes to define the object. For $\mathbb{Z}_2\mathbb{Z}_4$-additive codes, a new type called Z2Z4Code is defined. It contains the following attributes: MinimumLeeWeightWord, MinimumLeeWeightLowerBound, MinimumLeeWeightUpperBound, LeeWeightDistribution, CoveringRadius, MinimumLeeWeight, Alpha, Code, and Length. However, only the last three

are always assigned when we define a $\mathbb{Z}_2\mathbb{Z}_4$-additive code. The attribute
`Alpha` contains the value of α, and `Code` contains the quaternary code associated to a $\mathbb{Z}_2\mathbb{Z}_4$-additive code after changing the ones by twos in the first α
binary coordinates. Finally, `Length` contains the length of the code stored
in `Code`, that is, $\alpha + \beta$.

Since to represent a $\mathbb{Z}_2\mathbb{Z}_4$-additive code, we use quaternary linear codes,
it is necessary to remember that for modules defined over rings with zero
divisors, it is not possible to talk about the concept of dimension (the modules
are not always free) [46, p. 5783]. However, in MAGMA, each code over
such a ring has a unique generator matrix corresponding to the Howell form
[96, 153]. The number of rows k in this generator matrix will be called the
pseudo-dimension of the code. It should be noted that this pseudo-dimension
is not invariant between equivalent codes, and so does not provide structural
information like the dimension of a code over a finite field.

In order to use some of the functions included in this package, it is necessary to install the package "Codes over Z4", which can also be downloaded
from the web page `http://ccsg.uab.cat`. The package "Codes over Z4"
contains functions that expand the previous functionality for codes over $\mathbb{Z}_4$
in MAGMA. These functions are included in the standard MAGMA distribution from version 2.15-15 or later [46, Chapter 163], and in this case it is not
necessary to install this package.

Chapter 2

$\mathbb{Z}_2\mathbb{Z}_4$-Additive and $\mathbb{Z}_2\mathbb{Z}_4$-Linear Codes

In this chapter, some basic definitions related to the parameters of a $\mathbb{Z}_2\mathbb{Z}_4$-additive code and its associated binary code through a generalized Gray map are given in Section 2.1. In Section 2.2, generator matrices for $\mathbb{Z}_2\mathbb{Z}_4$-additive codes are defined along with the generator matrices in standard form. In Section 2.3, the binary linear residue and torsion codes of a $\mathbb{Z}_2\mathbb{Z}_4$-additive code are defined. Finally, in Section 2.4, some trivial families of $\mathbb{Z}_2\mathbb{Z}_4$-additive codes are also described. Examples by using the MAGMA package on $\mathbb{Z}_2\mathbb{Z}_4$-additive codes [29] developed by the authors are also included.

2.1 Basic Parameters

For a vector $\mathbf{u} \in \mathbb{Z}_2^\alpha \times \mathbb{Z}_4^\beta$, we write $\mathbf{u} = (u \mid u')$, where $u \in \mathbb{Z}_2^\alpha$ and $u' \in \mathbb{Z}_4^\beta$. The weight of $\mathbf{u} = (u \mid u')$ is defined as

$$\mathrm{wt}(\mathbf{u}) = \mathrm{wt}_H(u) + \mathrm{wt}_L(u'). \tag{2.1}$$

Given two elements $\mathbf{u} = (u \mid u'), \mathbf{v} = (v \mid v') \in \mathbb{Z}_2^\alpha \times \mathbb{Z}_4^\beta$, we define the distance between $\mathbf{u}$ and $\mathbf{v}$ as

$$d(\mathbf{u}, \mathbf{v}) = d_H(u, v) + d_L(u', v'). \tag{2.2}$$

The usual Gray map $\phi : \mathbb{Z}_4 \longrightarrow \mathbb{Z}_2^2$ is defined by

$$\phi(0) = (0,0), \ \ \phi(1) = (0,1), \ \ \phi(2) = (1,1), \ \ \phi(3) = (1,0).$$

There are parts of this chapter that have been previously published. Reprinted by permission from Springer Nature Customer Service Centre GmbH: J. Borges, C. Fernández-Córdoba, J. Pujol, J. Rifà, and M. Villanueva. "$\mathbb{Z}_2\mathbb{Z}_4$-linear codes: generator matrices and duality". *Designs, Codes and Cryptography*, v. 54, 2, pp. 167–179. ©(2010).

© Springer Nature Switzerland AG 2022
J. Borges et al., *Z₂Z₄-Linear Codes*, https://doi.org/10.1007/978-3-031-05441-9_2

If $a = (a_1, \ldots, a_\beta) \in \mathbb{Z}_4^\beta$, then the Gray map of a is the coordinate-wise extended map $\phi(a) = (\phi(a_1), \ldots, \phi(a_\beta)) \in \mathbb{Z}_2^{2\beta}$. We naturally extend the Gray map for vectors $\mathbf{u} = (u \mid u') \in \mathbb{Z}_2^\alpha \times \mathbb{Z}_4^\beta$ so that $\Phi(\mathbf{u}) = (u \mid \phi(u')) \in \mathbb{Z}_2^{\alpha+2\beta}$. Clearly, $\mathrm{wt}(\mathbf{u}) = \mathrm{wt}_H(\Phi(\mathbf{u}))$ and $d(\mathbf{u}, \mathbf{v}) = d_H(\Phi(\mathbf{u}), \Phi(\mathbf{v}))$, for vectors $\mathbf{u}, \mathbf{v} \in \mathbb{Z}_2^\alpha \times \mathbb{Z}_4^\beta$. Hence, the Gray map Φ is an isometry which transforms distance defined in $\mathbb{Z}_2^\alpha \times \mathbb{Z}_4^\beta$ to Hamming distance defined in $\mathbb{Z}_2^n$, where $n = \alpha + 2\beta$.

Recall that a $\mathbb{Z}_2\mathbb{Z}_4$-additive code $\mathcal{C}$ is a subgroup of $\mathbb{Z}_2^\alpha \times \mathbb{Z}_4^\beta$ for some non-negative integers α and β. Therefore, it is isomorphic to an abelian structure $\mathbb{Z}_2^\gamma \times \mathbb{Z}_4^\delta$ for some non-negative integers γ and δ. We have that the number of codewords is $|\mathcal{C}| = 2^\gamma 4^\delta$. Note that $2^{\gamma+\delta}$ of them have order at most two.

Let X (respectively Y) be the set of $\mathbb{Z}_2$ (respectively $\mathbb{Z}_4$) coordinate positions, so $|X| = \alpha$ and $|Y| = \beta$. Unless otherwise stated, the set X corresponds to the first α coordinates and Y corresponds to the last β coordinates. Call $\mathcal{C}_X$ (respectively $\mathcal{C}_Y$) the projection of $\mathcal{C}$ onto X (respectively Y). Let $\mathcal{C}_b$ be the subcode of $\mathcal{C}$ which contains exactly all codewords of order at most two, and let κ be the dimension of $(\mathcal{C}_b)_X$, which is a binary linear code. Figure 2.1 shows the general structure of a $\mathbb{Z}_2\mathbb{Z}_4$-additive code with $\mathcal{C}_X$, $\mathcal{C}_Y$, $\mathcal{C}_b$ and $(\mathcal{C}_b)_X$. For the case $\alpha = 0$, we write $\kappa = 0$. Taking into account all these parameters, we say that $\mathcal{C}$ (or equivalently $C = \Phi(\mathcal{C})$) is of type $(\alpha, \beta; \gamma, \delta; \kappa)$, $(\alpha, \beta; \gamma, \delta)$, or $2^\gamma 4^\delta$ when the other parameters are clear from the context or are not necessary.

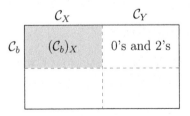

Figure 2.1: Structure of a $\mathbb{Z}_2\mathbb{Z}_4$-additive code

Definition 2.1. *Let $\mathcal{C}$ be a $\mathbb{Z}_2\mathbb{Z}_4$-additive code, that is, a subgroup of $\mathbb{Z}_2^\alpha \times \mathbb{Z}_4^\beta$. The binary image $C = \Phi(\mathcal{C})$ is called $\mathbb{Z}_2\mathbb{Z}_4$-linear code of length $n = \alpha + 2\beta$ and type $(\alpha, \beta; \gamma, \delta; \kappa)$, where γ, δ and κ are defined as above.*

Example 2.2. Let $\mathcal{C}$ be the $\mathbb{Z}_2\mathbb{Z}_4$-additive code with the following codewords:

$$\mathcal{C} = \{(0,0 \mid 0,0), (0,0 \mid 0,1), (0,0 \mid 0,2), (0,0 \mid 0,3),$$
$$(0,0 \mid 2,0), (0,0 \mid 2,1), (0,0 \mid 2,2), (0,0 \mid 2,3),$$
$$(1,1 \mid 0,0), (1,1 \mid 0,1), (1,1 \mid 0,2), (1,1 \mid 0,3),$$
$$(1,1 \mid 2,0), (1,1 \mid 2,1), (1,1 \mid 2,2), (1,1 \mid 2,3)\}.$$

We have that $\mathcal{C} \subset \mathbb{Z}_2^2 \times \mathbb{Z}_4^2$, so $\alpha = 2$ and $\beta = 2$. We also have that the code has 16 codewords and 8 of them have order at most two. Therefore, $\gamma = 2$ and $\delta = 1$. The subcode of $\mathcal{C}$ containing all codewords of order at most two is

$$\mathcal{C}_b = \{(0,0 \mid 0,0), (0,0 \mid 0,2), (0,0 \mid 2,0), (0,0 \mid 2,2),$$
$$(1,1 \mid 0,0), (1,1 \mid 0,2), (1,1 \mid 2,0), (1,1 \mid 2,2)\}.$$

Then, $(\mathcal{C}_b)_X = \{(0,0),(1,1)\}$ and $\kappa = 1$. Therefore, we have that $\mathcal{C}$ is of type $(2,2;2,1;1)$.

The binary code $C = \Phi(\mathcal{C})$ is a $\mathbb{Z}_2\mathbb{Z}_4$-linear code of length $n = 6$ and type $(2,2;2,1;1)$ with the following codewords:

$$C = \{(0,0,0,0,0,0), (0,0,0,0,0,1), (0,0,0,0,1,1), (0,0,0,0,1,0),$$
$$(0,0,1,1,0,0), (0,0,1,1,0,1), (0,0,1,1,1,1), (0,0,1,1,1,0),$$
$$(1,1,0,0,0,0), (1,1,0,0,0,1), (1,1,0,0,1,1), (1,1,0,0,1,0),$$
$$(1,1,1,1,0,0), (1,1,1,1,0,1), (1,1,1,1,1,1), (1,1,1,1,1,0)\}.$$

$\triangle$

Let $\mathcal{C}$ be a $\mathbb{Z}_2\mathbb{Z}_4$-additive code of type $(\alpha, \beta; \gamma, \delta; \kappa)$. The minimum distance of $\mathcal{C}$ is defined as $d(\mathcal{C}) = \min\{d(\mathbf{u},\mathbf{v}) : \mathbf{u}, \mathbf{v} \in \mathcal{C} \text{ and } \mathbf{u} \neq \mathbf{v}\}$. The $\mathbb{Z}_2\mathbb{Z}_4$-linear code $C = \Phi(\mathcal{C})$ is a binary code which is not necessarily linear in general. As a binary code, the minimum Hamming distance of C is $d_H(C) = \min\{d_H(\Phi(\mathbf{u}), \Phi(\mathbf{v})) : \mathbf{u}, \mathbf{v} \in \mathcal{C} \text{ and } \mathbf{u} \neq \mathbf{v}\}$. Note that $d(\mathcal{C}) = d_H(\Phi(\mathcal{C}))$ since $d(\mathbf{u}, \mathbf{v}) = d_H(\Phi(\mathbf{u}), \Phi(\mathbf{v}))$. Therefore, C is a $(\alpha + 2\beta, 2^\gamma 4^\delta, d)$ binary code, where $d = d(\mathcal{C})$.

The $\mathbb{Z}_2\mathbb{Z}_4$-linear codes are a generalization of binary linear codes and $\mathbb{Z}_4$-linear codes. On the one hand, when $\beta = 0$, the binary code $C = \mathcal{C}$ corresponds to a binary linear code. In fact, if C is a binary linear $[\alpha, k]$ code, then C is a $\mathbb{Z}_2\mathbb{Z}_4$-additive code of type $(\alpha, 0; k, 0; k)$. On the other hand, when $\alpha = 0$, the $\mathbb{Z}_2\mathbb{Z}_4$-additive code $\mathcal{C}$ is a quaternary linear code and the corresponding $\mathbb{Z}_2\mathbb{Z}_4$-linear code $C = \Phi(\mathcal{C})$ is a $\mathbb{Z}_4$-linear code. Any quaternary linear code of length β and type $2^\gamma 4^\delta$ is a $\mathbb{Z}_2\mathbb{Z}_4$-additive code of type $(0, \beta; \gamma, \delta; 0)$.

Note that changing ones by twos in the binary coordinates, we can see $\mathbb{Z}_2\mathbb{Z}_4$-additive codes as quaternary linear codes. Also, applying the modulo two map in the quaternary coordinates or considering the binary coordinates as elements in $\mathbb{Z}_4$, we can derive binary and quaternary codes. We define the following maps. First, let χ be the map from $\mathbb{Z}_2$ to $\mathbb{Z}_4$, which is the usual inclusion from the additive structure in $\mathbb{Z}_2$ to $\mathbb{Z}_4$: $\chi(0) = 0$, $\chi(1) = 2$. Let ξ be the natural modulo two map from $\mathbb{Z}_4$ to $\mathbb{Z}_2$, that is $\xi(0) = 0$, $\xi(1) = 1$, $\xi(2) = 0$, $\xi(3) = 1$. Finally, let ι be the identity map from $\mathbb{Z}_2$ to $\mathbb{Z}_4$, that is $\iota(0) = 0$, $\iota(1) = 1$. Let I denote the identity map. These maps can be extended to the maps $(\chi, I) : \mathbb{Z}_2^\alpha \times \mathbb{Z}_4^\beta \longrightarrow \mathbb{Z}_4^\alpha \times \mathbb{Z}_4^\beta$, $(I, \xi) : \mathbb{Z}_2^\alpha \times \mathbb{Z}_4^\beta \longrightarrow \mathbb{Z}_2^\alpha \times \mathbb{Z}_2^\beta$ and $(\iota, I) : \mathbb{Z}_2^\alpha \times \mathbb{Z}_4^\beta \longrightarrow \mathbb{Z}_4^\alpha \times \mathbb{Z}_4^\beta$, which are also denoted by χ, ξ and ι, respectively.

Let $\mathcal{C}$ be a $\mathbb{Z}_2\mathbb{Z}_4$-additive code of type $(\alpha, \beta; \gamma, \delta; \kappa)$. Note that $\mathcal{C}_Y$ is a quaternary linear code of length β and type $2^{\gamma_y} 4^\delta$, where $0 \leq \gamma_y \leq \gamma$, and $\mathcal{C}_X$ is a binary linear $[\alpha, \gamma_x]$ code, where $\kappa \leq \gamma_x \leq \kappa + \delta$.

Definition 2.3. *A $\mathbb{Z}_2\mathbb{Z}_4$-additive code $\mathcal{C}$ is said to be separable if $\mathcal{C} = \mathcal{C}_X \times \mathcal{C}_Y$.*

Let $\kappa_1 \leq \kappa$ and $\delta_2 \leq \delta$ be the integers such that the subcodes $\{(u \mid \mathbf{0}) \in \mathcal{C}\}$ and $\langle \{(\mathbf{0} \mid u') \in \mathcal{C} : u' = \mathbf{0}$ or the order of u' is four$\} \rangle$ are of type $(\alpha, \beta; \kappa_1, 0; \kappa_1)$ and $(\alpha, \beta; \gamma', \delta_2; 0)$ for an integer $\gamma' \leq \gamma$, respectively. With the definition of κ_1 and δ_2, consider the values κ_2 and δ_1 such that

$$\kappa = \kappa_1 + \kappa_2 \quad \text{and} \quad \delta = \delta_1 + \delta_2. \tag{2.3}$$

Remark 2.4. Let $\mathcal{C}$ be a $\mathbb{Z}_2\mathbb{Z}_4$-additive code of type $(\alpha, \beta; \gamma, \delta; \kappa)$. By the definition of the parameters κ_1 and δ_2, we have that $\mathcal{C}_Y$ is a quaternary linear code of length β and type $2^{\gamma - \kappa_1} 4^\delta$, and $\mathcal{C}_X$ is a binary linear $[\alpha, \kappa + \delta_1]$ code. Moreover, it is clear that $\mathcal{C}$ is separable if and only if κ_2 and δ_1 are zero; that is, $\kappa = \kappa_1$ and $\delta = \delta_2$.

Remark 2.5. Let $\mathcal{C}$ be a $\mathbb{Z}_2\mathbb{Z}_4$-additive code of type $(\alpha, \beta; \gamma, \delta; \kappa)$, with $\kappa = \kappa_1 + \kappa_2$ and $\delta = \delta_1 + \delta_2$ given in (2.3), and let $\mathcal{C}_b$ the subcode of $\mathcal{C}$ containing all codewords of order two. Then, it is easy to check that $\kappa_2 = 0$ if and only if $\mathcal{C}_b$ is separable.

From binary and quaternary linear codes, it is possible to construct separable $\mathbb{Z}_2\mathbb{Z}_4$-additive codes, as we can see through the following result that can be easily proven.

Proposition 2.6. *Let $\mathcal{D}_1$ be a binary linear $[\alpha, \kappa]$ code, and let $\mathcal{D}_2$ be a quaternary linear code of length β and type $2^\gamma 4^\delta$. Then, $\mathcal{C} = \mathcal{D}_1 \times \mathcal{D}_2$ is a separable $\mathbb{Z}_2\mathbb{Z}_4$-additive code of type $(\alpha, \beta; \gamma + \kappa, \delta; \kappa)$.*

Note that the separable $\mathbb{Z}_2\mathbb{Z}_4$-additive code $\mathcal{C}$ constructed by using Proposition 2.6 satisfies $\mathcal{C}_X = \mathcal{D}_1$ and $\mathcal{C}_Y = \mathcal{D}_2$. In Example 2.7, we give two different $\mathbb{Z}_2\mathbb{Z}_4$-additive codes, $\mathcal{C}$ and $\mathcal{D}$ with $\mathcal{C}_X = \mathcal{D}_X$ and $\mathcal{C}_Y = \mathcal{D}_Y$, one being separable and the other one non-separable.

Example 2.7. Consider the $\mathbb{Z}_2\mathbb{Z}_4$-additive code $\mathcal{C}$ of type $(2,2;2,1;1)$ given in Example 2.2. We have that $\mathcal{C}_X = \{(1,1),(0,0)\}$ and $\mathcal{C}_Y = \{(0,0),(2,0),(0,1),(0,2),(0,3),(2,1),(2,2),(2,3)\}$. Moreover, since $\mathcal{C} = \mathcal{C}_X \times \mathcal{C}_Y$, the code $\mathcal{C}$ is separable. Note that, $\{(0,0 \mid 0,0),(1,1 \mid 0,0)\}$ is a subcode of type $(2,2;1,0;1)$, so $\kappa_1 = 1$; and the subcode $\{(0,0 \mid 0,0),(0,0 \mid 0,1),(0,0 \mid 0,2),(0,0 \mid 0,3),(0,0 \mid 2,0),(0,0 \mid 2,1),(0,0 \mid 2,2),(0,0 \mid 2,3)\}$ is of type $(2,2;1,1;0)$, thus $\delta_2 = 1$. Therefore, $\kappa_1 = \kappa = 1$ and $\delta_2 = \delta = 1$.

Now, consider the $\mathbb{Z}_2\mathbb{Z}_4$-additive code $\mathcal{D}$ of type $(2,2;1,1;1)$ with the following codewords:

$$\mathcal{D} = \{(0,0 \mid 0,0),(0,0 \mid 0,1),(0,0 \mid 0,2),(0,0 \mid 0,3),$$
$$(1,1 \mid 2,0),(1,1 \mid 2,1),(1,1 \mid 2,2),(1,1 \mid 2,3)\}.$$

In this case, we also have that $\mathcal{D}_X = \mathcal{C}_X$ and $\mathcal{D}_Y = \mathcal{C}_Y$. However, since $\mathcal{D} \neq \mathcal{D}_X \times \mathcal{D}_Y$, the code $\mathcal{D}$ is non-separable. We have that $\kappa_1 = 0$ and $\delta_2 = 1$, since $\{(u \mid 0,0) \in \mathcal{D}\}$ is the subcode with only the all-zero codeword and the subcode $\{(0,0 \mid 0,0),(0,0 \mid 0,1),(0,0 \mid 0,2),(0,0 \mid 0,3)\}$ is of type $(2,2;0,1;0)$. Therefore, $\kappa_1 \neq \kappa$. △

Example 2.8. The $\mathbb{Z}_2\mathbb{Z}_4$-additive code $\mathcal{C}$ given in Example 2.2 can be generated by using the MAGMA package [29]. Then, we check that it is separable. We also compute the parameters κ_1 and δ_2, which are equal to κ and δ, respectively.

```
> Z4 := IntegerRing(4);
> V := RSpace(Z4, 4);
> C := Z2Z4AdditiveCode(
                [V![0,0,0,0], V![0,0,0,1], V![0,0,0,2], V![0,0,0,3],
                 V![0,0,2,0], V![0,0,2,1], V![0,0,2,2], V![0,0,2,3],
                 V![1,1,0,0], V![1,1,0,1], V![1,1,0,2], V![1,1,0,3],
                 V![1,1,2,0], V![1,1,2,1], V![1,1,2,2], V![1,1,2,3]], 2);
> Z2Z4Type(C);
[ 2, 2, 2, 1, 1 ]
> delta := Z2Z4Type(C)[4];
> kappa := Z2Z4Type(C)[5];
>
> CX := LinearBinaryCode(C);
> CY := LinearQuaternaryCode(C);
> Cnew := DirectSum(Z2Z4AdditiveCode(CX), Z2Z4AdditiveCode(CY));
> C eq Cnew;
```

```
true
> IsSeparable(C);
true
>
> subsetA := [c : c in Set(C) | (c[3] eq 0) and (c[4] eq 0) ];
> A := Z2Z4AdditiveCode(subsetA, 2);
> kappa1 := Z2Z4Type(A)[5];
> kappa1 eq kappa;
true
> subsetB := [c : c in Set(C) | (c[1] eq 0) and (c[2] eq 0)
                                  and 2*c ne V!0 ];
> B := Z2Z4AdditiveCode(subsetB, 2);
> delta2 := Z2Z4Type(B)[4];
> delta2 eq delta;
true
```
$\triangle$

Two $\mathbb{Z}_2\mathbb{Z}_4$-additive codes $\mathcal{C}_1$ and $\mathcal{C}_2$ of the same type $(\alpha, \beta; \gamma, \delta; \kappa)$ are said to be *monomially equivalent* (or simply, *equivalent*), if one can be obtained from the other by permuting the coordinates and (if necessary) changing the sign of some $\mathbb{Z}_4$ coordinates. Two $\mathbb{Z}_2\mathbb{Z}_4$-additive codes are said to be *permutation equivalent* if they only differ by a permutation of coordinates. Recall that a monomial matrix is a square matrix with exactly one non-zero entry in each row and column. Therefore, we can also say that $\mathcal{C}_1$ and $\mathcal{C}_2$ are monomially equivalent provided there is a monomial matrix M of the following form:

$$M = \begin{pmatrix} P_X & \mathbf{0} \\ \mathbf{0} & DP_Y \end{pmatrix}, \tag{2.4}$$

such that $\mathcal{C}_2 = \mathcal{C}_1 M$, where P_X and P_Y are permutation matrices of size α and β, respectively; and D is a diagonal matrix with entries in $\{1, 3\}$. The codes are permutation equivalent if there is a matrix M where D is the identity matrix. If two $\mathbb{Z}_2\mathbb{Z}_4$-additive codes $\mathcal{C}_1$ and $\mathcal{C}_2$ are monomially equivalent, then, after the Gray map, the corresponding $\mathbb{Z}_2\mathbb{Z}_4$-linear codes $C_1 = \Phi(\mathcal{C}_1)$ and $C_2 = \Phi(\mathcal{C}_2)$ are permutation equivalent as binary codes. Note that the inverse statement is not always true.

The *monomial automorphism group* of a $\mathbb{Z}_2\mathbb{Z}_4$-additive code $\mathcal{C}$, denoted by MAut($\mathcal{C}$), is the group generated by all permutations and sign-changes of the $\mathbb{Z}_4$ coordinates that preserves the set of codewords of $\mathcal{C}$. In other words, is the group of all monomial matrices of the form (2.4) that map $\mathcal{C}$ to itself. The *permutation automorphism group* of $\mathcal{C}$, denoted by PAut($\mathcal{C}$), is the group generated by all permutations that preserves the set of codewords of $\mathcal{C}$ [97].

Example 2.9. Let $\mathcal{C}$ be the $\mathbb{Z}_2\mathbb{Z}_4$-additive code of type $(4, 6; 1, 2; 1)$ with generator matrix

$$\mathcal{G} = \begin{pmatrix} 1 & 1 & 1 & 1 & 0 & 2 & 0 & 2 & 0 & 0 \\ 0 & 1 & 0 & 1 & 2 & 1 & 1 & 1 & 1 & 0 \\ 0 & 0 & 1 & 1 & 1 & 1 & 2 & 3 & 0 & 1 \end{pmatrix}.$$

The permutation automorphism group of $\mathcal{C}$ is the subgroup $\mathrm{PAut}(\mathcal{C}) = \langle (6,8)(7,9), (1,3)(2,4)(6,8), (1,2)(3,4)(5,10) \rangle \subset \mathcal{S}_{10}$ of order $8 = 2^3$. On the other hand, the permutation automorphism group of the corresponding $\mathbb{Z}_2\mathbb{Z}_4$-linear code $C = \Phi(\mathcal{C})$ is the subgroup

$$\mathrm{PAut}(C) = \langle (1,16,10,12,3,15,14,8)(2,5,9,11,4,6,13,7),$$
$$(1,7,13,4,11,14)(2,8,9,3,12,10)(6,15,16),$$
$$(1,15,3,5)(2,16,4,6)(7,14,12,10)(8,9,11,13) \rangle \subset \mathcal{S}_{16}$$

of order $9216 = 2^{10}3^2$. Note that any permutation in $\mathrm{PAut}(\mathcal{C})$ gives a permutation in $\mathrm{PAut}(C)$. For example, the permutation $(1,3)(2,4)(6,8) \in \mathrm{PAut}(\mathcal{C})$ gives the permutation $(1,3)(2,4)(7,11)(8,12) \in \mathrm{PAut}(C)$, considering that the binary coordinate positions $1,2,3,4$ in $\mathcal{C}$ correspond to the same coordinates in C; the quaternary coordinate position 6 in $\mathcal{C}$ corresponds to the pair of coordinate positions 7, 8 in C; and 8 corresponds to the pair 11, 12. Finally, in this case, the monomial automorphism group of $\mathcal{C}$ is $\mathrm{MAut}(\mathcal{C}) = \mathrm{PAut}(\mathcal{C})$.   △

2.2   Generator Matrices

In this section, mainly, we define and prove some results related to generator matrices and generator matrices in standard form for $\mathbb{Z}_2\mathbb{Z}_4$-additive codes, generalizing the already known results for quaternary linear codes.

A $\mathbb{Z}_2\mathbb{Z}_4$-additive code $\mathcal{C}$ of type $(\alpha, \beta; \gamma, \delta; \kappa)$ can also be seen as a submodule of the $\mathbb{Z}_4$-module $\mathbb{Z}_2^\alpha \times \mathbb{Z}_4^\beta$. As a submodule, $\mathcal{C}$ is free if and only if $\gamma = 0$. Although $\mathcal{C}$ is not a free module in general, there exist $\{\mathbf{u}_i\}_{i=1}^\gamma$ and $\{\mathbf{v}_j\}_{j=1}^\delta$ such that every codeword in $\mathcal{C}$ can be uniquely expressible in the form

$$\sum_{i=1}^{\gamma} \lambda_i \mathbf{u}_i + \sum_{j=1}^{\delta} \mu_j \mathbf{v}_j,$$

where $\lambda_i \in \mathbb{Z}_2$ for all $1 \le i \le \gamma$, $\mu_j \in \mathbb{Z}_4$ for all $1 \le j \le \delta$ and $\mathbf{u}_i, \mathbf{v}_j$ are codewords of order two and order four, respectively. The matrix $\mathcal{G}$ of size $(\gamma + \delta) \times (\alpha + \beta)$ that has as rows the vectors $\{\mathbf{u}_i\}_{i=1}^\gamma$ and $\{\mathbf{v}_j\}_{j=1}^\delta$ is a generator matrix for $\mathcal{C}$. Moreover, we can write $\mathcal{G}$ as

$$\mathcal{G} = \left(\begin{array}{c|c} B_1 & 2B_3 \\ B_2 & Q \end{array} \right), \tag{2.5}$$

where B_1, B_2 are matrices over $\mathbb{Z}_2$ of size $\gamma \times \alpha$ and $\delta \times \alpha$, respectively; B_3 is a matrix over $\mathbb{Z}_4$ of size $\gamma \times \beta$ with all entries in $\{0, 1\} \subset \mathbb{Z}_4$; and Q is a matrix over $\mathbb{Z}_4$ of size $\delta \times \beta$ with quaternary row vectors of order four.

Although it is also possible to generate $\mathcal{C}$ with more than $\gamma + \delta$ vectors, we usually consider generator matrices with minimum number of rows, $\gamma + \delta$.

For any integer $k > 0$, let I_k be the identity matrix of size $k \times k$. Let $\mathbf{0}$ be the all-zero matrix. In [92, 157], it is shown that any quaternary linear code of length β and type $2^\gamma 4^\delta$ is permutation equivalent to a quaternary linear code with a generator matrix of the form

$$\mathcal{G}_S = \left(\begin{array}{ccc} 2T & 2I_\gamma & \mathbf{0} \\ S & R & I_\delta \end{array} \right), \tag{2.6}$$

where R, T are matrices over $\mathbb{Z}_4$ with all entries in $\{0, 1\} \subset \mathbb{Z}_4$, of size $\delta \times \gamma$ and $\gamma \times (\beta - \gamma - \delta)$, respectively; and S is a matrix over $\mathbb{Z}_4$ of size $\delta \times (\beta - \gamma - \delta)$. We say that this matrix $\mathcal{G}_S$ is in standard form. Next, we generalize this result for $\mathbb{Z}_2\mathbb{Z}_4$-additive codes. As for binary and quaternary linear codes, there is a standard form for the generator matrix of a $\mathbb{Z}_2\mathbb{Z}_4$-additive code.

Theorem 2.10 ([37]). *Let $\mathcal{C}$ be a $\mathbb{Z}_2\mathbb{Z}_4$-additive code of type $(\alpha, \beta; \gamma, \delta; \kappa)$. Then, the code $\mathcal{C}$ is permutation equivalent to a $\mathbb{Z}_2\mathbb{Z}_4$-additive code with a generator matrix in standard form, that is, of the form*

$$\mathcal{G}_S = \left(\begin{array}{cc|ccc} I_\kappa & T_b & 2T_2 & \mathbf{0} & \mathbf{0} \\ \mathbf{0} & \mathbf{0} & 2T_1 & 2I_{\gamma-\kappa} & \mathbf{0} \\ \mathbf{0} & S_b & S_q & R & I_\delta \end{array} \right), \tag{2.7}$$

where T_b, S_b are matrices over $\mathbb{Z}_2$; T_1, T_2, R are matrices over $\mathbb{Z}_4$ with all entries in $\{0, 1\} \subset \mathbb{Z}_4$; and S_q is a matrix over $\mathbb{Z}_4$.

Proof. Since κ is the dimension of the matrix B_1 over $\mathbb{Z}_2$ given in (2.5), the code $\mathcal{C}$ has a generator matrix of the form

$$\left(\begin{array}{cc|cc} I_\kappa & \bar{B}_1 & 2\bar{B}_3 \\ \mathbf{0} & \mathbf{0} & 2\bar{B}_4 \\ \mathbf{0} & \bar{B}_2 & \bar{Q} \end{array} \right),$$

where $\bar{B}_1, \bar{B}_2$ are matrices over $\mathbb{Z}_2$ of size $\kappa \times (\alpha - \kappa)$ and $\delta \times (\alpha - \kappa)$, respectively; $\bar{B}_3, \bar{B}_4$ are matrices over $\mathbb{Z}_4$ with all entries in $\{0, 1\} \subset \mathbb{Z}_4$ of size $\kappa \times \beta$ and $(\gamma - \kappa) \times \beta$, respectively; and $\bar{Q}$ is a matrix over $\mathbb{Z}_4$ of size $\delta \times \beta$.

The quaternary linear code $\mathcal{C}^-$ of type $(0, \alpha - \kappa + \beta; \gamma - \kappa, \delta; 0)$ generated by the matrix

$$\begin{pmatrix} 0 & 2\bar{B}_4 \\ 2\bar{B}_2 & \bar{Q} \end{pmatrix}$$

is permutation equivalent to a quaternary linear code with generator matrix of the form

$$\mathcal{G}^- = \begin{pmatrix} 0 & 2T_1 & 2I_{\gamma - \kappa} & 0 \\ 2S_b & S_q & R & I_\delta \end{pmatrix}, \tag{2.8}$$

where the permutation of coordinates fixes the first $\alpha - \kappa$ coordinates (see [92] or (2.6)). Therefore, the quaternary linear code $\chi(\mathcal{C})$ generated by the matrix

$$\begin{pmatrix} 2I_\kappa & 2\bar{B}_1 & 2\bar{B}_3 \\ 0 & 0 & 2\bar{B}_4 \\ 0 & 2\bar{B}_2 & \bar{Q} \end{pmatrix}$$

is permutation equivalent to a quaternary linear code with generator matrix of the form

$$\mathcal{G}_\chi = \begin{pmatrix} 2I_\kappa & 2T_b & 2T_2 & 0 & 0 \\ 0 & 0 & 2T_1 & 2I_{\gamma - \kappa} & 0 \\ 0 & 2S_b & S_q & R & I_\delta \end{pmatrix}. \tag{2.9}$$

Finally, we have that $\mathcal{C}$ is permutation equivalent to a $\mathbb{Z}_2\mathbb{Z}_4$-additive code with generator matrix $\chi^{-1}(\mathcal{G}_\chi) = \mathcal{G}_S$. $\qquad \square$

Example 2.11. Let $\mathcal{C}$ denote the $\mathbb{Z}_2\mathbb{Z}_4$-additive code of type $(1, 3; 1, 2; 1)$ with generator matrix

$$\mathcal{G} = \begin{pmatrix} 1 & 2 & 2 & 2 \\ 0 & 1 & 1 & 0 \\ 1 & 1 & 2 & 3 \end{pmatrix}.$$

The code $\mathcal{C}$ can also be generated by the matrix

$$\begin{pmatrix} 1 & 2 & 2 & 2 \\ 0 & 1 & 1 & 0 \\ 0 & 1 & 0 & 3 \end{pmatrix}.$$

The quaternary linear code $\mathcal{C}^-$ generated by $\begin{pmatrix} 1 & 1 & 0 \\ 1 & 0 & 3 \end{pmatrix}$ is permutation equivalent to the one with generator matrix $\mathcal{G}^- = \begin{pmatrix} 3 & 1 & 0 \\ 1 & 0 & 1 \end{pmatrix}$. Therefore, the quaternary linear code $\chi(\mathcal{C})$ generated by

$$\begin{pmatrix} 2 & 2 & 2 & 2 \\ 0 & 1 & 1 & 0 \\ 0 & 1 & 0 & 3 \end{pmatrix}$$

is permutation equivalent to a quaternary linear code with generator matrix

$$\mathcal{G}_\chi = \begin{pmatrix} 2 & 2 & 0 & 0 \\ 0 & 3 & 1 & 0 \\ 0 & 1 & 0 & 1 \end{pmatrix}.$$

Finally, the code $\mathcal{C}$ is permutation equivalent to a $\mathbb{Z}_2\mathbb{Z}_4$-additive code with the following generator matrix in standard form:

$$\mathcal{G}_S = \chi^{-1}(\mathcal{G}_\chi) = \left(\begin{array}{c|ccc} 1 & 2 & 0 & 0 \\ 0 & 3 & 1 & 0 \\ 0 & 1 & 0 & 1 \end{array} \right).$$

$\triangle$

Example 2.12. The $\mathbb{Z}_2\mathbb{Z}_4$-additive code $\mathcal{C}$ given in Example 2.11 can be generated by using the MAGMA package [29]. Then, the type of $\mathcal{C}$ and a permutation equivalent code $\mathcal{C}_S$ with generator matrix in standard form is found. In this case, $\mathcal{C}_S$ is equal to $\mathcal{C}$ since the permutation is the identity. Note that the permutation equivalent code obtained in Example 2.11 is generated by a different matrix in standard form. Actually, this code is equal to $\mathcal{C}$ after applying the permutation $(3,4) \in \mathcal{S}_4$.

```
> Z4 := Integers(4);
> G := Matrix(Z4, [[1,2,2,2],
                    [0,1,1,0],
                    [1,1,2,3]]);
> C := Z2Z4AdditiveCode(G, 1);
> typeC := Z2Z4Type(C);
> typeC;
[ 1, 3, 1, 2, 1 ]
> Cs, _, Gs, p := StandardForm(C);
> Gs;
[2 2 0 0]
[0 1 1 0]
[0 3 0 1]
> p;
Id($)
> IsStandardFormMatrix(G, typeC);
false
```

```
> IsStandardFormMatrix(Gs, typeC);
true
> C eq Cs;
true
>
> Gs2 := Matrix(Z4, [[2,2,0,0],
                     [0,3,1,0],
                     [0,1,0,1]]);
> IsStandardFormMatrix(Gs2, typeC);
true
> Cs2 := Z2Z4AdditiveCode(Gs2, 1);
> C eq Cs2;
false
> P := PermutationMatrix(Z4, Sym(4)!(3,4));
> Cs2 eq Z2Z4AdditiveCode(G*P, 1);
true                                                    △
```

Corollary 2.13 ([79]). *There exists a $\mathbb{Z}_2\mathbb{Z}_4$-additive code C of type $(\alpha, \beta; \gamma, \delta; \kappa)$ as long as*

$$\alpha, \beta, \gamma, \delta, \kappa \geq 0, \quad \alpha + \beta > 0,$$
$$0 \leq \delta + \gamma \leq \beta + \kappa \quad and \quad \kappa \leq \min\{\alpha, \gamma\}. \tag{2.10}$$

Corollary 2.14. *Let C be a $\mathbb{Z}_2\mathbb{Z}_4$-additive code of type $(\alpha, \beta; \gamma, \delta; \kappa)$. Let $\mathcal{G}_S$ be a generator matrix of a permutation equivalent code in standard form (2.7). Then, C is separable if and only if S_b and T_2 are all-zero matrices.*

Example 2.15. Consider the $\mathbb{Z}_2\mathbb{Z}_4$-additive codes C and D given in Example 2.7. These codes have generator matrices

$$\mathcal{G}_C = \begin{pmatrix} 1 & 1 & 0 & 0 \\ 0 & 0 & 2 & 0 \\ 0 & 0 & 0 & 1 \end{pmatrix} \quad and \quad \mathcal{G}_D = \begin{pmatrix} 1 & 1 & 2 & 0 \\ 0 & 0 & 0 & 1 \end{pmatrix},$$

respectively. Note that both generator matrices are in standard form (2.7). Moreover, since the code C is separable, we have that the matrices S_b and T_2 in $\mathcal{G}_C$ are all-zero matrices, as it is shown in Corollary 2.14. △

Let C be a $\mathbb{Z}_2\mathbb{Z}_4$-additive code of type $(\alpha, \beta; \gamma, \delta; \kappa)$, with $\kappa = \kappa_1 + \kappa_2$ and $\delta = \delta_1 + \delta_2$ given in (2.3). By applying convenient column permutations and linear combinations of rows to a generator matrix in standard form (2.7), we can obtain a matrix of the form

$$\mathcal{G} = \begin{pmatrix} I_{\kappa_1} & T_{b_1} & T_{b_2} & T_{b_3} & 0 & 0 & 0 & 0 & 0 \\ 0 & I_{\kappa_2} & T_{b_4} & T_{b_5} & 2T_2 & 2T_2' & 0 & 0 & 0 \\ 0 & 0 & 0 & 0 & 2T_1 & 2T_1' & 2I_{\gamma-\kappa} & 0 & 0 \\ 0 & 0 & S_{b_1} & S_{b_2} & S_{q_1} & S_{q_2} & R_1 & I_{\delta_1} & 0 \\ 0 & 0 & 0 & 0 & S_{q_3} & S_{q_4} & R_2 & R_3 & I_{\delta_2} \end{pmatrix}, \tag{2.11}$$

where the matrices T_{b_i}, S_{b_j} are over $\mathbb{Z}_2$ for $1 \leq i \leq 5$, $1 \leq j \leq 2$; the matrices T_i, T_i', R_j are over $\mathbb{Z}_4$ with all entries in $\{0,1\} \subset \mathbb{Z}_4$ for $1 \leq i \leq 2$, $1 \leq j \leq 3$; and S_{q_i} are matrices over $\mathbb{Z}_4$ for $1 \leq i \leq 4$. The matrices S_{b_1} and T_2' are square matrices of full rank δ_1 and κ_2, respectively. Therefore, we have that $\mathcal{C}$ is permutation equivalent to a $\mathbb{Z}_2\mathbb{Z}_4$-additive code with generator matrix $\mathcal{G}$ of the form given in (2.11). Note that this generator matrix $\mathcal{G}$ in general is not in standard form, except when T_{b_1} and R_3 are all-zero matrices.

Example 2.16. Let $\mathcal{C}$ be the $\mathbb{Z}_2\mathbb{Z}_4$-additive code of type $(3, 5; 3, 3; 2)$ with generator matrix in standard form

$$\mathcal{G}_S = \left(\begin{array}{ccc|ccccc} 1 & 0 & 1 & 2 & 0 & 0 & 0 & 0 \\ 0 & 1 & 1 & 2 & 0 & 0 & 0 & 0 \\ 0 & 0 & 0 & 2 & 2 & 0 & 0 & 0 \\ 0 & 0 & 1 & 2 & 1 & 1 & 0 & 0 \\ 0 & 0 & 1 & 0 & 1 & 0 & 1 & 0 \\ 0 & 0 & 1 & 1 & 1 & 0 & 0 & 1 \end{array} \right).$$

By applying some linear combinations of rows, we have that a generator matrix of $\mathcal{C}$ of the form given in (2.11) is

$$\mathcal{G} = \left(\begin{array}{ccc|ccccc} 1 & 1 & 1 & 0 & 0 & 0 & 0 & 0 \\ 0 & 1 & 1 & 2 & 0 & 0 & 0 & 0 \\ 0 & 0 & 0 & 2 & 2 & 0 & 0 & 0 \\ 0 & 0 & 1 & 2 & 1 & 1 & 0 & 0 \\ 0 & 0 & 0 & 2 & 0 & 3 & 1 & 0 \\ 0 & 0 & 0 & 3 & 0 & 3 & 0 & 1 \end{array} \right),$$

which is not in standard form. From this last matrix, we can get the parameters $\kappa_1 = 1$ and $\delta_2 = 2$, so $\kappa_2 = 1$ and $\delta_1 = 1$.   △

Let $\mathcal{C}$ be a $\mathbb{Z}_2\mathbb{Z}_4$-additive code of type $(\alpha, \beta; \gamma, \delta; \kappa)$ and $\mathcal{G}$ a generator matrix of $\mathcal{C}$ of the form

$$\mathcal{G} = \left(\begin{array}{c|c} B_1 & 2B_3 \\ B_2 & Q \end{array} \right),$$

where B_1, B_2 are matrices over $\mathbb{Z}_2$ of size $\gamma \times \alpha$ and $\delta \times \alpha$, respectively; B_3 is a matrix over $\mathbb{Z}_4$ of size $\gamma \times \beta$ with all entries in $\{0,1\} \subset \mathbb{Z}_4$; and Q is a matrix over $\mathbb{Z}_4$ of size $\delta \times \beta$ with row vectors of order four. Note that $\mathcal{G}$ defines in fact a one-to-one morphism from $\mathbb{Z}_2^\gamma \times \mathbb{Z}_4^\delta$ to $\mathbb{Z}_2^\alpha \times \mathbb{Z}_4^\beta$. For a given $\mathbf{u} = (u \mid u') \in \mathbb{Z}_2^\gamma \times \mathbb{Z}_4^\delta$, the image under this morphism is denoted by $\mathbf{u}\mathcal{G}$, is called the product of the vector $\mathbf{u}$ by the matrix $\mathcal{G}$, and can be obtained as follows:

$$\mathbf{u}\mathcal{G} = \chi^{-1}(\iota(\mathbf{u})\chi(\mathcal{G})) \in \mathbb{Z}_2^\alpha \times \mathbb{Z}_4^\beta, \tag{2.12}$$

where χ and ι are defined on page 18.

2.3 Residue and Torsion Codes

In this section, the binary linear residue and torsion codes of a $\mathbb{Z}_2\mathbb{Z}_4$-additive code are defined.

Let $\mathcal{C}$ be a $\mathbb{Z}_2\mathbb{Z}_4$-additive code of type $(\alpha, \beta; \gamma, \delta; \kappa)$. We define the residue and torsion codes of $\mathcal{C}$, denoted respectively by $\mathrm{Res}(\mathcal{C})$ and $\mathrm{Tor}(\mathcal{C})$, as the following binary linear codes:

$$\mathrm{Res}(\mathcal{C}) = \{\xi(\mathbf{c}) : \mathbf{c} \in \mathcal{C}\},$$

$$\mathrm{Tor}(\mathcal{C}) = \{c \in \mathbb{Z}_2^{\alpha+\beta} : 2\iota(c) \in \mathcal{C}\},$$

where ξ and ι are defined on page 18. Consider now the map ξ projected onto the elements of the $\mathbb{Z}_2\mathbb{Z}_4$-additive code $\mathcal{C}$. Note that $\mathrm{Im}(\xi) = \mathrm{Res}(\mathcal{C})$ and it is easy to see that $\mathrm{Ker}(\xi) \cong \mathrm{Tor}(\mathcal{C})$. Therefore, by the first group isomorphism theorem, we have $\mathcal{C}/\mathrm{Tor}(\mathcal{C}) \cong \mathrm{Res}(\mathcal{C})$, so

$$|\mathcal{C}| = |\mathrm{Res}(\mathcal{C})||\mathrm{Tor}(\mathcal{C})|.$$

Lemma 2.17. *Let $\mathcal{C}$ be a $\mathbb{Z}_2\mathbb{Z}_4$-additive code with a generator matrix as in (2.7). Then, $\mathrm{Res}(\mathcal{C})$ and $\mathrm{Tor}(\mathcal{C})$ are binary linear $[\alpha + \beta, \kappa + \delta]$ and $[\alpha + \beta, \gamma - \kappa + \delta]$ codes, respectively, with generator matrices, respectively,*

$$G_{Res} = \begin{pmatrix} I_\kappa & T_b & \mathbf{0} & \mathbf{0} & \mathbf{0} \\ \mathbf{0} & S_b & \xi(S_q) & \xi(R) & \xi(I_\delta) \end{pmatrix}, \text{ and}$$

$$G_{Tor} = \begin{pmatrix} \mathbf{0} & \mathbf{0} & \xi(T_1) & \xi(I_{\gamma-\kappa}) & \mathbf{0} \\ \mathbf{0} & \mathbf{0} & \xi(S_q) & \xi(R) & \xi(I_\delta) \end{pmatrix}.$$

Note that, in general, $\mathrm{Tor}(\mathcal{C})_Y \neq \mathrm{Tor}(\mathcal{C}_Y)$, since there may be codewords in $\mathcal{C}$ of the form $\mathbf{u} = (u \mid 2u')$ with $u \neq \mathbf{0}$, having all non-zero coordinate in u' equal to 1 and such that $(\mathbf{0} \mid 2u') \notin \mathcal{C}$. Therefore, $u' \in \mathrm{Tor}(\mathcal{C}_Y)$ and $(u \mid u') \notin \mathrm{Tor}(\mathcal{C})$ that implies $u' \notin \mathrm{Tor}(\mathcal{C})_Y$.

Example 2.18. Let $\mathcal{C}$ be the $\mathbb{Z}_2\mathbb{Z}_4$-additive code generated by the matrix

$$\mathcal{G}_S = \begin{pmatrix} 1 & 0 & 2 & 0 & 0 \\ 0 & 1 & 0 & 2 & 0 \\ 1 & 1 & 0 & 0 & 1 \end{pmatrix}.$$

The binary linear code $\mathrm{Res}(\mathcal{C})$ is generated by

$$G_{Res} = \begin{pmatrix} 1 & 0 & 0 & 0 & 0 \\ 0 & 1 & 0 & 0 & 0 \\ 1 & 1 & 0 & 0 & 1 \end{pmatrix}.$$

In this case, we have that $\text{Tor}(\mathcal{C})_Y \neq \text{Tor}(\mathcal{C}_Y)$ because the binary linear code $\text{Tor}(\mathcal{C})$ is $\{(0,0,0,0,0), (0,0,0,0,1)\}$, and hence $\text{Tor}(\mathcal{C})_Y = \{(0,0,0),(0,0,1)\}$, whereas $\text{Tor}(\mathcal{C}_Y)$ is the linear code generated by the matrix

$$\begin{pmatrix} 1 & 0 & 0 \\ 0 & 1 & 0 \\ 0 & 0 & 1 \end{pmatrix}.$$

$\triangle$

2.4 Some Basic Families of $\mathbb{Z}_2\mathbb{Z}_4$-Linear Codes

In this section, we show that several basic and well-known families of binary linear codes can be viewed as $\mathbb{Z}_2\mathbb{Z}_4$-linear codes. Given a binary code C of length $n > 0$, in anyone of these families, C is always the binary image of a $\mathbb{Z}_2\mathbb{Z}_4$-additive code $\mathcal{C} \subseteq \mathbb{Z}_2^{\alpha} \times \mathbb{Z}_4^{\beta}$, for all non-negative values α and β such that $\alpha + 2\beta = n$.

i) *The zero code*

Let $C = \{\mathbf{0}\}$ be the binary linear $[n,0]$ code. Let α and β be non-negative integers such that $\alpha + 2\beta = n$. We have that $C = \Phi(\mathcal{C})$ is a $\mathbb{Z}_2\mathbb{Z}_4$-linear code of type $(\alpha, \beta; 0, 0; 0)$, where $\mathcal{C} = \{(\mathbf{0} \mid \mathbf{0})\}$.

ii) *The universe code*

Let $C = \mathbb{Z}_2^n$ be the binary linear $[n,n]$ code. A generator matrix for C is $G = I_n$. Let α and β be non-negative integers such that $\alpha + 2\beta = n$. Then, we have that $C = \Phi(\mathcal{C})$ is a $\mathbb{Z}_2\mathbb{Z}_4$-linear code of type $(\alpha, \beta; \alpha, \beta; \alpha)$, where $\mathcal{C}$ is generated by the matrix

$$\mathcal{G} = \left(\begin{array}{c|c} I_\alpha & \mathbf{0} \\ \mathbf{0} & I_\beta \end{array} \right).$$

iii) *The repetition code*

Let $C = \{\mathbf{0}, \mathbf{1}\}$ be a binary linear $[n,1]$ code. Let α and β be non-negative integers such that $\alpha + 2\beta = n$. We have that $C = \Phi(\mathcal{C})$ is a $\mathbb{Z}_2\mathbb{Z}_4$-linear code of type $(\alpha, \beta; 1, 0; 1)$, where $\mathcal{C} = \{(\mathbf{0} \mid \mathbf{0}), (\mathbf{1} \mid \mathbf{2})\}$.

iv) *The even code*

Let C be the binary linear $[n, n-1]$ code containing all the vectors in $\mathbb{Z}_2^n$ of even weight. Let α and β be non-negative integers such that

$\alpha + 2\beta = n$. Then, we have that $C = \Phi(\mathcal{C})$ is a $\mathbb{Z}_2\mathbb{Z}_4$-linear code of type $(\alpha, \beta; \alpha - 1, \beta; \alpha - 1)$, where $\mathcal{C}$ is generated by the matrix

$$\mathcal{G} = \left(\begin{array}{c|cc} I_\alpha & 0 & 1 \\ 0 & I_{\beta-1} & 1 \\ 0 & 0 & 2 \end{array} \right),$$

if $\beta > 0$, and $\mathcal{G} = \left(\begin{array}{cc} I_{\alpha-1} & 1 \end{array} \right)$, if $\beta = 0$ and $\alpha > 1$.

These basic $\mathbb{Z}_2\mathbb{Z}_4$-additive codes can be generated by using the MAGMA package for $\mathbb{Z}_2\mathbb{Z}_4$-additive codes [29], as we show in the following example.

Example 2.19. For $\alpha = 2$ and $\beta = 3$, it is checked that the zero $\mathbb{Z}_2\mathbb{Z}_4$-additive code of type $(\alpha, \beta; 0, 0; 0)$ is contained in a random $\mathbb{Z}_2\mathbb{Z}_4$-additive code of type $(\alpha, \beta; \gamma, \delta; \kappa)$, and similarly a random $\mathbb{Z}_2\mathbb{Z}_4$-additive code is contained in the universe $\mathbb{Z}_2\mathbb{Z}_4$-additive code of type $(\alpha, \beta; \alpha, \beta; \alpha)$. Finally, it is also checked that the image under the Gray map of the even $\mathbb{Z}_2\mathbb{Z}_4$-additive code is the $[\alpha + 2\beta, \alpha + 2\beta - 1]$ binary even code.

```
> alpha := 2; beta := 3;
> U := Z2Z4AdditiveUniverseCode(alpha, beta);
> Z := Z2Z4AdditiveZeroCode(alpha, beta);
> R := RandomZ2Z4AdditiveCode(alpha, beta);
> E := Z2Z4AdditiveEvenWeightCode(alpha, beta);
> (Z subset R) and (R subset U);
true
> (Z subset E) and (E subset U);
true
> _, binaryE := HasLinearGrayMapImage(E);
> binaryE eq EvenWeightCode(alpha + 2*beta);
true                                                        △
```

Chapter 3

Duality of $\mathbb{Z}_2\mathbb{Z}_4$-Additive Codes

In this chapter, first of all, a standard inner product for $\mathbb{Z}_2\mathbb{Z}_4$-additive codes is defined in Section 3.1. From this inner product, the additive dual code is defined in the usual way. Finally, it is also shown that the corresponding $\mathbb{Z}_2\mathbb{Z}_4$-linear codes of a $\mathbb{Z}_2\mathbb{Z}_4$-additive code and its additive dual code are not dual as binary codes in general, but their weight enumerator polynomials always satisfy the MacWilliams transform. In Section 3.2, the type of the additive dual code of a $\mathbb{Z}_2\mathbb{Z}_4$-additive code is computed. In addition, a parity check matrix of a $\mathbb{Z}_2\mathbb{Z}_4$-additive code from a generator matrix is provided.

3.1 Additive Dual Codes

For linear codes over finite fields or finite rings there exists the well-known concept of duality. In this section, we study this concept for $\mathbb{Z}_2\mathbb{Z}_4$-additive codes. We show that the images under the Gray map of a $\mathbb{Z}_2\mathbb{Z}_4$-additive code and its additive dual code are not necessarily binary dual codes. However, their weight enumerator polynomials satisfy the MacWilliams transform.

The *Fundamental Theorem of Finite Abelian Groups* says that any finite abelian group G is isomorphic to a direct product of cyclic groups, each one of order a prime power. Say $G = \langle a_1 \rangle \times \cdots \times \langle a_k \rangle$, where a_i is of order a prime power $p_i^{\alpha_i}$, for any $i \in \{1, \ldots, k\}$, where the p_i's are not necessarily distinct. The set of elements $a_i \in G$ is called a basis of G and fully determines the algebraic structure of G.

There are parts of this chapter that have been previously published. Reprinted by permission from Springer Nature Customer Service Centre GmbH: J. Borges, C. Fernández-Córdoba, J. Pujol, J. Rifà, and M. Villanueva. "$\mathbb{Z}_2\mathbb{Z}_4$-linear codes: generator matrices and duality". *Designs, Codes and Cryptography*, v. 54, 2, pp. 167–179. ©(2010).

Given a basis of G, every element $u \in G$ can be represented by a k-tuple of integers, $u = (u_1, u_2, \ldots, u_k)$ with $u = \sum_{i=1}^{k} u_i a_i$. This expression is unique in the sense that any other expression $u = \sum_{i=1}^{k} u_i' a_i$ means that, for all indices $i \in \{1, \ldots, k\}$, $u_i \equiv u_i' \pmod{p_i^{\alpha_i}}$. Note that the exponent m of the group G can be computed as $m = \operatorname{lcm}\{p_i^{\alpha_i} : i = 1, \ldots, k\}$ and is divisible by any $p_i^{\alpha_i}$. Therefore, $m = s_i p_i^{\alpha_i}$ for a positive integer s_i, which is of order $p_i^{\alpha_i}$ in $\mathbb{Z}_m$. Given a basis of G and fixed elements $\bar{s}_i$ of order $p_i^{\alpha_i}$ in $\mathbb{Z}_m$ (e.g., the above s_i), the inner product of elements $u = (u_1, u_2, \ldots, u_k), v = (v_1, v_2, \ldots, v_k) \in G$ is uniquely defined as the equivalence class of $\sum_{i=1}^{k} \bar{s}_i u_i v_i$ in $\mathbb{Z}_m$ and denoted by $(u \cdot v)_m$. That is,

$$(u \cdot v)_m = \sum_{i=1}^{k} \bar{s}_i u_i v_i \in \mathbb{Z}_m. \tag{3.1}$$

Now, consider the finite Abelian group $\mathbb{Z}_2^{\alpha} \times \mathbb{Z}_4^{\beta}$, whose elements are vectors of $\alpha + \beta$ coordinates (the first α ones in $\mathbb{Z}_2$ and the last β ones in $\mathbb{Z}_4$). We fix a generator for each component $\mathbb{Z}_2$ and $\mathbb{Z}_4$. Since $m = 4$, we have that $4 = \bar{s}_i \cdot 2 = \bar{s}_j \cdot 2^2$, for $1 \leq i \leq \alpha$ and $\alpha + 1 \leq j \leq \alpha + \beta$. Then, $\bar{s}_i = 2$ for $1 \leq i \leq \alpha$, since 2 is the only element of order two in $\mathbb{Z}_4$; and $\bar{s}_j \in \{1, 3\}$ for $\alpha + 1 \leq j \leq \alpha + \beta$. We consider $\bar{s}_j = 1$ for $\alpha + 1 \leq j \leq \alpha + \beta$, so that when $\alpha = 0$ the inner product coincides with the usual inner product defined for vectors in $\mathbb{Z}_4^{\beta}$. Then, we can write the inner product given by (3.1) in the following way:

$$\mathbf{u} \cdot \mathbf{v} = 2\left(\sum_{i=1}^{\alpha} u_i v_i\right) + \sum_{j=1}^{\beta} u_j' v_j' = 2(u \cdot v)_2 + (u' \cdot v')_4 \in \mathbb{Z}_4, \tag{3.2}$$

where $\mathbf{u} = (u \mid u') = (u_1, \ldots, u_{\alpha} \mid u_1', \ldots, u_{\beta}')$, $\mathbf{v} = (v \mid v') = (v_1, \ldots, v_{\alpha} \mid v_1', \ldots, v_{\beta}') \in \mathbb{Z}_2^{\alpha} \times \mathbb{Z}_4^{\beta}$. Note that the computations are made considering the zeros and ones in the α binary coordinates as quaternary zeros and ones, respectively. We refer to this product as the *standard inner product*, that can also be written as

$$\mathbf{u} \cdot \mathbf{v} = \mathbf{u} \left(\begin{array}{c|c} 2I_{\alpha} & \mathbf{0} \\ \hline \mathbf{0} & I_{\beta} \end{array} \right) \mathbf{v}^t.$$

Note that when $\alpha = 0$ the inner product is the usual one for quaternary vectors, and when $\beta = 0$ it is twice the usual one for binary vectors.

Let $\mathcal{C}$ be a $\mathbb{Z}_2\mathbb{Z}_4$-additive code of type $(\alpha, \beta; \gamma, \delta; \kappa)$ and let $C = \Phi(\mathcal{C})$ be the corresponding $\mathbb{Z}_2\mathbb{Z}_4$-linear code. The *additive orthogonal code* of $\mathcal{C}$, denoted by $\mathcal{C}^{\perp}$, is defined in the standard way as

$$\mathcal{C}^{\perp} = \{\mathbf{v} \in \mathbb{Z}_2^{\alpha} \times \mathbb{Z}_4^{\beta} : \mathbf{u} \cdot \mathbf{v} = 0 \text{ for all } \mathbf{u} \in \mathcal{C}\}.$$

We also call $C^\perp$ the *additive dual code* of $\mathcal{C}$. The corresponding binary code $\Phi(\mathcal{C}^\perp)$ is denoted by $C_\perp$ and called $\mathbb{Z}_2\mathbb{Z}_4$-*dual code* of $\mathcal{C}$. In the case that $\alpha = 0$, that is, when $\mathcal{C}$ is a quaternary linear code, $\mathcal{C}^\perp$ is also called the *quaternary dual code* of $\mathcal{C}$ and $C_\perp$ the $\mathbb{Z}_4$-*dual code* of $\mathcal{C}$. The additive dual code $\mathcal{C}^\perp$ is also a $\mathbb{Z}_2\mathbb{Z}_4$-additive code, that is, a subgroup of $\mathbb{Z}_2^\alpha \times \mathbb{Z}_4^\beta$.

The codes C and $C_\perp$ are not necessarily linear. Moreover, in general, they are not dual codes in the binary linear sense. We have the following diagram:

$$
\begin{array}{ccc}
\mathcal{C} & \xrightarrow{\;\Phi\;} & C = \Phi(\mathcal{C}) \\
{\scriptstyle \perp}\downarrow & & \\
\mathcal{C}^\perp & \xrightarrow{\;\Phi\;} & C_\perp = \Phi(\mathcal{C}^\perp).
\end{array}
\tag{3.3}
$$

Note that, in general, $C_\perp = \Phi(\mathcal{C}^\perp)$ is not the dual of $C = \Phi(\mathcal{C})$, so we cannot add an arrow on the right side to produce a commutative diagram. If we have a $\mathbb{Z}_2\mathbb{Z}_4$-additive code $\mathcal{C}$ such that the image under the Gray map is not linear, then clearly $\Phi(\mathcal{C})^\perp \neq \Phi(\mathcal{C}^\perp)$. Now, we see three examples where we consider $\mathbb{Z}_2\mathbb{Z}_4$-additive codes $\mathcal{C}$, with $\Phi(\mathcal{C})$ linear, having different duality relationship between $\Phi(\mathcal{C})$ and $\Phi(\mathcal{C}^\perp)$. In the first one, $\Phi(\mathcal{C})$ and $\Phi(\mathcal{C}^\perp)$ are both linear, and $\Phi(\mathcal{C})^\perp = \Phi(\mathcal{C}^\perp)$. In the second one, even though $\Phi(\mathcal{C})$ and $\Phi(\mathcal{C}^\perp)$ are both linear, we have that $\Phi(\mathcal{C})^\perp \neq \Phi(\mathcal{C}^\perp)$. Finally, in the third one, $\Phi(\mathcal{C}^\perp)$ is not linear, and therefore, $\Phi(\mathcal{C})^\perp \neq \Phi(\mathcal{C}^\perp)$.

Example 3.1. Let $\mathcal{C}$ be the $\mathbb{Z}_2\mathbb{Z}_4$-additive code of type $(2, 2; 2, 1; 1)$ given in Example 2.2 with generator matrix $\mathcal{G}_\mathcal{C}$ given in Example 2.15. The additive dual code of $\mathcal{C}$ is $\mathcal{C}^\perp = \{(0, 0 \mid 0, 0), (0, 0 \mid 2, 0), (1, 1 \mid 0, 0), (1, 1 \mid 2, 0)\}$. In this case, the corresponding $\mathbb{Z}_2\mathbb{Z}_4$-linear codes $C = \Phi(\mathcal{C})$ and $C_\perp = \Phi(\mathcal{C}^\perp)$ are linear. Actually, they are generated by the matrices

$$
G = \begin{pmatrix} 1 & 1 & 0 & 0 & 0 & 0 \\ 0 & 0 & 1 & 1 & 0 & 0 \\ 0 & 0 & 0 & 0 & 1 & 0 \\ 0 & 0 & 0 & 0 & 0 & 1 \end{pmatrix} \quad \text{and} \quad H = \begin{pmatrix} 1 & 1 & 0 & 0 & 0 & 0 \\ 0 & 0 & 1 & 1 & 0 & 0 \end{pmatrix},
$$

respectively. Moreover, it is easy to see that $C_\perp$ is the dual code of C in the binary linear sense, that is, $C^\perp = C_\perp$. $\triangle$

Example 3.2. Let $\mathcal{C}$ be the $\mathbb{Z}_2\mathbb{Z}_4$-additive code generated by

$$
\mathcal{G} = \begin{pmatrix} 1 & 0 & \vline & 0 & 0 & 0 \\ 0 & 1 & \vline & 0 & 0 & 0 \\ 0 & 0 & \vline & 0 & 2 & 0 \\ 0 & 0 & \vline & 1 & 0 & 1 \end{pmatrix}.
$$

The additive dual code of $\mathcal{C}$ is

$$\mathcal{C}^\perp = \{(0,0 \mid 0,0,0),(0,0 \mid 1,0,3),(0,0 \mid 2,0,2),(0,0 \mid 3,0,1),$$
$$(0,0 \mid 0,2,0),(0,0 \mid 1,2,3),(0,0 \mid 2,2,2),(0,0 \mid 3,2,1)\}.$$

Both codes, $C = \Phi(\mathcal{C})$ and $C_\perp = \Phi(\mathcal{C}^\perp)$, are linear and have generator matrices

$$G = \begin{pmatrix} 1 & 0 & 0 & 0 & 0 & 0 & 0 & 0 \\ 0 & 1 & 0 & 0 & 0 & 0 & 0 & 0 \\ 0 & 0 & 1 & 0 & 0 & 0 & 1 & 0 \\ 0 & 0 & 0 & 1 & 0 & 0 & 0 & 1 \\ 0 & 0 & 0 & 0 & 1 & 1 & 0 & 0 \end{pmatrix} \quad \text{and} \quad H = \begin{pmatrix} 0 & 0 & 1 & 0 & 0 & 0 & 0 & 1 \\ 0 & 0 & 0 & 1 & 0 & 0 & 1 & 0 \\ 0 & 0 & 0 & 0 & 1 & 1 & 0 & 0 \end{pmatrix},$$

respectively. Note that the inner product of $(0,0,1,0,0,0,1,0) \in C$ and $(0,0,1,0,0,0,0,1) \in C_\perp$ is not zero, so C and $C_\perp$ are not dual codes, that is, $C^\perp \neq C_\perp$.   △

Example 3.3. Let $\mathcal{C}$ be the $\mathbb{Z}_2\mathbb{Z}_4$-additive code generated by

$$\mathcal{G} = \begin{pmatrix} 1 & 1 & 2 & 0 & 0 \\ 0 & 1 & 1 & 1 & 1 \end{pmatrix},$$

It is easy to check that $\Phi(\mathcal{C})$ is linear. Its additive dual code $\mathcal{C}^\perp$ is generated by the matrix

$$\mathcal{H} = \begin{pmatrix} 1 & 1 & 2 & 0 & 0 \\ 0 & 1 & 1 & 1 & 0 \\ 0 & 1 & 1 & 0 & 1 \end{pmatrix}.$$

We check that $C_\perp = \Phi(\mathcal{C}^\perp)$ is not linear. Note that $\Phi(0,1 \mid 1,1,0), \Phi(0,1 \mid 1,0,1) \in C_\perp$. Since $\Phi^{-1}(\Phi(0,1 \mid 1,1,0) + \Phi(0,1 \mid 1,0,1)) = (0,0 \mid 0,1,1) \notin \mathcal{C}^\perp$, $\Phi(0,1 \mid 1,1,0) + \Phi(0,1 \mid 1,0,1) \notin C_\perp$ and $C_\perp$ is not linear. In this case, since $C_\perp$ is not linear, $C^\perp \neq C_\perp$.   △

Example 3.4. For the code $\mathcal{C}$ given in Example 3.2, by using MAGMA, we check that the images of $\mathcal{C}$ and $\mathcal{C}^\perp$ under the Gray map are linear and that $C^\perp = \Phi(\mathcal{C})^\perp \neq \Phi(\mathcal{C}^\perp) = C_\perp$.

```
> Z4 := IntegerRing(4);
> V := RSpace(Z4, 5);
> C := Z2Z4AdditiveCode([V![1,0,0,0,0], V![0,1,0,0,0],
                         V![0,0,1,0,1], V![0,0,0,2,0]], 2);
> isLinear, Cbin := HasLinearGrayMapImage(C);
> isLinearDual, Cdualbin := HasLinearGrayMapImage(Dual(C));
> isLinear and isLinearDual;
true
> Dual(Cbin) eq Cdualbin;
false
```
                                                                        △

One could think on $\mathbb{Z}_2\mathbb{Z}_4$-additive codes just as quaternary linear codes, replacing the ones with twos in the binary coordinates. However, in this case, the corresponding quaternary linear codes of a $\mathbb{Z}_2\mathbb{Z}_4$-additive code $\mathcal{C}$ and its additive dual code $\mathcal{C}^\perp$ are not necessarily quaternary dual codes. Take, for example, $\alpha = \beta = 1$ and the vectors $\mathbf{v} = (1 \mid 3)$ and $\mathbf{w} = (1 \mid 2)$. It is easy to check that $\mathbf{v} \cdot \mathbf{w} = 0$, so $\mathbf{v}$ and $\mathbf{w}$ are orthogonal. However, if we replace the ones with twos in the binary coordinates of these vectors, we obtain the vectors $(2,3)$ and $(2,2)$, which are not orthogonal in the quaternary sense.

Let $\mathcal{C}$ be a $\mathbb{Z}_2\mathbb{Z}_4$-additive code of type $(\alpha, \beta; \gamma, \delta; \kappa)$. The *weight enumerator polynomial* of $\mathcal{C}$ is

$$W_{\mathcal{C}}(x,y) = \sum_{\mathbf{c} \in \mathcal{C}} x^{n - \mathrm{wt}(\mathbf{c})} y^{\mathrm{wt}(\mathbf{c})}, \tag{3.4}$$

where $n = \alpha + 2\beta$. Note that the weight enumerator polynomial of a $\mathbb{Z}_2\mathbb{Z}_4$-additive code $\mathcal{C}$, $W_{\mathcal{C}}(x,y)$, refers to the weight defined in (2.1), and the weight enumerator polynomial of the $\mathbb{Z}_2\mathbb{Z}_4$-linear code $C = \Phi(\mathcal{C})$, $W_C(x,y)$, refers to the Hamming weight. Then, $W_{\mathcal{C}}(x,y)$ coincides with $W_C(x,y)$.

The following lemma is easily proven.

Lemma 3.5. *Let $\mathcal{C}$ be a separable $\mathbb{Z}_2\mathbb{Z}_4$-additive code. Let $W_{\mathcal{C}_X}(x,y)$ and $W_{\mathcal{C}_Y}(x,y)$ be the weight enumerator of $\mathcal{C}_X$ and $\mathcal{C}_Y$ considering the Hamming and Lee weight, respectively. Then, $W_{\mathcal{C}}(x,y) = W_{\mathcal{C}_X}(x,y) W_{\mathcal{C}_Y}(x,y)$.*

In [134], it is first stated that $\mathbb{Z}_2\mathbb{Z}_4$-additive codes satisfy the MacWilliams identity.

Proposition 3.6 ([58, 134]). *Let $\mathcal{C}$ be a $\mathbb{Z}_2\mathbb{Z}_4$-additive code. If $W_{\mathcal{C}}(x,y)$ (respectively, $W_{\mathcal{C}^\perp}(x,y)$) is the weight enumerator polynomial of $\mathcal{C}$ (respectively, $\mathcal{C}^\perp$), then*

$$W_{\mathcal{C}^\perp}(x,y) = \frac{1}{|\mathcal{C}|} W_{\mathcal{C}}(x + y, x - y). \tag{3.5}$$

In addition, taking $x = y$, we obtain the following result.

Corollary 3.7 ([37]). *Let $\mathcal{C}$ be a $\mathbb{Z}_2\mathbb{Z}_4$-additive code of type $(\alpha, \beta; \gamma, \delta; \kappa)$ and $\mathcal{C}^\perp$ its additive dual code. Then, $|\mathcal{C}||\mathcal{C}^\perp| = 2^n$, where $n = \alpha + 2\beta$.*

As we have said, in general $C = \Phi(\mathcal{C})$ and $C_\perp = \Phi(\mathcal{C}^\perp)$ are not dual codes in the binary linear sense. However, the weight enumerator polynomial of $C_\perp$ is the MacWilliams transform of the weight enumerator polynomial of C. This remarkable relationship was first established for the specific case of $\mathbb{Z}_4$-linear codes in [92, 157], where it is pointed out that the Kerdock code is the $\mathbb{Z}_4$-dual of some Preparata-like code.

Example 3.8. Let $\mathcal{C}$ be the $\mathbb{Z}_2\mathbb{Z}_4$-additive code given in Example 3.3. The types of $\mathcal{C}$ and $\mathcal{C}^\perp$ are $(2,3;1,2;1)$ and $(2,3;1,1;1)$, respectively. We have that $|\mathcal{C}| = 2 \cdot 4^2 = 2^5$ and $|\mathcal{C}^\perp| = 2 \cdot 4 = 2^3$. Note that $|\mathcal{C}||\mathcal{C}^\perp| = 2^5 \cdot 2^3 = 2^8 = 2^n$.

From the generator matrix of $\mathcal{C}$, we have that

$$
\begin{aligned}
\mathcal{C} = \{ & (0,0 \mid 0,0,0), (0,1 \mid 1,0,1), (0,0 \mid 2,0,2), (0,1 \mid 3,0,3), \\
& (0,1 \mid 1,1,0), (0,0 \mid 2,1,1), (0,1 \mid 3,1,2), (0,0 \mid 0,1,3), \\
& (1,1 \mid 2,0,0), (1,0 \mid 3,0,1), (1,1 \mid 0,0,2), (1,0 \mid 1,0,3), \\
& (1,0 \mid 3,1,0), (1,0 \mid 1,2,1), (1,1 \mid 2,2,2), (1,0 \mid 3,2,3), \\
& (0,0 \mid 2,3,3), (0,1 \mid 3,3,0), (0,0 \mid 0,3,1), (0,1 \mid 1,3,2), \\
& (1,0 \mid 3,3,2), (1,1 \mid 0,3,3), (1,0 \mid 1,3,0), (1,1 \mid 2,3,1), \\
& (0,1 \mid 3,2,1), (0,0 \mid 0,2,2), (0,1 \mid 1,2,3), (0,0 \mid 2,2,0), \\
& (1,1 \mid 0,1,1), (1,0 \mid 1,1,2), (1,1 \mid 2,1,3), (1,1 \mid 0,2,0) \}.
\end{aligned}
$$

From this set of codewords, it is easy to see that $W_\mathcal{C}(x,y) = x^8 + 2x^6y^2 + 8x^5y^3 + 10x^4y^4 + 8x^3y^5 + 2x^2y^6 + y^8$. The codewords of the additive dual code of $\mathcal{C}$ are

$$
\begin{aligned}
\mathcal{C}^\perp = \{ & (0,0 \mid 0,0,0), (0,1 \mid 1,1,1), (0,0 \mid 2,2,2), (0,0 \mid 3,3,3), \\
& (1,1 \mid 2,0,0), (1,0 \mid 3,1,1), (1,0 \mid 0,2,2), (1,0 \mid 1,3,3) \}.
\end{aligned}
$$

Therefore, the weight enumerator polynomial of $\mathcal{C}^\perp$ is $W_{\mathcal{C}^\perp}(x,y) = x^8 + 5x^4y^4 + 2x^2y^6$. Note that it can be computed as $W_{\mathcal{C}^\perp}(x,y) = \frac{1}{2^5}W_\mathcal{C}(x+y, x-y) = x^8 + 5x^4y^4 + 2x^2y^6$. $\triangle$

Example 3.9. By using the MAGMA package for $\mathbb{Z}_2\mathbb{Z}_4$-additive codes [29], we compute the weight enumerator of $\mathcal{C}$ and $\mathcal{C}^\perp$ given in Example 3.8. They coincide with the weight enumerator of $C = \Phi(\mathcal{C})$ and $C_\perp = \Phi(\mathcal{C}^\perp)$, respectively.

```
> Z4 := Integers(4);
> G := Matrix(Z4, [[1,0,1,0,3],
                    [0,1,1,0,1],
                    [0,0,2,0,2],
                    [0,0,0,1,3]]);
> C := Z2Z4AdditiveCode(G, 2);
> WC<x,y> := LeeWeightEnumerator(C); WC;
x^8 + 2*x^6*y^2 + 8*x^5*y^3 + 10*x^4*y^4 + 8*x^3*y^5 + 2*x^2*y^6 + y^8
> WCdual<x,y> := LeeWeightEnumerator(Dual(C)); WCdual;
x^8 + 5*x^4*y^4 + 2*x^2*y^6
> DualLeeWeightDistribution(C) eq LeeWeightDistribution(Dual(C));
true
```
   $\triangle$

3.2 Parity Check Matrices

In this section, we show two methods to construct a parity check matrix of a $\mathbb{Z}_2\mathbb{Z}_4$-additive code, which is a generator matrix of its additive dual code. First, we define some maps and give statements that relate these maps with the inner product and duality. We also establish the type of an additive dual code from the type of its corresponding $\mathbb{Z}_2\mathbb{Z}_4$-additive code.

Let $\mathcal{C}$ be a $\mathbb{Z}_2\mathbb{Z}_4$-additive code of type $(\alpha, \beta; \gamma, \delta; \kappa)$. Since $\mathcal{C}$ is a subgroup of $\mathbb{Z}_2^\alpha \times \mathbb{Z}_4^\beta$, the code $\mathcal{C}$ could be seen as the kernel of a group homomorphism onto $\mathbb{Z}_2^{\bar{\gamma}} \times \mathbb{Z}_4^{\bar{\delta}}$, that is, $\mathcal{C} = \mathrm{Ker}(\vartheta)$, where

$$\vartheta : \ \mathbb{Z}_2^\alpha \times \mathbb{Z}_4^\beta \ \longrightarrow \ \mathbb{Z}_2^{\bar{\gamma}} \times \mathbb{Z}_4^{\bar{\delta}}.$$

The additive dual code $\mathcal{C}^\perp$ is also the kernel of another group homomorphism onto $\mathbb{Z}_2^\gamma \times \mathbb{Z}_4^\delta$, that is, $\mathcal{C}^\perp = \mathrm{Ker}(\bar{\vartheta})$, where

$$\bar{\vartheta} : \ \mathbb{Z}_2^\alpha \times \mathbb{Z}_4^\beta \ \longrightarrow \ \mathbb{Z}_2^\gamma \times \mathbb{Z}_4^\delta.$$

The homomorphism ϑ can be represented by a matrix $\mathcal{H}$, which can be seen as a parity check matrix for the $\mathbb{Z}_2\mathbb{Z}_4$-additive code $\mathcal{C}$ or as a generator matrix for its additive dual code $\mathcal{C}^\perp$. Vice versa, the homomorphism $\bar{\vartheta}$ can be represented by a matrix $\mathcal{G}$, which can be seen as a parity check matrix for the additive dual code $\mathcal{C}^\perp$ or as a generator matrix for the $\mathbb{Z}_2\mathbb{Z}_4$-additive code $\mathcal{C}$.

Example 3.10. We continue with the $\mathbb{Z}_2\mathbb{Z}_4$-additive code $\mathcal{C}$ given in Example 2.11, which has been generated by using MAGMA in Example 2.12. Now, we check that a parity check matrix of $\mathcal{C}$ generates the additive dual code $\mathcal{C}^\perp$, and a parity check matrix of $\mathcal{C}^\perp$ generates $\mathcal{C}$. Finally, the equality $|\mathcal{C}||\mathcal{C}^\perp| = 2^n$, where $n = \alpha + 2\beta$, is also checked in MAGMA.

```
> Z4 := Integers(4);
> G := Matrix(Z4, [[1,2,2,2],
                    [0,1,1,0],
                    [1,1,2,3]]);
> C := Z2Z4AdditiveCode(G, 1);
> typeC := Z2Z4Type(C);
> alpha := typeC[1];
> beta := typeC[2];
>
> Hc := MinRowsParityCheckMatrix(C);
> D := Dual(C);
> D eq  Z2Z4AdditiveCode(Hc, alpha);
true
> Hd := MinRowsParityCheckMatrix(D);
```

```
> C eq Z2Z4AdditiveCode(Hd, alpha);
true
> #C * #D eq 2^(alpha + 2*beta);
true
```
$\triangle$

Recall that ξ is the natural modulo two map from $\mathbb{Z}_4$ to $\mathbb{Z}_2$, ι is the identity map from $\mathbb{Z}_2$ to $\mathbb{Z}_4$, and χ is the natural inclusion from $\mathbb{Z}_2$ to $\mathbb{Z}_4$ $\chi(0) = 0$ and $\chi(1) = 2$, defined on page 18.

Lemma 3.11 ([37]). *If* $\mathbf{u} \in \mathbb{Z}_2^\alpha \times \mathbb{Z}_4^\beta$ *and* $\mathbf{v} \in \mathbb{Z}_4^{\alpha+\beta}$, *then* $(\chi(\mathbf{u}) \cdot \mathbf{v})_4 = \mathbf{u} \cdot \xi(\mathbf{v})$.

Proof. Let $\mathbf{u} = (u_1, \ldots, u_\alpha \mid u_1', \ldots, u_\beta'), \mathbf{v} = (v_1, \ldots, v_\alpha \mid v_1', \ldots, v_\beta') \in \mathbb{Z}_2^\alpha \times \mathbb{Z}_4^\beta$. Then, $(\chi(\mathbf{u}) \cdot \mathbf{v})_4 = \sum_{i=1}^\alpha (2u_i)v_i + \sum_{j=1}^\beta u_j' v_j' = \sum_{i=1}^\alpha (2u_i)(v_i \bmod 2) + \sum_{j=1}^\beta u_j' v_j' = \mathbf{u} \cdot \xi(\mathbf{v})$. $\square$

Corollary 3.12 ([37]). *If* $\mathbf{u}, \mathbf{v} \in \mathbb{Z}_2^\alpha \times \mathbb{Z}_4^\beta$, *then* $(\chi(\mathbf{u}) \cdot \iota(\mathbf{v}))_4 = \mathbf{u} \cdot \mathbf{v}$.

Proof. By Lemma 3.11, $(\chi(\mathbf{u}) \cdot \iota(\mathbf{v}))_4 = \mathbf{u} \cdot \xi(\iota(\mathbf{v})) = \mathbf{u} \cdot \mathbf{v}$. $\square$

Proposition 3.13 ([37]). *Let* $\mathcal{C}$ *be a* $\mathbb{Z}_2\mathbb{Z}_4$-*additive code of type* $(\alpha, \beta; \gamma, \delta; \kappa)$. *Then,*

$$\mathcal{C}^\perp = \xi(\chi(\mathcal{C})^\perp). \tag{3.6}$$

Proof. We know that if $\mathbf{v} \in \mathcal{C}^\perp$, then $\mathbf{u} \cdot \mathbf{v} = 0$ for all $\mathbf{u} \in \mathcal{C}$. By Corollary 3.12, $\mathbf{u} \cdot \mathbf{v} = (\chi(\mathbf{u}) \cdot \iota(\mathbf{v}))_4 = 0$. Therefore, $\xi(\iota(\mathbf{v})) = \mathbf{v} \in \xi(\chi(\mathcal{C})^\perp)$ and $\mathcal{C}^\perp \subseteq \xi(\chi(\mathcal{C})^\perp)$. On the other hand, if $\mathbf{v} \in \chi(\mathcal{C})^\perp$, then $(\chi(\mathbf{u}) \cdot \mathbf{v})_4 = 0$ for all $\mathbf{u} \in \mathcal{C}$. By Lemma 3.11, $(\chi(\mathbf{u}) \cdot \mathbf{v})_4 = \mathbf{u} \cdot \xi(\mathbf{v}) = 0$. Thus, $\xi(\chi(\mathcal{C})^\perp) \subseteq \mathcal{C}^\perp$ and we obtain the equality. $\square$

Proposition 3.14 ([37]). *Let* $\mathcal{C}$ *be a* $\mathbb{Z}_2\mathbb{Z}_4$-*additive code of type* $(\alpha, \beta; \gamma, \delta; \kappa)$. *Then,*

$$\mathcal{C}^\perp = \chi^{-1}(\xi^{-1}(\mathcal{C})^\perp). \tag{3.7}$$

Proof. Let $\mathcal{G}$ be a generator matrix of the $\mathbb{Z}_2\mathbb{Z}_4$-additive code $\mathcal{C}$ as in (2.5). Then, the quaternary linear code $\xi^{-1}(\mathcal{C})$ has a generator matrix of the form

$$\begin{pmatrix} 2I_\alpha & \mathbf{0} \\ B_1 & 2B_3 \\ B_2 & Q \end{pmatrix}. \tag{3.8}$$

We show that $\mathbf{v} \in \mathcal{C}^\perp$ if and only if $\chi(\mathbf{v}) \in \xi^{-1}(\mathcal{C})^\perp$. In fact, for each row vector w in the matrix $(2I_\alpha \ \mathbf{0})$, we have $(\chi(\mathbf{v}) \cdot w)_4 = \sum_{i=1}^\alpha w_i 2v_i = 0$ because there is only one index i such that $w_i = 2$. Moreover, by Corollary 3.12, $0 = \mathbf{u} \cdot \mathbf{v} = (\chi(\mathbf{v}) \cdot \iota(\mathbf{u}))_4$, for all $\mathbf{u} \in \mathcal{C}$. $\square$

Let $\mathcal{C}$ be a $\mathbb{Z}_2\mathbb{Z}_4$-additive code of type $(\alpha, \beta; \gamma, \delta; \kappa)$. Let $\mathcal{H}$ be a matrix representing the group homomorphism $\vartheta : \mathbb{Z}_2^\alpha \times \mathbb{Z}_4^\beta \longrightarrow \mathbb{Z}_2^{\bar\gamma} \times \mathbb{Z}_4^{\bar\delta}$ such that $\mathcal{C} = \mathrm{Ker}(\vartheta)$. Let $\mathbf{v} = (v \mid v') \in \mathbb{Z}_2^\alpha \times \mathbb{Z}_4^\beta$. We write $\mathbf{v}\mathcal{H}^T$ to refer to $\vartheta(\mathbf{v})$, which can be obtained as follows:

$$\mathbf{v}\mathcal{H}^T = \chi^{-1}(\iota(\mathbf{v})\chi(\mathcal{H})^T) \in \mathbb{Z}_2^{\bar\gamma} \times \mathbb{Z}_4^{\bar\delta}. \tag{3.9}$$

The next question we settle is how to compute the type of an additive dual code $\mathcal{C}^\perp$ from the type of the $\mathbb{Z}_2\mathbb{Z}_4$-additive code $\mathcal{C}$. The next proposition gives the result for the particular case of $\mathbb{Z}_2\mathbb{Z}_4$-additive codes with $\alpha = 0$, that is, for quaternary linear codes. Then, we generalize it for any $\mathbb{Z}_2\mathbb{Z}_4$-additive code.

Proposition 3.15 ([92]). *If $\mathcal{C}$ is a quaternary linear code of type $(0, \beta; \gamma, \delta; 0)$, then the quaternary dual code $\mathcal{C}^\perp$ is of type $(0, \beta; \gamma, \beta - \gamma - \delta; 0)$.*

Theorem 3.16 ([37]). *Let $\mathcal{C}$ be a $\mathbb{Z}_2\mathbb{Z}_4$-additive code of type $(\alpha, \beta; \gamma, \delta; \kappa)$. The additive dual code $\mathcal{C}^\perp$ is of type $(\alpha, \beta; \bar\gamma, \bar\delta; \bar\kappa)$, where*

$$
\begin{aligned}
\bar\gamma &= \alpha + \gamma - 2\kappa, \\
\bar\delta &= \beta - \gamma - \delta + \kappa, \\
\bar\kappa &= \alpha - \kappa.
\end{aligned}
$$

Proof. Let $\mathcal{G}$ be a generator matrix of the $\mathbb{Z}_2\mathbb{Z}_4$-additive code $\mathcal{C}$ as in (2.5). Then, the matrix (3.8) is a generator matrix for the quaternary linear code $\xi^{-1}(\mathcal{C})$, which is of type $(0, \alpha + \beta; \gamma', \delta'; 0)$, where $\gamma' = \alpha + \gamma - 2\kappa$ and $\delta' = \delta + \kappa$. The value of δ' comes from the fact that the κ independent binary vectors of $(\mathcal{C}_b)_X$ are in the submatrix B_1 in (2.5) and, so, the number of independent quaternary vectors of order four becomes $\delta + \kappa$. The value of γ' comes from the fact that the cardinality of the quaternary linear code $\xi^{-1}(\mathcal{C})$ is $2^{\gamma'+2\delta'} = 2^{\gamma+2\delta+\alpha}$.

By Proposition 3.15, the quaternary dual code $\xi^{-1}(\mathcal{C})^\perp$ is of type $(0, \alpha + \beta; \bar\gamma, \bar\delta; 0)$, where $\bar\gamma = \gamma'$ and $\bar\delta = \alpha + \beta - \gamma' - \delta' = \alpha + \beta - (\gamma + \alpha - 2\kappa) - (\delta + \kappa) = \beta - \gamma - \delta + \kappa$. Note that the $\bar\delta$ independent vectors in $\xi^{-1}(\mathcal{C})^\perp$, projected onto the first α coordinates, are vectors of order two, because in $\xi^{-1}(\mathcal{C})$ there are the row vectors of the matrix $(2I_\alpha \; \mathbf{0})$. Finally, applying χ^{-1} we obtain the additive dual code of $\mathcal{C}$. For this additive dual code $\mathcal{C}^\perp$, the value of $\bar\kappa$ can be easily computed from the fact that, again, the additive dual coincides with $\mathcal{C}$. $\qquad\square$

From Theorem 3.16, we obtain the following result about the number of codewords in $\mathcal{C}$, $\mathcal{C}_X$, $\mathcal{C}_Y$ and in their duals.

Proposition 3.17 ([39]). *Let $\mathcal{C}$ be a $\mathbb{Z}_2\mathbb{Z}_4$-additive code of type $(\alpha, \beta; \gamma, \delta; \kappa)$. Let κ_1 and δ_1 be the integers defined in (2.3). Then,*

$$|\mathcal{C}| = 2^\gamma 4^\delta, \quad |\mathcal{C}^\perp| = 2^{\alpha+\gamma-2\kappa} 4^{\beta-\gamma-\delta+\kappa},$$
$$|\mathcal{C}_X| = 2^{\kappa+\delta_1}, \quad |(\mathcal{C}_X)^\perp| = 2^{\alpha-\kappa-\delta_1},$$
$$|\mathcal{C}_Y| = 2^{\gamma-\kappa_1} 4^\delta, \quad |(\mathcal{C}_Y)^\perp| = 2^{\gamma-\kappa_1} 4^{\beta-\gamma-\delta+\kappa_1}.$$

When the $\mathbb{Z}_2\mathbb{Z}_4$-additive code $\mathcal{C}$ is separable, that is, $\mathcal{C} = \mathcal{C}_X \times \mathcal{C}_Y$, by Remark 2.4, we have that κ_2 and δ_1 are zero. Therefore, we obtain the following result.

Corollary 3.18. *Let $\mathcal{C}$ be a $\mathbb{Z}_2\mathbb{Z}_4$-additive code. Then, $\mathcal{C}$ is separable if and only if $\mathcal{C}^\perp$ is separable. Moreover, if $\mathcal{C}$ is separable, then $\mathcal{C}^\perp = (\mathcal{C}_X)^\perp \times (\mathcal{C}_Y)^\perp$.*

Example 3.19. Let $\mathcal{C}$ and $\mathcal{D}$ be the codes given in Example 2.7, with generator matrices

$$\mathcal{G}_\mathcal{C} = \begin{pmatrix} 1 & 1 & 0 & 0 \\ 0 & 0 & 2 & 0 \\ 0 & 0 & 0 & 1 \end{pmatrix} \quad \text{and} \quad \mathcal{G}_\mathcal{D} = \begin{pmatrix} 1 & 1 & 2 & 0 \\ 0 & 0 & 0 & 1 \end{pmatrix},$$

given in Example 2.15, and types $(2, 2; 2, 1; 1)$ and $(2, 2; 1, 1; 1)$, respectively. By Theorem 3.16, the type of $\mathcal{C}^\perp$ is $(2, 2; 2, 0; 1)$ and the type of $\mathcal{D}^\perp$ is $(2, 2; 1, 1; 1)$. Moreover, we have that their additive dual codes are

$$\mathcal{C}^\perp = \{(0, 0 \mid 0, 0), (0, 0 \mid 2, 0), (1, 1 \mid 0, 0), (1, 1 \mid 2, 0)\}, \text{ and}$$
$$\mathcal{D}^\perp = \{(0, 0 \mid 0, 0), (1, 0 \mid 1, 0), (0, 0 \mid 2, 0), (1, 0 \mid 3, 0),$$
$$(1, 1 \mid 0, 0), (0, 1 \mid 1, 0), (1, 1 \mid 2, 0), (0, 1 \mid 3, 0)\},$$

respectively. We have seen that $\mathcal{C}$ is separable in Example 2.7. Since $(\mathcal{C}^\perp)_X = \{(0, 0), (1, 1)\}$, $(\mathcal{C}^\perp)_Y = \{(0, 0), (2, 0)\}$ and $\mathcal{C}^\perp = (\mathcal{C}^\perp)_X \times (\mathcal{C}^\perp)_Y$, the code $\mathcal{C}^\perp$ is also separable. Moreover, it is easy to check that $(\mathcal{C}^\perp)_X = (\mathcal{C}_X)^\perp$ and $(\mathcal{C}^\perp)_Y = (\mathcal{C}_Y)^\perp$, so $\mathcal{C}^\perp = (\mathcal{C}_X)^\perp \times (\mathcal{C}_Y)^\perp$ as shown in Corollary 3.18. In Example 2.7, we have also seen that $\mathcal{D}$ is not separable. Note that $\mathcal{D}^\perp$ is not separable either.   △

Example 3.20. We continue with the $\mathbb{Z}_2\mathbb{Z}_4$-additive code $\mathcal{C}$ given in Example 3.19, which has been generated by using MAGMA in Example 2.8. Now, we check that the type of $\mathcal{C}^\perp$ can be computed from the type of $\mathcal{C}$. Moreover, since $\mathcal{C}$ is separable, we also see that $\mathcal{C}^\perp$ can be computed from the dual codes of $\mathcal{C}_X$ and $\mathcal{C}_Y$.

```
> Z4 := IntegerRing(4);
> V := RSpace(Z4, 4);
> C := Z2Z4AdditiveCode([V![0,0,0,1], V![0,0,2,0], V![1,1,0,0]], 2);
> typeC := Z2Z4Type(C);
> Cdual := Dual(C);
> typeCdual := Z2Z4Type(Cdual);
>
> typeCdual[1] eq typeC[1];
true
> typeCdual[2] eq typeC[2];
true
> typeCdual[3] eq typeC[1] + typeC[3] - 2*typeC[5];
true
> typeCdual[4] eq typeC[2] - typeC[3] - typeC[4] + typeC[5];
true
> typeCdual[5] eq typeC[1] - typeC[5];
>
> CX := LinearBinaryCode(C);
> CY := LinearQuaternaryCode(C);
> Cdualnew := DirectSum(Z2Z4AdditiveCode(Dual(CX)),
                        Z2Z4AdditiveCode(Dual(CY)));
> Cdual eq Cdualnew;
true                                                              △
```

There are two different methods to obtain the additive dual code $C^\perp$, one given by Proposition 3.13 and another one by Proposition 3.14. In both cases, we construct a generator matrix of $C^\perp$ starting from a generator matrix of C. However, the process to obtain the additive dual code, or equivalently the parity check matrix, is different as we see in Example 3.21 for a generator matrix in standard form.

Example 3.21. Let C_S denote the $\mathbb{Z}_2\mathbb{Z}_4$-additive code of type $(1, 3; 1, 2; 1)$ with generator matrix in standard form

$$
\mathcal{G}_S = \left(\begin{array}{c|ccc} 1 & 2 & 0 & 0 \\ 0 & 3 & 1 & 0 \\ 0 & 1 & 0 & 1 \end{array} \right),
$$

obtained in Example 2.11. By Theorem 3.16, the additive dual code $C_S^\perp$ is of type $(1, 3; 0, 1; 0)$.

The first method uses Proposition 3.13. We know that if C' is a quaternary linear code with generator matrix $\mathcal{G}' = \left(\begin{array}{cccc} 2 & 2 & 0 & 0 \\ 3 & 0 & 1 & 0 \\ 1 & 0 & 0 & 1 \end{array} \right)$, the quaternary dual code $(C')^\perp$ has generator matrix $\mathcal{H}' = \left(\begin{array}{cccc} 0 & 2 & 0 & 0 \\ 1 & 1 & 1 & 3 \end{array} \right)$. Therefore, the generator matrix of $\chi(\mathcal{C}_S)^{\perp}$ is $\begin{pmatrix} 2 & 0 & 0 & 0 \\ 1 & 1 & 1 & 3 \end{pmatrix}$ and finally, applying ξ, the generator matrix of $\mathcal{C}_S^{\perp} = \xi(\chi(\mathcal{C}_S)^{\perp})$ is

$$\mathcal{H}_S = \left(\, 1 \mid 1 \quad 1 \quad 3 \,\right).$$

The second method uses Proposition 3.14. We know that the quaternary linear code $\xi^{-1}(\mathcal{C}_S)$ with generator matrix

$$\begin{pmatrix} 2 & 0 & 0 & 0 \\ 1 & 2 & 0 & 0 \\ 0 & 3 & 1 & 0 \\ 0 & 1 & 0 & 1 \end{pmatrix},$$

or equivalently $\begin{pmatrix} 1 & 2 & 0 & 0 \\ 0 & 3 & 1 & 0 \\ 0 & 1 & 0 & 1 \end{pmatrix}$, has parity check matrix $\left(\, 2 \quad 1 \quad 1 \quad 3 \,\right)$.

Therefore, applying χ^{-1}, the generator matrix of $\mathcal{C}_S^{\perp} = \chi^{-1}(\xi^{-1}(\mathcal{C}_S)^{\perp})$ is

$$\mathcal{H}_S = \left(\, 1 \mid 1 \quad 1 \quad 3 \,\right).$$

△

The result in Theorem 3.23 shows how to construct a parity check matrix of a $\mathbb{Z}_2\mathbb{Z}_4$-additive code generated by a generator matrix in standard form as in (2.7) and it is proved using the method given by Proposition 3.13. Note also that we can apply both methods to any generator matrix, not necessary a generator matrix in standard form, to get a parity check matrix.

Lemma 3.22 ([92]). *If $\mathcal{C}$ is a quaternary linear code of type $(0, \beta; \gamma, \delta; 0)$ with generator matrix in standard form (2.6), then a generator matrix of $\mathcal{C}^{\perp}$ is*

$$\mathcal{H}_S = \begin{pmatrix} \mathbf{0} & 2I_{\gamma} & 2R^t \\ I_{\beta-\gamma-\delta} & T^t & -(S + RT)^t \end{pmatrix}, \tag{3.10}$$

where R, T are matrices over $\mathbb{Z}_4$ with all entries in $\{0, 1\} \subset \mathbb{Z}_4$ of size $\delta \times \gamma$ and $\gamma \times (\beta-\gamma-\delta)$, respectively; and S is a matrix over $\mathbb{Z}_4$ of size $\delta \times (\beta-\gamma-\delta)$.

Theorem 3.23 ([37]). *Let $\mathcal{C}$ be a $\mathbb{Z}_2\mathbb{Z}_4$-additive code of type $(\alpha, \beta; \gamma, \delta; \kappa)$ with generator matrix in standard form (2.7). Then, a generator matrix of $\mathcal{C}^{\perp}$ is*

$$\mathcal{H}_S = \begin{pmatrix} T_b^t & I_{\alpha-\kappa} & \mathbf{0} & \mathbf{0} & 2S_b^t \\ \mathbf{0} & \mathbf{0} & \mathbf{0} & 2I_{\gamma-\kappa} & 2R^t \\ T_2^t & \mathbf{0} & I_{\beta+\kappa-\gamma-\delta} & T_1^t & -\left(S_q + RT_1\right)^t \end{pmatrix}, \tag{3.11}$$

where T_b, T_2 are matrices over $\mathbb{Z}_2$; T_1, R, S_b are matrices over $\mathbb{Z}_4$ with all entries in $\{0, 1\} \subset \mathbb{Z}_4$; and S_q is a matrix over $\mathbb{Z}_4$.

Proof. By Lemma 3.22, if $\mathcal{C}'$ is a quaternary linear code with generator matrix

$$\mathcal{G}' = \begin{pmatrix} 2T_b & 2T_2 & 2I_\kappa & 0 & 0 \\ 0 & 2T_1 & 0 & 2I_{\gamma-\kappa} & 0 \\ 2S_b & S_q & 0 & R & I_\delta \end{pmatrix},$$

then the quaternary dual code $(\mathcal{C}')^\perp$ has generator matrix

$$\mathcal{H}' = \begin{pmatrix} 0 & 0 & 2I_\kappa & 0 & 0 \\ 0 & 0 & 0 & 2I_{\gamma-\kappa} & 2R^t \\ I_{\alpha-\kappa} & 0 & T_b^t & 0 & 2S_b^t \\ 0 & I_{\beta-\gamma-\delta+\kappa} & T_2^t & T_1^t & -(S_q+RT_1)^t \end{pmatrix}.$$

Hence, if $\mathcal{C}$ is a $\mathbb{Z}_2\mathbb{Z}_4$-additive code with generator matrix (2.7), then the generator matrix of $\chi(\mathcal{C})^\perp$ is

$$\mathcal{H}_\xi = \begin{pmatrix} 2I_\kappa & 0 & 0 & 0 & 0 \\ 0 & 0 & 0 & 2I_{\gamma-\kappa} & 2R^t \\ T_b^t & I_{\alpha-\kappa} & 0 & 0 & 2S_b^t \\ T_2^t & 0 & I_{\beta-\gamma-\delta+\kappa} & T_1^t & -(S_q+RT_1)^t \end{pmatrix}.$$

Finally, by Proposition 3.13, $\mathcal{H}_S = \xi(\mathcal{H}_\xi)$ is the generator matrix of $\mathcal{C}^\perp$. $\quad\square$

Note that by Theorem 3.16 and Theorem 3.23, if $\mathcal{C}$ is a $\mathbb{Z}_2\mathbb{Z}_4$-additive code of type $(\alpha,\beta;\gamma,\delta;\kappa)$ with generator matrix in standard form (2.7), then $\mathcal{C}^\perp$ is permutation equivalent to a $\mathbb{Z}_2\mathbb{Z}_4$-additive code with generator matrix in standard form

$$\begin{pmatrix} I_{\bar\kappa} & T_b^t & 2S_b^t & 0 & 0 \\ 0 & 0 & 2R^t & 2I_{\bar\gamma-\bar\kappa} & 0 \\ 0 & T_2^t & -(S_q+RT_1)^t & T_1^t & I_{\bar\delta} \end{pmatrix}, \tag{3.12}$$

where T_b, T_2 are matrices over $\mathbb{Z}_2$; T_1, R, S_b are matrices over $\mathbb{Z}_4$ with all entries in $\{0,1\} \subset \mathbb{Z}_4$; and S_q is a matrix over $\mathbb{Z}_4$. Moreover, $\bar\gamma = \alpha+\gamma-2\kappa$, $\bar\delta = \beta-\gamma-\delta+\kappa$ and $\bar\kappa = \alpha-\kappa$.

Finally, recall that a generator matrix of $\mathcal{C}^\perp$ can be seen as a parity check matrix for $\mathcal{C}$. Analogously, we can use a generator matrix of $\mathcal{C}$ as a parity check matrix of $\mathcal{C}^\perp$. Therefore, Theorem 3.23 also shows how to construct a parity check matrix of a $\mathbb{Z}_2\mathbb{Z}_4$-additive code generated by a generator matrix in standard form as in (2.7).

Example 3.24. Let $\mathcal{C}$ be the $\mathbb{Z}_2\mathbb{Z}_4$-additive code of type $(4, 6; 1, 2; 1)$ given in Example 2.9 with generator matrix in standard form

$$\mathcal{G}_S = \left(\begin{array}{cccc|cccccc} 1 & 1 & 1 & 1 & 0 & 2 & 0 & 2 & 0 & 0 \\ 0 & 1 & 0 & 1 & 2 & 1 & 1 & 1 & 1 & 0 \\ 0 & 0 & 1 & 1 & 1 & 1 & 2 & 3 & 0 & 1 \end{array} \right).$$

By Theorem 3.16 and Theorem 3.23, the additive dual code $\mathcal{C}^\perp$ is of type $(4, 6; 3, 4; 3)$ and has generator matrix

$$\mathcal{H}_S = \left(\begin{array}{cccc|cccccc} 1 & 1 & 0 & 0 & 0 & 0 & 0 & 0 & 2 & 0 \\ 1 & 0 & 1 & 0 & 0 & 0 & 0 & 0 & 0 & 2 \\ 1 & 0 & 0 & 1 & 0 & 0 & 0 & 0 & 2 & 2 \\ 0 & 0 & 0 & 0 & 1 & 0 & 0 & 0 & 2 & 3 \\ 1 & 0 & 0 & 0 & 0 & 1 & 0 & 0 & 3 & 3 \\ 0 & 0 & 0 & 0 & 0 & 0 & 1 & 0 & 3 & 2 \\ 1 & 0 & 0 & 0 & 0 & 0 & 0 & 1 & 3 & 1 \end{array} \right).$$

Moreover, $\mathcal{G}_S$ and $\mathcal{H}_S$ can be seen as parity check matrices of $\mathcal{C}^\perp$ and $\mathcal{C}$, respectively.

In this case, the corresponding $\mathbb{Z}_2\mathbb{Z}_4$-linear codes $C = \Phi(\mathcal{C})$ and $C_\perp = \Phi(\mathcal{C}^\perp)$ are not linear, so $C_\perp$ is not the dual code of C in the binary sense. Actually, $C_\perp$ has length 16 and 2^{11} codewords, but $C^\perp = \langle C \rangle^\perp$ is a binary linear $[16, 10]$ code. $\triangle$

Chapter 4

$\mathbb{Z}_2\mathbb{Z}_4$-Additive Self-Dual Codes

Self-duality for binary and quaternary linear codes has been extensively studied, see [125, 132] for a complete description and an extensive bibliography. Quaternary self-dual codes are also interesting, since their binary Gray map images are formally self-dual in the sense that their Hamming weight enumerators are invariant under the MacWilliams transform [92].

In this chapter, we study $\mathbb{Z}_2\mathbb{Z}_4$-additive self-dual codes and their binary Gray map images. First, we characterize $\mathbb{Z}_2\mathbb{Z}_4$-additive self-dual codes in terms of some properties such us separability, antipodality and weight parity. Then, we determine all the possible values for the parameters α and β of a $\mathbb{Z}_2\mathbb{Z}_4$-additive self-dual code of type $(\alpha, \beta; \gamma, \delta; \kappa)$, as well as the weight enumerator of such codes. We also show some constructions of $\mathbb{Z}_2\mathbb{Z}_4$-additive self-dual codes starting from a given $\mathbb{Z}_2\mathbb{Z}_4$-additive self-dual code. Finally, we establish similar results for $\mathbb{Z}_2\mathbb{Z}_4$-additive formally self-dual codes.

4.1   Properties of $\mathbb{Z}_2\mathbb{Z}_4$-Additive Self-Dual Codes

Let $\mathcal{C}$ be a $\mathbb{Z}_2\mathbb{Z}_4$-additive code. We say that $\mathcal{C}$ is a $\mathbb{Z}_2\mathbb{Z}_4$-additive self-orthogonal code if $\mathcal{C} \subseteq \mathcal{C}^\perp$ and $\mathcal{C}$ is a $\mathbb{Z}_2\mathbb{Z}_4$-additive self-dual code if $\mathcal{C} = \mathcal{C}^\perp$. Let $C = \Phi(\mathcal{C})$ be the corresponding $\mathbb{Z}_2\mathbb{Z}_4$-linear code. We say that C is a self $\mathbb{Z}_2\mathbb{Z}_4$-orthogonal code if $C \subseteq C_\perp$ and C is a self $\mathbb{Z}_2\mathbb{Z}_4$-dual code if $C = C_\perp$, where $C_\perp = \Phi(\mathcal{C}^\perp)$.

There are parts of this chapter that have been previously published. Reprinted by permission from American Institute of Mathematical Sciences: J. Borges, S. T. Dougherty, and C. Fernández-Córdoba. "Characterization and constructions of self-dual codes over $\mathbb{Z}_2 \times \mathbb{Z}_4$". *Advances in Mathematics of Communications*, v. 6, 3, pp. 287-303 ©(2012). Reprinted by permission from Springer Nature Customer Service Centre GmbH: S. T. Dougherty and C. Fernández-Córdoba. "$\mathbb{Z}_2\mathbb{Z}_4$-additive formally self-dual codes". *Designs, Codes and Cryptography*, v. 72, 2, pp. 435–453. ©(2014).

Note that in the case that $\beta = 0$, so when $\mathcal{C} = C$ is a binary linear code, we also say that $\mathcal{C}$ is *binary self-dual* (or *binary self-orthogonal*) if $\mathcal{C} = \mathcal{C}^\perp$ (or $\mathcal{C} \subseteq \mathcal{C}^\perp$). In the case that $\alpha = 0$, so when $\mathcal{C}$ is a quaternary linear code, we also say that $\mathcal{C}$ is *quaternary self-dual* (or *quaternary self-orthogonal*) if $\mathcal{C} = \mathcal{C}^\perp$ (or $\mathcal{C} \subseteq \mathcal{C}^\perp$).

Example 4.1. Consider the following $\mathbb{Z}_2\mathbb{Z}_4$-additive code $\mathcal{C}$ of type $(2, 2; 1, 1; 1)$:

$$\mathcal{C} = \{(0,0 \mid 0,0), (0,0 \mid 2,2), (1,1 \mid 0,2), (1,1 \mid 2,0)$$
$$(0,1 \mid 1,1), (0,1 \mid 3,3), (1,0 \mid 1,3), (1,0 \mid 3,1)\}.$$

For any $\mathbf{u}, \mathbf{v} \in \mathcal{C}$, we have that $\mathbf{u} \cdot \mathbf{v} = 0$. Therefore, $\mathcal{C} \subseteq \mathcal{C}^\perp$, that is, $\mathcal{C}$ is self-orthogonal. Since $|\mathcal{C}| = 2^3$ and, by Corollary 3.7, $|\mathcal{C}||\mathcal{C}^\perp| = 2^{\alpha+2\beta} = 2^6$, we have that $|\mathcal{C}^\perp| = 2^3$ and hence $\mathcal{C}$ is self-dual.

The binary code $C = \Phi(\mathcal{C})$, with codewords

$$C = \{(0,0,0,0,0,0), (0,0,1,1,1,1), (1,1,0,0,1,1), (1,1,1,1,0,0)$$
$$(0,1,0,1,0,1), (0,1,1,0,1,0), (1,0,0,1,1,0), (1,0,1,0,0,1)\},$$

is a self $\mathbb{Z}_2\mathbb{Z}_4$-dual code. However, note that C is not a binary self-orthogonal code since, for example, $(0,1,0,1,0,1) \cdot (0,1,1,0,1,0) \neq 0$, and then it is not binary self-dual. $\triangle$

As shown in Chapter 3, if $\mathcal{C}$ is a $\mathbb{Z}_2\mathbb{Z}_4$-additive code with generator matrix $\mathcal{G}$, then the parity check matrix of $\mathcal{C}$, $\mathcal{H}$, is the generator matrix of $\mathcal{C}^\perp$ and vice versa. Therefore, we have that $\mathcal{C}$ is self-dual if and only if $\mathcal{H}$ is a generator matrix for $\mathcal{C}$. In the following example, we obtain a family of $\mathbb{Z}_2\mathbb{Z}_4$-additive self-dual codes by considering their generator matrices.

Example 4.2. Consider the $\mathbb{Z}_2\mathbb{Z}_4$-additive code $\mathcal{C}$ of type $(2\kappa, \beta; \beta + \kappa - 2\delta, \delta; \kappa)$, for $\kappa, \beta, \delta \geq 0$ and $\delta \leq \beta$, with generator matrix in standard form

$$\mathcal{G} = \begin{pmatrix} I_\delta & \mathbf{0} & I_\delta & \mathbf{0} & 2I_\delta & \mathbf{0} & \mathbf{0} \\ \mathbf{0} & I_{\kappa-\delta} & \mathbf{0} & I_{\kappa-\delta} & \mathbf{0} & \mathbf{0} & \mathbf{0} \\ \mathbf{0} & \mathbf{0} & \mathbf{0} & \mathbf{0} & \mathbf{0} & 2I_{\beta-2\delta} & \mathbf{0} \\ \mathbf{0} & \mathbf{0} & I_\delta & \mathbf{0} & I_\delta & \mathbf{0} & I_\delta \end{pmatrix}.$$

By using Theorem 3.23 to construct the parity check matrix of $\mathcal{C}$, we can see that we obtain again the matrix $\mathcal{G}$. Hence, $\mathcal{C}^\perp = \mathcal{C}$ and $\mathcal{C}$ is a $\mathbb{Z}_2\mathbb{Z}_4$-additive self-dual code. $\triangle$

Now, we study some properties of $\mathbb{Z}_2\mathbb{Z}_4$-additive self-dual codes. The following lemma establish some conditions on the parameters of a $\mathbb{Z}_2\mathbb{Z}_4$-additive self-dual code. It can be easily proven by considering the parameters of the dual of a $\mathbb{Z}_2\mathbb{Z}_4$-additive code shown in Theorem 3.16.

Lemma 4.3 ([32]). *If $\mathcal{C}$ is a $\mathbb{Z}_2\mathbb{Z}_4$-additive self-dual code, then $\mathcal{C}$ is of type $(2\kappa, \beta; \beta + \kappa - 2\delta, \delta; \kappa)$, $|\mathcal{C}| = 2^{\beta+\kappa}$ and $|\mathcal{C}_b| = 2^{\beta+\kappa-\delta}$.*

Let $\mathcal{C}$ be a $\mathbb{Z}_2\mathbb{Z}_4$-additive self-dual code of type $(2\kappa, \beta; \beta + \kappa - 2\delta, \delta; \kappa)$. It is easy to check that codewords in $(\mathcal{C}_X)^\perp \times (\mathcal{C}_Y)^\perp$ are orthogonal to $\mathcal{C}$ and, hence, $(\mathcal{C}_X)^\perp \times (\mathcal{C}_Y)^\perp \subseteq \mathcal{C}^\perp$, where $|\mathcal{C}^\perp| = |\mathcal{C}| = 2^{\beta+\kappa}$ by Lemma 4.3. Moreover, if $\mathcal{C}$ is separable $\mathcal{C}^\perp = (\mathcal{C}_X)^\perp \times (\mathcal{C}_Y)^\perp$ by Corollary 3.18.

Proposition 4.4 ([71]). *Let $\mathcal{C}$ be a separable $\mathbb{Z}_2\mathbb{Z}_4$-additive code. Then, $\mathcal{C}$ is a $\mathbb{Z}_2\mathbb{Z}_4$-additive self-dual code of type $(2\kappa, \beta; \beta + \kappa - 2\delta, \delta; \kappa)$ if and only if $\mathcal{C}_X$ is a $[2\kappa, \kappa]$ binary self-dual code and $\mathcal{C}_Y$ is a quaternary self-dual code of type $(0, \beta; \beta - 2\delta, \delta; 0)$.*

Proof. Let $\mathcal{C}$ be a separable $\mathbb{Z}_2\mathbb{Z}_4$-additive code. For all $u \in \mathcal{C}_X$, we have that $(u \mid \mathbf{0}) \in \mathcal{C}$. If $\mathcal{C}$ is self-dual, $(u \mid \mathbf{0}) \in \mathcal{C}^\perp$ and $(u \mid \mathbf{0}) \cdot (v \mid \mathbf{0}) = 0$ for all $v \in \mathcal{C}_X$. Since $(u \mid \mathbf{0}) \cdot (v \mid \mathbf{0}) = 0$, we have that $(u \cdot v)_2 = 0$ and $u \in (\mathcal{C}_X)^\perp$. Therefore, $\mathcal{C}_X$ is a binary self-dual code. Similarly, if $u' \in \mathcal{C}_Y$, then $(\mathbf{0} \mid u') \in \mathcal{C}$ and for all $v' \in \mathcal{C}_Y$ we have $(\mathbf{0} \mid v') \in \mathcal{C}^\perp$ and $(u' \cdot v')_4 = (\mathbf{0} \mid u') \cdot (\mathbf{0} \mid v') = 0$. Hence, $\mathcal{C}_Y$ is a quaternary self-dual code.

Assume now that $\mathcal{C}_X$ and $\mathcal{C}_Y$ are binary and quaternary self-dual codes, respectively. Then, by Corollary 3.18, $\mathcal{C}^\perp = (\mathcal{C}_X)^\perp \times (\mathcal{C}_Y)^\perp = \mathcal{C}_X \times \mathcal{C}_Y = \mathcal{C}$, and $\mathcal{C}$ is self-dual. $\qquad\square$

For a vector $z' \in \mathbb{Z}_4^\beta$, denote by $p(z')$ the number of order four coordinates of z', that is, those coordinates in $\{1, 3\}$. The following lemma is easy to prove.

Lemma 4.5 ([32]). *Let $\mathcal{C}$ be a $\mathbb{Z}_2\mathbb{Z}_4$-additive self-dual code, and $(u \mid u') \in \mathcal{C}$.*

i) If $\mathrm{wt}_H(u)$ is even, then $p(u') \equiv 0 \pmod{4}$.

ii) If $\mathrm{wt}_H(u)$ is odd, then $p(u') \equiv 2 \pmod{4}$.

iii) The vector $(\mathbf{0} \mid \mathbf{2})$ is a codeword in $\mathcal{C}$.

Lemma 4.6 ([32]). *If $\mathcal{C}$ is a $\mathbb{Z}_2\mathbb{Z}_4$-additive self-dual code, then the subcode $(\mathcal{C}_b)_X$ is a binary self-dual code.*

Proof. By Lemma 4.3, the code $\mathcal{C}$ is of type $(2\kappa, \beta; \beta + \kappa - 2\delta, \delta; \kappa)$. Since for any pair of codewords $(u \mid u'), (v \mid v') \in \mathcal{C}_b$ we have that u' and v' are orthogonal quaternary vectors, $(\mathcal{C}_b)_X \subseteq (\mathcal{C}_b)_X^\perp$. Moreover, since $(\mathcal{C}_b)_X$ has dimension κ (by definition) and length 2κ, we have that $(\mathcal{C}_b)_X$ is binary self-dual. $\qquad\square$

Lemma 4.7 ([32]). *Let C be a $\mathbb{Z}_2\mathbb{Z}_4$-additive self-dual code of type $(2\kappa, \beta; \beta + \kappa - 2\delta, \delta; \kappa)$. There is an integer number r, $0 \le r \le \kappa$, such that each codeword in C_Y appears 2^r times in C and $|C_Y| \ge 2^\beta$.*

Proof. Consider the subcode $\bar{C} = \{(u \mid \mathbf{0}) \in C\}$. Clearly, $\bar{C}_X$ is a binary linear code. Let $r = \dim(\bar{C}_X)$. Thus, any vector in C_Y appears 2^r times in C. Note that $\bar{C}_X$ is also a subcode of $(C_b)_X$, hence $r \le \kappa$. Finally, we have that $|C| = 2^{\beta+\kappa} = |C_Y| \cdot 2^r$, therefore $|C_Y| \ge 2^\beta$. $\qquad\square$

In the following statement, we give some relationships between the properties of self-orthogonality and self-duality for the binary linear code C_X as well as for the quaternary linear code C_Y of a $\mathbb{Z}_2\mathbb{Z}_4$-additive self-dual code C.

Theorem 4.8 ([32]). *Let C be a $\mathbb{Z}_2\mathbb{Z}_4$-additive self-dual code of type $(2\kappa, \beta; \beta + \kappa - 2\delta, \delta; \kappa)$. The following statements are equivalent:*

i) C_X is a binary self-orthogonal code.

ii) C_X is a binary self-dual code.

iii) $|C_X| = 2^\kappa$.

iv) C_Y is a quaternary self-orthogonal code.

v) C_Y is a quaternary self-dual code.

vi) $|C_Y| = 2^\beta$.

vii) C is separable.

Proof. $(i) \Leftrightarrow (ii)$: By Lemma 4.6, $|C_X| \ge 2^\kappa$, thus (i) and (ii) are equivalent statements.

$(ii) \Leftrightarrow (iii)$: Clearly, (ii) implies (iii). The statement (iii) implies $C_X = (C_b)_X$, so C_X is binary self-dual by Lemma 4.6.

$(ii) \Leftrightarrow (v)$: Straightforward.

$(iv) \Leftrightarrow (v)$: By Lemma 4.7, $|C_Y| \ge 2^\beta$, thus (iv) and (v) are equivalent statements.

$(ii) \Leftrightarrow (vii)$: If C_X is binary self-dual, then C_Y is quaternary self-dual, $|C_X| = 2^\kappa$ and $|C_Y| = 2^\beta$. Since $|C| = 2^{\beta+\kappa}$, we have that the set of codewords in C is $C_X \times C_Y$. Reciprocally, if $C = C_X \times C_Y$, then C_X is a binary self-dual code and C_Y is a quaternary self-dual code by Proposition 4.4.

$(v) \Rightarrow (vi)$: Trivial.

$(vi) \Rightarrow (iii)$: By Lemma 4.7, each vector in C_Y appears 2^κ times in C. Thus, for any vector $u_b \in (C_b)_X$, the vector $(u_b \mid \mathbf{0})$ is a codeword in C_b.

This means that given any codeword $(u \mid u') \in \mathcal{C}$, we have that u and u_b are orthogonal binary vectors, for all $u_b \in (\mathcal{C}_b)_X$, since $(u \mid u') \cdot (u_b \mid \mathbf{0}) = 0$. Therefore, for all $u \in \mathcal{C}_X$, $u \in (\mathcal{C}_b)_X^\perp$ and $\mathcal{C}_X \subseteq (\mathcal{C}_b)_X^\perp$. By Lemma 4.6, $(\mathcal{C}_b)_X^\perp = (\mathcal{C}_b)_X$, which implies $\mathcal{C}_X = (\mathcal{C}_b)_X$, hence $|\mathcal{C}_X| = 2^\kappa$. $\qquad\square$

From Theorem 4.8, if $\mathcal{C}$ is a $\mathbb{Z}_2\mathbb{Z}_4$-additive self-dual code, then $\mathcal{C}_X$ is binary self-dual if and only if $\mathcal{C}_Y$ is quaternary self-dual. If $\mathcal{C}$ is a separable $\mathbb{Z}_2\mathbb{Z}_4$-additive code, $\mathcal{C}_X$ is binary self-dual and $\mathcal{C}_Y$ is quaternary self-dual, then $\mathcal{C}$ is also $\mathbb{Z}_2\mathbb{Z}_4$-additive self-dual as it is proven in Proposition 4.4.

We say that a binary code C is *antipodal* if, for any codeword $z \in C$, we have that $z + \mathbf{1} \in C$. If $\mathcal{C}$ is a $\mathbb{Z}_2\mathbb{Z}_4$-additive code, we say that $\mathcal{C}$ is antipodal if $\Phi(\mathcal{C})$ is antipodal. Clearly, a $\mathbb{Z}_2\mathbb{Z}_4$-additive code $\mathcal{C}$ is antipodal if and only if $(\mathbf{1} \mid \mathbf{2}) \in \mathcal{C}$.

Let $\mathcal{C}$ be a $\mathbb{Z}_2\mathbb{Z}_4$-additive code. If all the codewords of $\mathcal{C}$ have even weights, then $\mathcal{C}$ is called an *even code*, otherwise $\mathcal{C}$ is called an *odd code*. If all the codewords of $\mathcal{C}$ have doubly-even weights (multiple of four weights), then $\mathcal{C}$ is called a *doubly-even code*. A $\mathbb{Z}_2\mathbb{Z}_4$-additive even self-dual code is said to be *Type II* if all the codewords have doubly-even weight, and *Type I*, otherwise. A $\mathbb{Z}_2\mathbb{Z}_4$-additive odd self-dual code is said to be *Type 0*.

Let $\mathcal{C}$ be a separable $\mathbb{Z}_2\mathbb{Z}_4$-additive self-dual code. By Proposition 4.4, $\mathcal{C}_X$ and $\mathcal{C}_Y$ are self-dual codes. Then, it is easy to see that $\mathcal{C}$ is Type II if and only if $\mathcal{C}_X$ and $\mathcal{C}_Y$ are both Type II codes, and $\mathcal{C}$ is Type I, otherwise.

Now, we give some relationships among weight parity (Type 0, Type I or Type II), separability and antipodality.

Proposition 4.9 ([32]). *Let $\mathcal{C}$ be a $\mathbb{Z}_2\mathbb{Z}_4$-additive self-dual code. If $\mathcal{C}_X$ has only doubly-even weights, then $\mathcal{C}$ is separable.*

Proof. Given any pair of vectors $u, v \in \mathcal{C}_X$, we have that $\mathrm{wt}_H(u + v) = w_H(u) + \mathrm{wt}_H(v) - 2\,\mathrm{wt}_H(u * v)$, where $u * v$ is the componentwise product of u and v. Since $\mathrm{wt}_H(u + v)$, $\mathrm{wt}_H(u)$ and $\mathrm{wt}_H(v)$ are 0 modulo 4, we conclude that $\mathrm{wt}_H(u * v)$ is even and then u and v are orthogonal. Therefore, $\mathcal{C}_X$ is self-orthogonal and, by Theorem 4.8, $\mathcal{C}$ is separable. $\qquad\square$

Proposition 4.10 ([32]). *Let $\mathcal{C}$ be a $\mathbb{Z}_2\mathbb{Z}_4$-additive self-dual code.*

  i) $\mathcal{C}$ is antipodal if and only if $\mathcal{C}$ is Type I or Type II.

  ii) If $\mathcal{C}$ is separable, then $\mathcal{C}$ is antipodal.

  iii) If $\mathcal{C}$ is Type 0, then $\mathcal{C}$ is non-separable and non-antipodal.

Proof. We have that $\mathcal{C}$ is antipodal if and only if $(\mathbf{1}\mid\mathbf{2})\in\mathcal{C}$. Since $(\mathbf{0}\mid\mathbf{2})\in\mathcal{C}$ by Lemma 4.5, we obtain that $\mathcal{C}$ is antipodal if and only if $(\mathbf{1}\mid\mathbf{0})\in\mathcal{C}$. However, this condition is equivalent to saying that all codewords of $\mathcal{C}_X$ are of even weight. Therefore, (i) is proven.

To prove (ii), if $\mathcal{C}$ is separable, then $\mathcal{C}=\mathcal{C}_X\times\mathcal{C}_Y$, where $\mathcal{C}_X$ is a binary self-dual code and $\mathcal{C}_Y$ is a quaternary self-dual code by Proposition 4.4. Hence, $\mathcal{C}_X$ contains the all-one vector and $\mathcal{C}_Y$ contains the all-two vector. Therefore, $(\mathbf{1}\mid\mathbf{2})\in\mathcal{C}$.

Finally, from (i) and (ii) we obtain (iii).   □

The following examples show the existence of all possible cases we have described.

Example 4.11 (Type 0). Let $\mathcal{C}_1$ be the $\mathbb{Z}_2\mathbb{Z}_4$-additive self-dual code given in Example 4.1. A generator matrix of $\mathcal{C}_1$ in standard form is

$$\mathcal{G}_1=\left(\begin{array}{cc|cc}1&1&2&0\\0&1&1&1\end{array}\right),$$

and the weight enumerator of this code is $W_{\mathcal{C}_1}(x,y)=x^6+4x^3y^3+3x^2y^4$. Note that it has codewords of odd weight, hence it is a Type 0 code and it is non-separable and non-antipodal by Proposition 4.10.   △

Example 4.12 (Type I, separable). A $\mathbb{Z}_2\mathbb{Z}_4$-additive self-dual code with $\alpha,\beta\geq1$ should have $\alpha\geq2$, since α must be even by Lemma 4.3. A $\mathbb{Z}_2\mathbb{Z}_4$-additive self-dual code with minimum length has $\alpha=2$, $\beta=1$ and $2^{\beta+\kappa}=2^{1+1}=4$ codewords. For example, $\mathcal{C}_2=\{(0,0\mid0),(0,0\mid2),(1,1\mid0),(1,1\mid2)\}$ is a $\mathbb{Z}_2\mathbb{Z}_4$-additive self-dual code of type $(2,1;2,0;1)$ with generator matrix

$$\mathcal{G}_2=\left(\begin{array}{cc|c}1&1&0\\0&0&2\end{array}\right),$$

and weight enumerator $W_{\mathcal{C}_2}(x,y)=x^4+2x^2y^2+y^4$. Note that for $\alpha=2$ and $\beta=1$, it is not possible to have odd weight codewords. Thus, the code must be Type I and antipodal. Also, we have that the code projected onto the quaternary coordinates is $\{0,2\}$, which is self-dual and hence, by Theorem 4.8, $\mathcal{C}_2$ is separable.   △

Example 4.13 (Type I, non-separable). Let $\mathcal{C}_3$ and $\mathcal{C}_4$ be the $\mathbb{Z}_2\mathbb{Z}_4$-additive codes generated by the matrices in standard form $\mathcal{G}_3$ and $\mathcal{G}_4$, respectively, where

$$
\mathcal{G}_3 = \begin{pmatrix}
1 & 1 & 1 & 1 & 0 & 0 & 0 & 0 \\
0 & 1 & 0 & 1 & 2 & 0 & 0 & 0 \\
0 & 1 & 0 & 1 & 0 & 2 & 0 & 0 \\
0 & 1 & 0 & 1 & 0 & 0 & 2 & 0 \\
0 & 0 & 1 & 1 & 1 & 1 & 1 & 1
\end{pmatrix}
\quad
\mathcal{G}_4 = \begin{pmatrix}
1 & 1 & 1 & 1 & 0 & 0 & 0 & 0 & 0 & 0 \\
0 & 1 & 0 & 1 & 2 & 2 & 0 & 0 & 0 & 0 \\
0 & 0 & 0 & 0 & 2 & 0 & 2 & 0 & 0 & 0 \\
0 & 1 & 0 & 1 & 0 & 0 & 0 & 2 & 0 & 0 \\
0 & 1 & 0 & 1 & 1 & 1 & 1 & 0 & 1 & 0 \\
0 & 0 & 1 & 1 & 1 & 0 & 1 & 1 & 0 & 1
\end{pmatrix}.
$$

It is easy to check that both codes are self-orthogonal. Since $|\mathcal{C}_3| = 2^6$ and $|\mathcal{C}_4| = 2^8$, it is also easy to see they are self-dual. They are non-separable by Corollary 2.14. Finally, since their weight enumerators are $W_{\mathcal{C}_3}(x,y) = x^{12} + 15x^8 y^4 + 32x^6 y^6 + 15x^4 y^8 + y^{12}$ and $W_{\mathcal{C}_4}(x,y) = x^{16} + 12x^{12} y^4 + 64x^{10} y^6 + 102x^8 y^8 + 64x^6 y^{10} + 12x^4 y^{12} + y^{16}$, they are Type I codes. $\triangle$

Example 4.14 (Type II, separable). Let C be the extended binary Hamming code of length 8 and let $\mathcal{D}$ be the quaternary linear code generated by

$$
\begin{pmatrix}
2 & 2 & 0 & 0 \\
2 & 0 & 2 & 0 \\
1 & 1 & 1 & 1
\end{pmatrix}.
$$

Then, $|\mathcal{D}| = 2^2 4^1 = 2^4$, which is the correct size to be self-dual. Clearly, $\mathcal{D}$ is quaternary self-orthogonal and hence self-dual. On the other hand, C is a binary self-dual code. Since both codes have only doubly-even weights, we conclude that $\mathcal{C}_5 = C \times \mathcal{D}$ is Type II and separable. Its weight enumerator is $W_{\mathcal{C}_5}(x,y) = x^{16} + 28x^{12} y^4 + 198x^8 y^8 + 28x^4 y^{12} + y^{16}$. $\triangle$

Example 4.15 (Type II, non-separable). The $\mathbb{Z}_2\mathbb{Z}_4$-additive code $\mathcal{C}_6$ generated by the matrix in standard form $\mathcal{G}_6$ is self-orthogonal, where

$$
\mathcal{G}_6 = \begin{pmatrix}
1 & 0 & 0 & 1 & 0 & 1 & 1 & 0 & 0 & 0 & 0 & 0 \\
0 & 1 & 0 & 0 & 1 & 1 & 1 & 0 & 0 & 0 & 0 & 0 \\
0 & 0 & 1 & 0 & 0 & 1 & 1 & 1 & 0 & 0 & 0 & 0 \\
0 & 0 & 0 & 0 & 0 & 1 & 1 & 0 & 2 & 0 & 0 & 0 \\
0 & 0 & 0 & 0 & 0 & 1 & 1 & 0 & 0 & 2 & 0 & 0 \\
0 & 0 & 0 & 0 & 0 & 1 & 1 & 0 & 0 & 0 & 2 & 0 \\
0 & 0 & 0 & 1 & 1 & 0 & 1 & 1 & 1 & 1 & 1 & 1
\end{pmatrix}.
$$

Since $|\mathcal{C}_6| = 2^8$, the code $\mathcal{C}_6$ is self-dual. Clearly, it is non-separable by Corollary 2.14. We can also see that it is non-separable by Theorem 4.8, since $(\mathcal{C}_6)_X$ is not self-orthogonal. On the other hand, it can be checked that the weight enumerator of $\mathcal{C}_6$ is $W_{\mathcal{C}_6}(x,y) = x^{16} + 28x^{12} y^4 + 198x^8 y^8 + 28x^4 y^{12} + y^{16}$. Therefore, $\mathcal{C}_6$ is Type II. $\triangle$

Example 4.16. By using the MAGMA package for $\mathbb{Z}_2\mathbb{Z}_4$-additive codes [29], we construct the code $\mathcal{C}_6$ given in Example 4.15 and check that it is a non-separable self-dual code. By computing its weight enumerator, we also see that $\mathcal{C}_6$ is a Type II code.

```
> Z4 := IntegerRing(4);
> V := RSpace(Z4, 12);
> C := Z2Z4AdditiveCode([V![1,0,0,1,0,1,1,0, 0,0,0,0],
                         V![0,1,0,0,1,1,1,0, 0,0,0,0],
                         V![0,0,1,0,0,1,1,1, 0,0,0,0],
                         V![0,0,0,0,0,1,1,0, 2,0,0,0],
                         V![0,0,0,0,0,1,1,0, 0,2,0,0],
                         V![0,0,0,0,0,1,1,0, 0,0,2,0],
                         V![0,0,0,1,1,0,1,1, 1,1,1,1]], 8);
> IsSelfDual(C);
true
> IsSeparable(C);
false
> W<x,y> := LeeWeightEnumerator(C); W;
x^16 + 28*x^12*y^4 + 198*x^8*y^8 + 28*x^4*y^12 + y^16          △
```

Now, we study the weight enumerator of a $\mathbb{Z}_2\mathbb{Z}_4$-additive self-dual code and we relate it to the weight parity of the code.

Let $\mathcal{C}$ be a $\mathbb{Z}_2\mathbb{Z}_4$-additive self-dual code of type $(\alpha, \beta; \gamma, \delta; \kappa)$, and $C = \Phi(\mathcal{C})$. Recall that the weight enumerator of $\mathcal{C}$ and C coincides, that is, $W_{\mathcal{C}}(x,y) = W_C(x,y)$ (see Section 3.1). By (3.5), the weight enumerator of C satisfies $W_C(x,y) = \frac{1}{|C|}W_C(x+y, x-y)$. Since $W_C(x,y)$ is an homogeneous polynomial of degree $n = \alpha + 2\beta$ and $|C| = 2^{n/2}$, the previous equation becomes

$$W_C(x,y) = W_C\left(\frac{x+y}{\sqrt{2}}, \frac{x-y}{\sqrt{2}}\right).$$

Thus, $W_C(x,y)$ is left invariant by the action of the matrix

$$M = \frac{1}{\sqrt{2}}\begin{pmatrix} 1 & 1 \\ 1 & -1 \end{pmatrix}. \tag{4.1}$$

That is, $W_C(x,y) = W_C((x,y)M)$. Note that the matrix M satisfies $M^2 = I_2$. We also know that the length of C, $n = \alpha + 2\beta$, is even so $W_C(-x, -y) = W_C(x,y)$. Therefore, the weight enumerator $W_C(x,y)$ is also held invariant by the matrix

$$B = \begin{pmatrix} -1 & 0 \\ 0 & -1 \end{pmatrix}. \tag{4.2}$$

According to the invariant theory (see e.g. [112]), $W_C(x,y)$ is an invariant of the group of matrices generated by M and B. In general, for a given group

G of complex $m \times m$ matrices, a polynomial function $f(x_1, \ldots, x_m)$ is said to be an invariant of G if $f(x_1, \ldots, x_m) = f((x_1, \ldots, x_m)A)$ for every matrix $A \in G$. If G is the group $\langle M, B \rangle$, then $|G| = 4$ with $G = \{M, -M, B, I_2\}$. In [112], the invariant theory is done for this group. By using this theory and Gleason's theorems [89], we obtain the following result.

Theorem 4.17 ([32]). *Let $\mathcal{C}$ be a $\mathbb{Z}_2\mathbb{Z}_4$-additive self-dual code. Then,*

$$W_{\mathcal{C}}(x, y) \in \begin{cases} \mathbb{C}[x^2 + y^2, y(x - y)], & \text{if } \mathcal{C} \text{ is Type 0,} \\ \mathbb{C}[x^2 + y^2, x^2y^2(x^2 - y^2)^2], & \text{if } \mathcal{C} \text{ is Type I,} \\ \mathbb{C}[x^8 + 14x^4y^4 + y^8, x^4y^4(x^4 - y^4)^4], & \text{if } \mathcal{C} \text{ is Type II.} \end{cases} \quad (4.3)$$

Note that, following the invariant theory, the weight enumerators in Theorem 4.17 are considered as polynomials with complex coefficients, regardless that, in our case, all such coefficients are integer values.

4.2 Allowable Values of α and β

In this section, we discuss the possible values that the parameters α and β can take, for $\mathbb{Z}_2\mathbb{Z}_4$-additive self-dual codes of type $(\alpha, \beta; \gamma, \delta; \kappa)$. Recall that for these codes, $\alpha = 2\kappa$ and $\gamma = \beta + \kappa - 2\delta$ by Lemma 4.3.

The following lemma is easily proven.

Lemma 4.18 ([32]). *If $\mathcal{C}$ is a $\mathbb{Z}_2\mathbb{Z}_4$-additive self-dual code of type $(\alpha, \beta; \gamma, \delta; \kappa)$ and $\mathcal{D}$ is a $\mathbb{Z}_2\mathbb{Z}_4$-additive self-dual code of type $(\alpha', \beta'; \gamma', \delta'; \kappa')$, then $\mathcal{C} \times \mathcal{D}$ is a $\mathbb{Z}_2\mathbb{Z}_4$-additive self-dual code of type $(\alpha + \alpha', \beta + \beta'; \gamma + \gamma', \delta + \delta'; \kappa + \kappa')$.*

Example 4.19. By using the MAGMA package for $\mathbb{Z}_2\mathbb{Z}_4$-additive codes [29], we show how to construct a $\mathbb{Z}_2\mathbb{Z}_4$-additive self-dual code of type $(8, 8; 8, 2; 4)$ from the $\mathbb{Z}_2\mathbb{Z}_4$-additive self-dual codes C and D of type $(0, 4; 2, 1; 0)$ and $(8, 4; 6, 1; 4)$ given in Examples 4.14 and 4.15, respectively. Moreover, we check that it is a Type II code by computing its weight enumerator, since C and D are also Type II codes.

```
> Z4 := IntegerRing(4);
> V := RSpace(Z4, 12);
> C := Z2Z4AdditiveCode(Matrix(Z4, [[2,2,0,0],
                                    [2,0,2,0],
                                    [1,1,1,1]]), 0);
> D := Z2Z4AdditiveCode([V![1,0,0,1,0,1,1,0, 0,0,0,0],
                         V![0,1,0,0,1,1,1,0, 0,0,0,0],
                         V![0,0,1,0,0,1,1,1, 0,0,0,0],
                         V![0,0,0,0,0,1,1,0, 2,0,0,0],
                         V![0,0,0,0,0,1,1,0, 0,2,0,0],
```

```
                              V![0,0,0,0,0,1,1,0, 0,0,2,0],
                              V![0,0,0,1,1,0,1,1, 1,1,1,1]], 8);
> C:Minimal;
(4, 16) Z2Z4-additive code of type (0, 4; 2, 1; 0)
> D:Minimal;
(12, 256) Z2Z4-additive code of type (8, 4; 6, 1; 4)
> IsSelfDual(C) and IsSelfDual(D);
true
> WC<x,y> := LeeWeightEnumerator(C); WC;
x^8 + 14*x^4*y^4 + y^8
> WD<x,y> := LeeWeightEnumerator(D); WD;
x^16 + 28*x^12*y^4 + 198*x^8*y^8 + 28*x^4*y^12 + y^16
>
> newCode := DirectSum(C, D);
> newCode:Minimal;
(16, 4096) Z2Z4-additive code of type (8, 8; 8, 2; 4)
> IsSelfDual(newCode);
true
> [Z2Z4Type(C)[i] + Z2Z4Type(D)[i] : i in [1..5]] eq Z2Z4Type(newCode);
true
> W<x,y> := LeeWeightEnumerator(newCode); W;
x^24+42*x^20*y^4+591*x^16*y^8+2828*x^12*y^12+591*x^8*y^16+42*x+y^24          △
```

Proposition 4.4 and Lemma 4.18 give the following statement.

Corollary 4.20 ([32]). *There exist $\mathbb{Z}_2\mathbb{Z}_4$-additive self-dual codes of type $(\alpha, \beta; \gamma, \delta; \kappa)$ for all even α and all β.*

The following corollary follows from Gleason's theorem.

Corollary 4.21 ([32]). *Let C be a $\mathbb{Z}_2\mathbb{Z}_4$-additive code of type $(\alpha, \beta; \gamma, \delta; \kappa)$. If C is Type II, then $n = \alpha + 2\beta \equiv 0 \pmod{8}$.*

Lemma 4.22 ([32]). *Let C be a binary Type I code. Then, it has a linear subcode C_0 containing all doubly-even codewords of C and $|C_0| = |C|/2$.*

Proof. Define the map $f : C \longrightarrow \mathbb{Z}_2$, such that $f(x) = 0$ if x is doubly-even and $f(x) = 1$ if x is single-even, for all $x \in C$. Taking into account that any pair of codewords must be orthogonal, it is easy to verify that f is a group homomorphism. Therefore, $C/f^{-1}(0)$ is isomorphic to the image of f. Since C contains single-even codewords, this image is $\mathbb{Z}_2$. It follows that $|C|/|f^{-1}(0)| = 2$ and, clearly, $f^{-1}(0) = C_0$. $\square$

Theorem 4.23 ([32]). *Let C be a $\mathbb{Z}_2\mathbb{Z}_4$-additive code of type $(\alpha, \beta; \gamma, \delta; \kappa)$. If C is Type II, then $\alpha \equiv 0 \pmod{8}$.*

Proof. We shall prove that if $\mathcal{C}$ is a $\mathbb{Z}_2\mathbb{Z}_4$-additive Type II code, then there exists a binary Type II code of length α. It is well known that binary Type II codes only exist for lengths a multiple of 8 (see [17] and also Corollary 4.21). If the code is separable, then we already know that the binary part is a Type II code. Therefore, we shall assume that $\mathcal{C}$ is not a separable code.

First, we consider the code $(\mathcal{C}_b)_X$. We know that this code is a binary self-dual code by Lemma 4.6. If it is Type II, then we are done. Hence, assume that it is Type I. Thus, by Lemma 4.22, $(\mathcal{C}_b)_X$ contains a codimension 1 linear subcode consisting of the doubly-even codewords. Call this subcode $((\mathcal{C}_b)_X)_0$.

For Type II codes and codewords $(v \mid v')$ in those codes, we have

$$\mathrm{wt}_H(v) + \mathrm{wt}_L(v') \equiv 0 \pmod 4$$
$$2\,\mathrm{wt}_H(v) + p(v') \equiv 0 \pmod 4,$$

where $p(v')$ is the number of order four coordinates in v' as in Lemma 4.5. The first congruence is because the codewords are doubly-even and the second congruence is because they are self-orthogonal (see also Lemma 4.5). These equations imply that every codeword either has

i) doubly-even binary part with doubly-even many order four coordinates in the quaternary part with evenly many twos, or

ii) singly-even binary part with doubly-even many order four coordinates in the quaternary part with oddly many twos.

Consider the vector $(\mathbf{0} \mid \mathbf{1})$. This vector is not in $\mathcal{C}$ since there are codewords of case (b) in $\mathcal{C}$, otherwise it would be separable by Proposition 4.9. Moreover, $(\mathbf{0} \mid \mathbf{1}) + (\mathbf{0} \mid \mathbf{1}) \in \mathcal{C}$. Hence if we let $\mathcal{D} = \{\mathbf{v} \in \mathcal{C} : \mathbf{v} \cdot (\mathbf{0} \mid \mathbf{1}) = 0\}$, since the index of $\mathcal{C}$ in $\langle \mathcal{C}, (\mathbf{0} \mid \mathbf{1}) \rangle$ is 2, we have that the index of $\mathcal{D}$ in $\mathcal{C}$ is 2. Moreover, $\mathcal{D}$ consists precisely of those vectors described above with doubly-even binary part, that is, case (a).

Note that the code $\mathcal{D}_X$ contains $((\mathcal{C}_b)_X)_0$. If $\mathcal{D}_X = ((\mathcal{C}_b)_X)_0$ then $\mathcal{C}_X = (\mathcal{C}_b)_X$ and the code is separable. So $((\mathcal{C}_b)_X)_0$ is strictly contained in $\mathcal{D}_X$ giving that the dimension of $\mathcal{D}_X$ is at least the dimension of $(\mathcal{C}_b)_X$. Now $\mathcal{D}_X$ is a binary linear code consisting of only doubly-even vectors. Hence, as in Proposition 4.9, $\mathcal{D}_X$ is a linear, doubly-even, self-orthogonal code with dimension at least the dimension of a self-dual code, hence it must be a Type II code giving that α must be a multiple of 8.   □

Lemma 4.24 ([32]). *If $\mathcal{C}$ is a non-separable $\mathbb{Z}_2\mathbb{Z}_4$-additive self-dual code, then there exist two codewords $\mathbf{v} = (v \mid v')$ and $\mathbf{w} = (w \mid w')$ with $(v \cdot w)_2 = 1$ and $(v' \cdot w')_4 = 2$.*

Proof. If $(v \cdot w)_2 = 0$, then $2(v \cdot w)_2 = 0$ and so $(v' \cdot w')_4 = 0$ as well. If $(v \cdot w)_2 = 1$, then $(v' \cdot w')_4 = 2$. This last case must occur. Otherwise, $\mathcal{C}_X$ and $\mathcal{C}_Y$ would be self-orthogonal and, by Theorem 4.8, $\mathcal{C}$ would be separable. □

Corollary 4.25 ([32]). *Let $\mathcal{C}$ be a $\mathbb{Z}_2\mathbb{Z}_4$-additive self-dual code of type $(\alpha, \beta; \gamma, \delta; \kappa)$. If $\mathcal{C}$ is non-separable, then $\delta \geq 1$.*

Proof. By Lemma 4.24, there are two codewords $(v \mid v'), (w \mid w') \in \mathcal{C}$ such that $(v \cdot w)_2 = 1$ and $(v' \cdot w')_4 = 2$. Therefore, either v' or w' has a quaternary coordinate of order four and $\delta \geq 1$. □

Theorem 4.26 ([32]). *Let $\mathcal{C}$ be a $\mathbb{Z}_2\mathbb{Z}_4$-additive self-dual code of type $(\alpha, \beta; \gamma, \delta; \kappa)$, with $\alpha, \beta > 0$.*

   i) *If $\mathcal{C}$ is Type 0, then $\alpha \geq 2$ and $\beta \geq 2$.*

   ii) *If $\mathcal{C}$ is Type I and separable, then $\alpha \geq 2$ and $\beta \geq 1$.*

   iii) *If $\mathcal{C}$ is Type I and non-separable, then $\alpha \geq 4$ and $\beta \geq 4$.*

   iv) *If $\mathcal{C}$ is Type II, then $\alpha \geq 8$ and $\beta \geq 4$.*

Proof. By Lemma 4.3, $\alpha = 2\kappa$. If $\mathcal{C}$ is Type 0, then there is a codeword $(x \mid x') \in \mathcal{C}$, where $\text{wt}_H(x)$ is odd and hence $p(x') \equiv 2 \pmod 4$ by Lemma 4.5. Then, $\alpha \geq 2$ and $\beta \geq 2$.

If $\mathcal{C} = \mathcal{C}_X \times \mathcal{C}_Y$ is a Type I separable code, then $\mathcal{C}_X$ is binary self-dual and $\mathcal{C}_Y$ is quaternary self-dual by Theorem 4.8. Therefore, $\alpha \geq 2$ and $\beta \geq 1$.

Assume $\mathcal{C}$ is Type I non-separable. By Lemma 4.6, $(\mathcal{C}_b)_X$ is self-dual, and all codewords of $\mathcal{C}_X$ have even weight by Lemma 4.5. Therefore, if $\alpha = 2$ then $(\mathcal{C}_b)_X = \mathcal{C}_X$, which is not possible. Hence α must be at least 4. Moreover, by Lemma 4.5, $p(y) \equiv 0 \pmod 4$ for all $y \in \mathcal{C}_Y$. Since there is a codeword $y \in \mathcal{C}_Y$ with $p(y) \geq 1$ by Corollary 4.25, $p(y) \geq 4$ and β must also be at least 4.

If $\mathcal{C}$ is Type II, $\alpha \geq 8$ since $\alpha \equiv 0 \pmod 8$ by Theorem 4.23. By Corollary 4.21, $\alpha + 2\beta \equiv 0 \pmod 8$, so $2\beta \equiv 0 \pmod 8$ and $\beta \geq 4$. □

Let α_{min} and β_{min} be the minimum values of α and β, respectively, given in Theorem 4.26 for each case (i)–(iv). Note that the codes $\mathcal{C}_1, \mathcal{C}_2, \mathcal{C}_3, \mathcal{C}_5$ and $\mathcal{C}_6$ in Examples 4.11–4.15, respectively, have α and β as the minimum values α_{min} and β_{min}.

Theorem 4.27 ([32]). *Let α_{min} and β_{min} be as defined above.*

   i) *There exist a Type 0 and a Type I code of type $(\alpha, \beta; \gamma, \delta; \kappa)$ if and only if $\alpha = \alpha_{min} + 2a$ for any integer $a \geq 0$.*

ii) *There exist a Type II code of type $(\alpha, \beta; \gamma, \delta; \kappa)$ if and only if $\alpha = \alpha_{min} +$ $8a$ and $\beta = \beta_{min} + 4b$ for any integers $a, b \geq 0$.*

Proof. The necessary conditions are given by Lemma 4.3, Corollary 4.21 and Theorem 4.23. Moreover, we know that there exist a $\mathbb{Z}_2\mathbb{Z}_4$-additive code of type $(\alpha_{min}, \beta_{min}; \gamma, \delta; \kappa)$ for each case (i) to (iv) in Theorem 4.26.

Let $\mathcal{C}$ be a Type 0 or Type I code of type $(\alpha, \beta; \gamma, \delta; \kappa)$, with $\alpha = \alpha_{min} + 2a$ and $a \geq 0$. Consider the Type I codes $\mathcal{C}'$ and $\mathcal{C}''$ generated by $\mathcal{G}'$ and $\mathcal{G}''$, respectively, where

$$\mathcal{G}' = \left(\, 1 \;\; 1 \, | \right) \quad \text{and} \quad \mathcal{G}'' = \left(|\, 2 \, \right). \tag{4.4}$$

Then, by Lemma 4.18, $\mathcal{C} \times \mathcal{C}'$ is a $\mathbb{Z}_2\mathbb{Z}_4$-additive self-dual code of type $(\alpha + 2, \beta; \gamma + 1, \delta; \kappa + 1)$ and $\mathcal{C} \times \mathcal{C}''$ is a $\mathbb{Z}_2\mathbb{Z}_4$-additive self-dual code of type $(\alpha, \beta + 1; \gamma + 1, \delta; \kappa)$. If $\mathcal{C}$ is a Type 0 (resp. Type I) code, then both codes $\mathcal{C} \times \mathcal{C}'$ and $\mathcal{C} \times \mathcal{C}''$ are Type 0 (resp. Type I).

Let $\mathcal{C}$ be a Type II code of type $(\alpha, \beta; \gamma, \delta; \kappa)$, with $\alpha = \alpha_{min} + 8a$, $\beta = \beta_{min} + 4b$, and $a, b \geq 0$. Consider the following two Type II codes: the extended Hamming code $\mathcal{C}'$ of length 8 and the code $\mathcal{C}''$ generated by

$$\mathcal{G}'' = \begin{pmatrix} | & 2 & 2 & 0 & 0 \\ | & 2 & 0 & 2 & 0 \\ | & 1 & 1 & 1 & 1 \end{pmatrix}. \tag{4.5}$$

By Lemma 4.18, $\mathcal{C} \times \mathcal{C}'$ is a Type II code of type $(\alpha + 8, \beta; \gamma + 8, \delta; \kappa + 8)$ and $\mathcal{C} \times \mathcal{C}''$ is a Type II code of type $(\alpha, \beta + 4; \gamma + 2, \delta + 1; \kappa)$. □

Finally, we remark a special case where the binary Gray map image is also a self-dual code.

Theorem 4.28 ([32]). *If $\mathcal{C}$ is a $\mathbb{Z}_2\mathbb{Z}_4$-additive Type II code and $\Phi(\mathcal{C})$ is linear, then $\Phi(\mathcal{C})$ is a binary Type II code.*

Proof. For any pair of codewords $v, w \in \Phi(\mathcal{C})$, we have that $v + w \in \Phi(\mathcal{C})$. Hence v, w and $v + w$ have doubly-even weight. Since $\text{wt}_H(v+w) = \text{wt}_H(v) + \text{wt}_H(w) - 2\,\text{wt}_H(v * w)$, where $v * w$ is the component-wise product, we have that $\text{wt}_H(v * w)$ is even. In other words, v and w are orthogonal. Therefore, $\Phi(\mathcal{C})$ is a self-dual code. □

In Table 4.1, we summarize all the results related with the separability, antipodality, weight enumerator and the allowable values of α and β for a $\mathbb{Z}_2\mathbb{Z}_4$-additive self-dual code depending on its weight parity.

	Type 0	Type I		Type II
separability	non-separable	separable or non-separable		separable or non-separable
antipodality	non-antipodal	antipodal		antipodal
allowable α, β	$\alpha = 2a \geq 2$ $\beta \geq 2$	sep. $\alpha = 2a \geq 2$ $\beta \geq 1$	non-sep. $\alpha = 2a \geq 4$ $\beta \geq 4$	$\alpha = 8a \geq 8$ $\beta = 4b \geq 4$
$W_{\mathcal{C}}(x,y)$	$\mathbb{C}[x^2 + y^2,$ $y(x-y)]$	$\mathbb{C}[x^2 + y^2,$ $x^2 y^2 (x^2 - y^2)^2]$		$\mathbb{C}[x^8 + 14x^4 y^4 + y^8,$ $x^4 y^4 (x^4 - y^4)^4]$

Table 4.1: Separability, antipodality, weight enumerator and allowable values of α and β for $\mathbb{Z}_2\mathbb{Z}_4$-additive self-dual codes with different weight parity

4.3   Some Constructions of $\mathbb{Z}_2\mathbb{Z}_4$-Additive Self-Dual Codes

In this section, we show different constructions of $\mathbb{Z}_2\mathbb{Z}_4$-additive self-dual codes starting from a given $\mathbb{Z}_2\mathbb{Z}_4$-additive self-dual code. The constructions that are included are the following: the building up construction, the neighbour construction and finally a construction using the shadow of the code.

4.3.1 Building up Construction

The building up construction is a technique for constructing larger self-dual codes from smaller ones by extending the length and, if necessary, adding some codewords. This technique was first given for binary codes in [48] and later in [67], [65], [66], and [32] for self-dual codes over different alphabets.

Let $\mathcal{C}$ be a $\mathbb{Z}_2\mathbb{Z}_4$-additive self-dual code in $\mathbb{Z}_2^\alpha \times \mathbb{Z}_4^\beta$ and let $\mathbf{v} \in \mathbb{Z}_2^\alpha \times \mathbb{Z}_4^\beta$ such that $\mathbf{v} \notin \mathcal{C}$. We define $\mathcal{C}_{\mathbf{v}} = \{\mathbf{u} \in \mathcal{C} : \mathbf{u} \cdot \mathbf{v} = 0\}$. It is immediate that $\mathcal{C}_{\mathbf{v}}$ is a subgroup of $\mathcal{C}$ and that the index $[\mathcal{C} : \mathcal{C}_{\mathbf{v}}]$ is either 2 or 4. In either case, we have that $[\mathcal{C} : \mathcal{C}_{\mathbf{v}}] = [\mathcal{C}_{\mathbf{v}}^\perp : \mathcal{C}]$ (indeed, $|\mathcal{C}||\mathcal{C}| = 2^{\alpha + 2\beta} = |\mathcal{C}_{\mathbf{v}}||\mathcal{C}_{\mathbf{v}}^\perp|$) and that $\mathcal{C}_{\mathbf{v}}^\perp = \langle \mathcal{C}, \mathbf{v} \rangle$. Let $\mathbf{w}$ be a vector such that $\mathcal{C} = \langle \mathcal{C}_{\mathbf{v}}, \mathbf{w} \rangle$. We can then write $\mathcal{C}_{\mathbf{v}}^\perp = \langle \mathcal{C}_{\mathbf{v}}, \mathbf{w}, \mathbf{v} \rangle$.

Example 4.29. Let $\mathcal{C}$ be the $\mathbb{Z}_2\mathbb{Z}_4$-additive self-dual code given in Examples 4.1 and 4.11, and let $\mathbf{v} = (0, 0 \mid 2, 0) \notin \mathcal{C}$. Then, $\mathcal{C}_{\mathbf{v}} = \{(0, 0 \mid 0, 0), (1, 1 \mid 2, 0), (0, 0 \mid 2, 2), (1, 1 \mid 0, 2)\}$, which is generated by

$$\mathcal{G}_{\mathbf{v}} = \begin{pmatrix} 1 & 1 & | & 2 & 0 \\ 0 & 0 & | & 2 & 2 \end{pmatrix},$$

and $\mathcal{C} = \langle \mathcal{C}_{\mathbf{v}}, \mathbf{w} \rangle$, where $\mathbf{w} = (0, 1 \mid 1, 1)$. Therefore, a generator matrix for the code $\mathcal{C}_{\mathbf{v}}^{\perp}$ is

$$
\mathcal{H}_{\mathbf{v}} = \begin{pmatrix} 1 & 1 & 2 & 0 \\ 0 & 0 & 2 & 2 \\ 0 & 0 & 2 & 0 \\ 0 & 1 & 1 & 1 \end{pmatrix}.
$$

$\triangle$

We can form a $\mathbb{Z}_2\mathbb{Z}_4$-additive self-dual code $\mathcal{D}$ by extending the code $\mathcal{C}_{\mathbf{v}}^{\perp}$ in the following way. For $\mathbf{u} = (u_X \mid u_Y) \in \mathcal{C}_{\mathbf{v}}^{\perp}$ we let $\mathbf{u}^* = (u'_X, u_X \mid u_Y, u'_Y)$, where u'_X is an extension of the binary part and u'_Y is an extension of the quaternary part. We choose u'_X and u'_Y so that $(\mathcal{C}_{\mathbf{v}}^{\perp})^* = \langle \{ \mathbf{u}^* : \mathbf{u} \in \mathcal{C}_{\mathbf{v}}^{\perp} \} \rangle$ is self-orthogonal. This is always possible by the proof of Theorem 4.30. For example, if $\mathbf{u} \in \mathcal{C}_{\mathbf{v}}$, we let u'_X and u'_Y be $\mathbf{0}$; and if $\mathbf{u}$ is $\mathbf{w}$ or $\mathbf{v}$, u'_X and u'_Y can be chosen according to Tables 4.2-4.5. Then, $(\mathcal{C}_{\mathbf{v}}^{\perp})^* = \langle \{ \mathbf{u}^* : \mathbf{u} \in \mathcal{C}_{\mathbf{v}} \}, \mathbf{w}^*, \mathbf{v}^* \rangle$. If $(\mathcal{C}_{\mathbf{v}}^{\perp})^*$ is self-dual, then $\mathcal{D} = (\mathcal{C}_{\mathbf{v}}^{\perp})^*$; otherwise, we may need to add additional vectors in order to obtain $\mathcal{D}$. We denote by α' the length of u'_X and by β' the length of u'_Y. We separate the construction into three cases, the first case when $\beta' = 0$, the second when $\alpha' = 0$, and the last one when neither α' nor β' are 0.

Theorem 4.30 ([32]). *Let $\mathcal{C}$ be a $\mathbb{Z}_2\mathbb{Z}_4$-additive self-dual code of type $(\alpha, \beta; \gamma, \delta; \kappa)$ and $\mathbf{v} \notin \mathcal{C}$. Let $\mathcal{C}_{\mathbf{v}}$ and $\mathbf{w}$ be defined as before and $\mathcal{C}_{\mathbf{v}}^{\perp} = \langle \mathcal{C}_{\mathbf{v}}, \mathbf{w}, \mathbf{v} \rangle$. Then, there exists a $\mathbb{Z}_2\mathbb{Z}_4$-additive self-dual code $\mathcal{D} = \langle (\mathcal{C}_{\mathbf{v}}^{\perp})^*, V \rangle$ of type $(\alpha + \alpha', \beta + \beta'; \gamma', \delta'; \kappa')$, for some set of vectors V, satisfying*

   i) $\alpha' \neq 0$ and $\beta' = 0$ only if $\mathbf{v} \cdot \mathbf{w} = 2$ and $\mathbf{v} \cdot \mathbf{v} \in \{0, 2\}$,

   ii) $\alpha' = 0$ and $\beta' \neq 0$ only if

      a) $\mathbf{v} \cdot \mathbf{w} = 2$, or

      b) $\mathbf{v} \cdot \mathbf{w} \in \{1, 3\}$ and $\mathbf{v} \cdot \mathbf{v} \in \{1, 3\}$,

   iii) $\alpha' \neq 0$ and $\beta' \neq 0$.

The proof of Theorem 4.30 can be found in [32]. Here, we describe the steps to obtain a $\mathbb{Z}_2\mathbb{Z}_4$-additive self-dual code $\mathcal{D}$ as an extension of a $\mathbb{Z}_2\mathbb{Z}_4$-additive self-dual code $\mathcal{C}$. First, select $\mathbf{v} \notin \mathcal{C}$ and $\mathbf{w} \in \mathcal{C} \backslash \mathcal{C}_{\mathbf{v}}$ such that $\mathcal{C} = \langle \mathcal{C}_{\mathbf{v}}, \mathbf{w} \rangle$, with the conditions on $\mathbf{v} \cdot \mathbf{v}$ and $\mathbf{v} \cdot \mathbf{w}$ described in Theorem 4.30. After that, determine the values of v'_X, v'_Y, w'_X, w'_Y and V from Tables 4.2-4.5.

Finally, if $\mathcal{G}_\mathbf{v}$ is a generator matrix of $\mathcal{C}_\mathbf{v}$, then a generator matrix of $\mathcal{D}$ is

$$
\mathcal{G}_\mathcal{D} = \begin{pmatrix} \mathbf{0} & \mathcal{G}_\mathbf{v} & \mathbf{0} \\ v'_X & \mathbf{v} & v'_Y \\ w'_X & \mathbf{w} & w'_Y \\ & V & \end{pmatrix} .
$$

$\mathbf{v} \cdot \mathbf{v}$	v'_X	w'_X	V
0	$(0,0,1,1)$	$(0,1,0,1)$	$\{(1,1,1,1,\mathbf{0})\}$
2	$(0,1)$	$(1,1)$	$\emptyset$

Table 4.2: Case $\alpha' \neq 0, \beta' = 0$

$\mathbf{v} \cdot \mathbf{v}$	v'_Y	w'_Y	V
0	$(1,1,1,1)$	$(2,0,0,0)$	$\{(\mathbf{0},0,2,2,0),(\mathbf{0},0,0,2,2)\}$
1	$(1,1,1)$	$(2,0,0)$	$\{(\mathbf{0},0,2,2)\}$
2	$(1,1)$	$(2,0)$	$\emptyset$
3	(1)	(2)	$\emptyset$

Table 4.3: Case $\alpha' = 0, \beta' \neq 0, \mathbf{v} \cdot \mathbf{w} = 2$

$\mathbf{v} \cdot \mathbf{v}$	v'_Y	w'_Y	V
1	$(1,1,1,0)$	$(1,1,1,1)$	$\{(\mathbf{0},0,2,2,0),(\mathbf{0},2,2,0,0)\}$
3	$(3,0,0,0)$	$(1,1,1,1)$	$\{(\mathbf{0},0,2,2,0),(\mathbf{0},0,0,2,2)\}$

Table 4.4: Case $\alpha' = 0, \beta' \neq 0, \mathbf{v} \cdot \mathbf{w} = 1$ or 3

$\mathbf{v} \cdot \mathbf{v}$	v'_X	v'_Y	w'_X	w'_Y	V
0	$(1,0)$	$(1,1)$	$(1,1)$	$(2,2)$	$\{(1,1,\mathbf{0},2,0)\}$
1	$(1,0)$	$(1,0)$	$(1,1)$	$(0,2)$	$\{(1,1,\mathbf{0},2,0)\}$
2	$(1,1)$	$(1,3)$	$(1,0)$	$(1,1)$	$\emptyset$
3	$(0,0)$	$(0,1)$	$(1,1)$	$(0,2)$	$\{(1,1,\mathbf{0},2,0)\}$

Table 4.5: Case $\alpha' \neq 0, \beta' \neq 0, \mathbf{v} \cdot \mathbf{w} = 2$

$\mathbf{v}\cdot\mathbf{v}$	v'_X	v'_Y	w'_X	w'_Y	V
0	$(1,0)$	$(1,0,1)$	$(1,0)$	$(1,1,0)$	$\{(1,1,\mathbf{0},2,0,0),(1,1,\mathbf{0},0,2,0)\}$
1	$(1,0)$	$(1,0)$	$(1,0)$	$(1,1)$	$\{(1,1,\mathbf{0},2,0)\}$
2	$(1,1)$	$(0,1,1)$	$(1,0)$	$(1,1,0)$	$\{(1,1,\mathbf{0},2,0,0),(1,1,\mathbf{0},0,2,2)\}$
3	$(1,1)$	$(1,0)$	$(1,0)$	$(1,1)$	$\{(1,1,\mathbf{0},0,2)\}$

Table 4.6: Case $\alpha'\neq 0,\ \beta'\neq 0,\ \mathbf{v}\cdot\mathbf{w}=1$ or 3

Example 4.31. Let $\mathcal{C}$ be the $\mathbb{Z}_2\mathbb{Z}_4$-additive self-dual code given in Example 4.1. Let $\mathbf{v}=(0,0\mid 2,0)$ and $\mathbf{w}=(0,1\mid 1,1)$ be the vectors given in Example 4.29, where it is also shown that

$$\mathcal{G}_{\mathbf{v}}=\begin{pmatrix} 1 & 1 & 2 & 0 \\ 0 & 0 & 2 & 2 \end{pmatrix}$$

is a generator matrix of $\mathcal{C}_{\mathbf{v}}$ and $\mathcal{C}=\langle\mathcal{C}_{\mathbf{v}},\mathbf{w}\rangle$. Now, we construct $\mathbb{Z}_2\mathbb{Z}_4$-additive self-dual codes starting from $\mathcal{C}$ and extending the length with the construction given in Theorem 4.30 and using the Tables 4.2-4.5. We have that $\mathbf{v}\cdot\mathbf{v}=0$ and $\mathbf{v}\cdot\mathbf{w}=2$. Therefore, we can give a construction for each case (i) to (iii) in Theorem 4.30.

i) By considering the values of v'_X, w'_X and V given in Table 4.2, we obtain a $\mathbb{Z}_2\mathbb{Z}_4$-additive self-dual code $\mathcal{D}$ from $\mathcal{C}$, extending the binary part in 4 coordinates, with generator matrix

$$\mathcal{G}_{\mathcal{D}}=\begin{pmatrix} 0 & 0 & 0 & 0 & 1 & 1 & 2 & 0 \\ 0 & 0 & 0 & 0 & 0 & 0 & 2 & 2 \\ 0 & 0 & 1 & 1 & 0 & 0 & 2 & 0 \\ 0 & 1 & 0 & 1 & 0 & 1 & 1 & 1 \\ 1 & 1 & 1 & 1 & 0 & 0 & 0 & 0 \end{pmatrix}.$$

ii) By considering the values of v'_Y, w'_Y and V given in Table 4.3, we obtain a $\mathbb{Z}_2\mathbb{Z}_4$-additive self-dual code $\mathcal{D}$ from $\mathcal{C}$, extending the quaternary part in 4 coordinates, with generator matrix

$$\mathcal{G}_{\mathcal{D}}=\begin{pmatrix} 1 & 1 & 2 & 0 & 0 & 0 & 0 & 0 \\ 0 & 0 & 2 & 2 & 0 & 0 & 0 & 0 \\ 0 & 0 & 2 & 0 & 1 & 1 & 1 & 1 \\ 0 & 1 & 1 & 1 & 2 & 0 & 0 & 0 \\ 0 & 0 & 0 & 0 & 0 & 2 & 2 & 0 \\ 0 & 0 & 0 & 0 & 0 & 0 & 2 & 2 \end{pmatrix}.$$

iii) By considering the values of v'_X, v'_Y, w'_X, w'_Y and V given in Table 4.5, we obtain a $\mathbb{Z}_2\mathbb{Z}_4$-additive self-dual code $\mathcal{D}$ from $\mathcal{C}$, extending the binary part in 2 coordinates and the quaternary part in 2 coordinates, with generator matrix

$$
\mathcal{G}_{\mathcal{D}} = \left(
\begin{array}{cccc|cccc}
0 & 0 & 1 & 1 & 2 & 0 & 0 & 0 \\
0 & 0 & 0 & 0 & 2 & 2 & 0 & 0 \\
1 & 0 & 0 & 0 & 2 & 0 & 1 & 1 \\
1 & 1 & 0 & 1 & 1 & 1 & 2 & 2 \\
1 & 1 & 0 & 0 & 0 & 0 & 2 & 0
\end{array}
\right).
$$

$\triangle$

4.3.2 Neighbour Construction

The following construction technique is a generalization of the neighbour construction used for codes over finite fields [125, Chapter 2].

Two $\mathbb{Z}_2\mathbb{Z}_4$-additive self-dual codes, $\mathcal{C}_1, \mathcal{C}_2 \subseteq \mathbb{Z}_2^\alpha \times \mathbb{Z}_4^\beta$, are called *neighbours* if $|\mathcal{C}_1 \cap \mathcal{C}_2| = 2^{\frac{\alpha+2\beta}{2}-1}$. Let $\mathcal{C}$ be a $\mathbb{Z}_2\mathbb{Z}_4$-additive self-dual code in $\mathbb{Z}_2^\alpha \times \mathbb{Z}_4^\beta$ and let $\mathbf{v}$ be a self-orthogonal vector such that $\mathbf{v} \notin \mathcal{C}$. As before, we denote by $\mathcal{C}_{\mathbf{v}}$ the subcode $\mathcal{C}_{\mathbf{v}} = \{\mathbf{u} \in \mathcal{C} : \mathbf{u} \cdot \mathbf{v} = 0\}$. Define the $\mathbb{Z}_2\mathbb{Z}_4$-additive code $N(\mathcal{C}, \mathbf{v})$ as

$$
N(\mathcal{C}, \mathbf{v}) = \langle \mathcal{C}_{\mathbf{v}}, \mathbf{v} \rangle.
$$

Theorem 4.32 ([32]). *Let $\mathcal{C}$ be a $\mathbb{Z}_2\mathbb{Z}_4$-additive self-dual code and let $\mathbf{v}$ be a self-orthogonal vector such that $\mathbf{v} \notin \mathcal{C}$. Then, $N(\mathcal{C}, \mathbf{v})$ is a $\mathbb{Z}_2\mathbb{Z}_4$-additive self-dual code.*

Proof. Clearly, $N(\mathcal{C}, \mathbf{v})$ is self-orthogonal. Moreover, it is also clear that $|N(\mathcal{C}, \mathbf{v})| = |\mathcal{C}|$. Hence, $N(\mathcal{C}, \mathbf{v})$ is a self-dual code.   □

Moreover, since $\mathcal{C} \cap N(\mathcal{C}_{\mathbf{v}}, \mathbf{v}) = \mathcal{C}_{\mathbf{v}}$, $|\mathcal{C}_{\mathbf{v}}| = |\mathcal{C}|/2$, and $|\mathcal{C}| = 2^{\frac{\alpha+2\beta}{2}}$, we have that $|\mathcal{C} \cap N(\mathcal{C}_{\mathbf{v}}, \mathbf{v})| = 2^{\frac{\alpha+2\beta}{2}-1}$ and $N(\mathcal{C}, \mathbf{v})$ is a neighbour of $\mathcal{C}$. We say that $N(\mathcal{C}, \mathbf{v})$ is the $\mathbb{Z}_2\mathbb{Z}_4$-additive self-dual code obtained from $\mathcal{C}$ and the self-orthogonal vector $\mathbf{v} \notin \mathcal{C}$, by the neighbour construction.

Example 4.33. Let $\mathcal{C}$ be the $\mathbb{Z}_2\mathbb{Z}_4$-additive self-dual code given in Example 4.1. The vector $\mathbf{v} = (0,0 \mid 2,0)$, considered in Example 4.29, is self-orthogonal and $\mathbf{v} \notin \mathcal{C}$. Moreover, the matrix

$$
\mathcal{G}_{\mathbf{v}} = \left(
\begin{array}{cc|cc}
1 & 1 & 2 & 0 \\
0 & 0 & 2 & 2
\end{array}
\right)
$$

is a generator matrix of $\mathcal{C}_{\mathbf{v}}$ as also shown in Examples 4.29 and 4.31. There-fore, $N(\mathcal{C}_{\mathbf{v}}, \mathbf{v})$ has generator matrix

$$\left(\begin{array}{cc|cc} 1 & 1 & 2 & 0 \\ 0 & 0 & 2 & 2 \\ 0 & 0 & 2 & 0 \end{array} \right) \quad \text{or} \quad \left(\begin{array}{cc|cc} 1 & 1 & 0 & 0 \\ 0 & 0 & 2 & 0 \\ 0 & 0 & 0 & 2 \end{array} \right)$$

in standard form, and it is a $\mathbb{Z}_2\mathbb{Z}_4$-additive self-dual code containing $\mathbf{v}$. $\triangle$

Proposition 4.34 ([32]). *Every* $\mathbb{Z}_2\mathbb{Z}_4$*-additive self-dual code can be obtained by repeated application of the neighbour construction starting from any* $\mathbb{Z}_2\mathbb{Z}_4$*-additive self-dual code of that length.*

Proof. Let $\mathcal{C}$ be a $\mathbb{Z}_2\mathbb{Z}_4$-additive self-dual code and let $\mathcal{D}$ be a $\mathbb{Z}_2\mathbb{Z}_4$-additive self-dual code, $\mathcal{D} \neq \mathcal{C}$. Let $\mathbf{v}_1, \mathbf{v}_2, \ldots, \mathbf{v}_k \in \mathcal{D}$ be a set of generators of $\mathcal{D}$ such that $\mathbf{v}_1, \mathbf{v}_2, \ldots, \mathbf{v}_s$ are not in $\mathcal{C}$. Set $\mathcal{D}_i = N(\mathcal{D}_{i-1}, \mathbf{v}_i)$ and $\mathcal{D}_0 = \mathcal{C}$, then $\mathcal{D}_s = \mathcal{D}$ and we have the result. $\square$

4.3.3 Using the Shadow of the Code

Let $\mathcal{C}$ be a $\mathbb{Z}_2\mathbb{Z}_4$-additive code and $\mathcal{C}_0$ be the additive subcode of $\mathcal{C}$ having all its even weight codewords. That is,

$$\mathcal{C}_0 = \{\mathbf{v} \in \mathcal{C} : \text{wt}(\mathbf{v}) \equiv 0 \pmod{2}\}.$$

Note that if $\mathcal{C}$ is an even code, then $\mathcal{C}_0 = \mathcal{C}$. Otherwise, $\mathcal{C}_0$ contains exactly half of the codewords of $\mathcal{C}$. Since $\mathbb{Z}_2\mathbb{Z}_4$-additive Type 0 codes have odd weight codewords, we have the following technical lemma.

Lemma 4.35 ([32]). *Let* $\mathcal{C}$ *be a* $\mathbb{Z}_2\mathbb{Z}_4$*-additive Type 0 code. Then, the subcode* $\mathcal{C}_0 = \{\mathbf{v} \in \mathcal{C} : \text{wt}(\mathbf{v}) \equiv 0 \pmod{2}\}$ *is an additive subcode with* $|\mathcal{C}| = 2|\mathcal{C}_0|$.

Example 4.36. Let $\mathcal{C}$ be the $\mathbb{Z}_2\mathbb{Z}_4$-additive Type 0 code given in Example 4.1. That is,

$$\mathcal{C} = \{(0, 0 \mid 0, 0), (0, 1 \mid 1, 1), (0, 0 \mid 2, 2), (0, 1 \mid 3, 3),$$
$$(1, 1 \mid 2, 0), (1, 0 \mid 3, 1), (1, 1 \mid 0, 2), (1, 0 \mid 1, 3)\}.$$

The vectors that have even weight are precisely the following ones:

$$\mathcal{C}_0 = \{(0, 0 \mid 0, 0), (1, 1 \mid 2, 0), (1, 1 \mid 0, 2), (0, 0 \mid 2, 2)\}, \tag{4.6}$$

which form the additive subcode $\mathcal{C}_0$ generated by

$$\left(\begin{array}{cc|cc} 1 & 1 & 2 & 0 \\ 0 & 0 & 2 & 2 \end{array} \right).$$

We have that $|\mathcal{C}_0| = 4$ and $2^3 = |\mathcal{C}| = 2|\mathcal{C}_0|$. $\triangle$

In the case a $\mathbb{Z}_2\mathbb{Z}_4$-additive code $\mathcal{C}$ is Type 0, we have seen that $\mathcal{C}_0$ is a proper subcode. In the following proposition, we see that the weight enumerator of $\mathcal{C}_0$ is related to the weight enumerator of $\mathcal{C}$.

Proposition 4.37 ([32]). *Let $\mathcal{C}$ be a $\mathbb{Z}_2\mathbb{Z}_4$-additive Type 0 code. Then,*

$$W_{\mathcal{C}_0}(x, y) = \frac{1}{2}(W_{\mathcal{C}}(x, -y) + W_{\mathcal{C}}(x, y)).$$

Proof. The monomials in the weight enumerator of $\mathcal{C}$ are of the form $x^{n-\text{wt}(\mathbf{v})}y^{\text{wt}(\mathbf{v})}$, for $\mathbf{v} \in \mathcal{C}$, where $n = \alpha + 2\beta$. For odd weight vectors, $x^{n-\text{wt}(\mathbf{v})}(-y)^{\text{wt}(\mathbf{v})} = -x^{n-\text{wt}(\mathbf{v})}y^{\text{wt}(\mathbf{v})}$ and, hence, the monomial in $W_{\mathcal{C}}(x, y)$ cancels with the monomial in $W_{\mathcal{C}}(x, -y)$. For even weight vectors, $x^{n-\text{wt}(\mathbf{v})}(-y)^{\text{wt}(\mathbf{v})} = x^{n-\text{wt}(\mathbf{v})}y^{\text{wt}(\mathbf{v})}$, so the vectors are counted twice and dividing by 2 gives the monomial that represents each even weight vector. $\quad\square$

We define the shadow of a $\mathbb{Z}_2\mathbb{Z}_4$-additive self-dual code $\mathcal{C}$ to be $S = \mathcal{C}_0^{\perp} \backslash \mathcal{C}$, and it consists of all the vectors $\mathbf{u} \in \mathbb{Z}_2^{\alpha} \times \mathbb{Z}_4^{\beta}$ satisfying

$$\mathbf{u} \cdot \mathbf{v} = 0 \text{ for all } \mathbf{v} \in \mathcal{C}_0, \text{ and}$$
$$\mathbf{u} \cdot \mathbf{v} \neq 0 \text{ for all } \mathbf{v} \in \mathcal{C} \backslash \mathcal{C}_0.$$

The shadow satisfies $|S| = |\mathcal{C}|$, but it is not an additive code. Shadows of binary self-dual codes first appeared in [158], but were first specifically labelled as a code in [54]. The shadow has been generalized to numerous alphabets, see [125] for a complete description. For a $\mathbb{Z}_2\mathbb{Z}_4$-additive Type 0 code $\mathcal{C}$, the weight enumerator of its shadow S is

$$W_S(x, y) = W_{\mathcal{C}_0^{\perp}}(x, y) - W_{\mathcal{C}}(x, y)$$
$$= \frac{1}{|\mathcal{C}|}W_{\mathcal{C}}(x + y, -(x - y)) = W_{\mathcal{C}}\left(\frac{x+y}{\sqrt{2}}, \frac{-(x-y)}{\sqrt{2}}\right), \qquad (4.7)$$

as can be seen in [32]. Note that the weight enumerator of a $\mathbb{Z}_2\mathbb{Z}_4$-additive Type 0 code is not necessarily a weight enumerator of any binary self-dual code. This is clear since a $\mathbb{Z}_2\mathbb{Z}_4$-additive Type 0 code may have odd weight codewords.

Example 4.38. Let $\mathcal{C}$ be the $\mathbb{Z}_2\mathbb{Z}_4$-additive Type 0 code given in Examples 4.11 and 4.36. The weight enumerator of $\mathcal{C}$, given in Example 4.11, is

$$W_{\mathcal{C}}(x, y) = x^6 + 4x^3y^3 + 3x^2y^4.$$

From the set of codewords of $\mathcal{C}_0$, given in Example 4.36, we have that $W_{\mathcal{C}_0}(x, y) = x^6 + 3x^2y^4$, which can also be computed from $W_{\mathcal{C}_0}(x, y) =$

$\frac{1}{2}(W_{\mathcal{C}}(x,-y) + W_{\mathcal{C}}(x,y))$ by Proposition 4.37. From the set of codewords of $\mathcal{C}_0^{\perp}$,

$$
\begin{aligned}
\mathcal{C}_0^{\perp} = \{ &(0,0 \mid 0,0), (1,0 \mid 1,1), (0,0 \mid 2,2), (1,0 \mid 3,3) \\
&(1,1 \mid 0,0), (0,1 \mid 1,1), (1,1 \mid 2,2), (0,1 \mid 3,3) \\
&(0,0 \mid 0,2), (1,0 \mid 1,3), (0,0 \mid 2,0), (1,0 \mid 3,1) \\
&(1,1 \mid 0,2), (0,1 \mid 1,3), (1,1 \mid 2,0), (0,1 \mid 3,1) \},
\end{aligned}
$$

we have that $W_{\mathcal{C}_0^{\perp}}(x,y) = x^6 + 3x^4y^2 + 8x^3y^3 + 3x^2y^4 + y^6$, which can also be computed from the MacWilliams transform $W_{\mathcal{C}_0^{\perp}}(x,y) = \frac{1}{|\mathcal{C}_0|} W_{\mathcal{C}_0}(x+y, x-y)$ given in Proposition 3.6. Finally, the set of codewords of the shadow of $\mathcal{C}$, $S = \mathcal{C}_0^{\perp} \setminus \mathcal{C}$, is

$$
\begin{aligned}
S = \{ &(1,1 \mid 0,0), (1,0 \mid 1,1), (0,1 \mid 1,3), (0,0 \mid 2,0), \\
&(1,0 \mid 3,3), (0,0 \mid 0,2), (1,1 \mid 2,2), (0,1 \mid 3,1) \}.
\end{aligned}
$$

Therefore, $W_S(x,y) = 3x^4y^2 + 4x^3y^3 + y^6$, which coincides with $W_S(x,y) = W_{\mathcal{C}_0^{\perp}}(x,y) - W_{\mathcal{C}}(x,y)$ and $W_S(x,y) = \frac{1}{|\mathcal{C}|} W_{\mathcal{C}}(x+y, -(x-y))$ by (4.7). △

Let $\mathcal{C}$ be a $\mathbb{Z}_2\mathbb{Z}_4$-additive Type 0 code. We see that the code $\mathcal{C}_0$ has 4 cosets in $\mathcal{C}_0^{\perp}$. Let $\mathcal{C}_{0,0} = \mathcal{C}_0$ and $\mathcal{C}_{1,0} = \mathcal{C} \setminus \mathcal{C}_{0,0}$. Then, we have that

$$
\mathcal{C}_0^{\perp} = \mathcal{C}_{0,0} \cup \mathcal{C}_{1,0} \cup \mathcal{C}_{0,1} \cup \mathcal{C}_{1,1},
$$

$$
\mathcal{C} = \mathcal{C}_{0,0} \cup \mathcal{C}_{1,0} \quad \text{and} \quad S = \mathcal{C}_{0,1} \cup \mathcal{C}_{1,1}.
$$

There are vectors $\mathbf{t}$ and $\mathbf{s}$ such that $\mathcal{C} = \langle \mathcal{C}_0, \mathbf{t} \rangle$ and $\mathcal{C}_0^{\perp} = \langle \mathcal{C}, \mathbf{s} \rangle$. Then, we have that $\mathcal{C}_{i,j} = \mathcal{C}_0 + i\mathbf{t} + j\mathbf{s}$. Note that $(\mathbf{1} \mid \mathbf{2}) \in \mathcal{C}_0^{\perp}$ and, by Proposition 4.10, $\mathcal{C}$ is non-antipodal. Then, $(\mathbf{1} \mid \mathbf{2}) \in \mathcal{C}_0^{\perp} \setminus \mathcal{C} = S$, and we can take $\mathbf{s} = (\mathbf{1} \mid \mathbf{2})$. We have that $\mathcal{C}_{0,0}$ and $\mathcal{C}_{0,1}$ consist of even weight vectors and $\mathcal{C}_{1,0}$ and $\mathcal{C}_{1,1}$ consist of odd weight vectors. Moreover, we have that $\mathbf{s} \cdot \mathbf{s} = \mathbf{t} \cdot \mathbf{t} = 0$ and $\mathbf{s} \cdot \mathbf{t} = 2$. Thus, we can define $\mathcal{C}_{i,j} \cdot \mathcal{C}_{i',j'}$ as the value of $\mathbf{x} \cdot \mathbf{y}$ for any $\mathbf{x} \in \mathcal{C}_{i,j}$ and any $\mathbf{y} \in \mathcal{C}_{i',j'}$ $(i,j,i',j' \in \{0,1\})$. Note that $\mathcal{C}_{i,j} \cdot \mathcal{C}_{i',j'} \in \{0,2\}$ and this product is well-defined. Hence, in Table 4.7, we show the orthogonality relations among the four cosets of $\mathcal{C}_0^{\perp}$. From these relationships, we obtain the following result.

Proposition 4.39 ([32]). *Let $\mathcal{C}$ be a $\mathbb{Z}_2\mathbb{Z}_4$-additive Type 0 code. Then, the codes $\mathcal{C}_{0,0} \cup \mathcal{C}_{0,1} = \langle \mathcal{C}_0, \mathbf{s} \rangle$ and $\mathcal{C}_{0,0} \cup \mathcal{C}_{1,1} = \langle \mathcal{C}_0, \mathbf{t}+\mathbf{s} \rangle$ are $\mathbb{Z}_2\mathbb{Z}_4$-additive self-dual codes. The code $\langle \mathcal{C}_0, \mathbf{t} + \mathbf{s} \rangle$ is Type 0 and the code $\langle \mathcal{C}_0, \mathbf{s} \rangle$ is not Type 0.*

The $\mathbb{Z}_2\mathbb{Z}_4$-additive self-dual codes $\mathcal{C} = \langle \mathcal{C}_0, \mathbf{t} \rangle$, $\langle \mathcal{C}_0, \mathbf{s} \rangle$ and $\langle \mathcal{C}_0, \mathbf{t} + \mathbf{s} \rangle$ considered in Proposition 4.39 are, in fact, neighbour codes of $\mathcal{C}_0$ and $\mathcal{C}_0^{\perp}$. These neighbour relations are shown in Figure 4.1.

	$\mathcal{C}_{0,0}$	$\mathcal{C}_{1,0}$	$\mathcal{C}_{0,1}$	$\mathcal{C}_{1,1}$
$\mathcal{C}_{0,0}$	$\perp$	$\perp$	$\perp$	$\perp$
$\mathcal{C}_{1,0}$	$\perp$	$\perp$	$\not\perp$	$\not\perp$
$\mathcal{C}_{0,1}$	$\perp$	$\not\perp$	$\perp$	$\not\perp$
$\mathcal{C}_{1,1}$	$\perp$	$\not\perp$	$\not\perp$	$\perp$

Table 4.7: Orthogonality relations given by the inner products

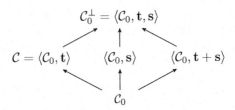

Figure 4.1: Neighbour codes from $\mathcal{C}_0$

Example 4.40. By using the MAGMA package for $\mathbb{Z}_2\mathbb{Z}_4$-additive codes [29], we construct the $\mathbb{Z}_2\mathbb{Z}_4$-additive Type 0 code given in Example 4.1 and explored in the subsequent examples. We also construct the partition of $\mathcal{C}_0^{\perp}$ into 4 cosets of $\mathcal{C}_0$, and check the properties of the $\mathbb{Z}_2\mathbb{Z}_4$-additive self-dual codes defined in Proposition 4.39.

```
> Z4 := IntegerRing(4);
> alpha := 2; beta := 2;
> V := RSpace(Z4, alpha + beta);
> C := Z2Z4AdditiveCode([V![1,1, 2,0], V![0,1, 1,1]], alpha);
> CO := Z2Z4AdditiveCode([c : c in Set(C) |
                    LeeWeight(c, alpha) mod 2 eq 0], alpha);
> COdual := Dual(CO);
> Shadow := {c : c in Set(COdual) | c notin C};
> P<x,y> := PolynomialRing(Integers(), 2);
> WC<x,y> := P!LeeWeightEnumerator(C);  WC;
x^6 + 4*x^3*y^3 + 3*x^2*y^4
> WCO<x,y> := P!LeeWeightEnumerator(CO);  WCO;
x^6 + 3*x^2*y^4
> WCOdual<x,y> := P!LeeWeightEnumerator(COdual); WCOdual;
x^6 + 3*x^4*y^2 + 8*x^3*y^3 + 3*x^2*y^4 + y^6
> WShadow := WCOdual - WC; WShadow;
3*x^4*y^2 + 4*x^3*y^3 + y^6
>
> COO := {c : c in Set(CO)};
> C10 := {c : c in Set(C) | c notin CO};
> t := Random(C10);
```

```
> s := V![2^^(alpha+beta)];
> C01 := {c + s : c in Set(C00)};
> C11 := {c + t + s : c in Set(C00)};
> Set(COdual) eq (C00 join C10 join C01 join C11);
true
>
> G0 := GeneratorMatrix(C0);
> Ct := Z2Z4AdditiveCode(Rows(G0) cat [t], alpha);
> Cs := Z2Z4AdditiveCode(Rows(G0) cat [s], alpha);
> Cts := Z2Z4AdditiveCode(Rows(G0) cat [s+t], alpha);
> Ct eq C;
true
> Set(Cs) eq (C00 join C01);
true
> Set(Cts) eq (C00 join C11);
true
> IsSelfDual(Ct) and (Ct subset COdual);
true
> IsSelfDual(Cs) and (Cs subset COdual);
true
> IsSelfDual(Cts) and (Cts subset COdual);
true
>
> WCt<x,y> := LeeWeightEnumerator(Ct); WCt;
x^6 + 4*x^3*y^3 + 3*x^2*y^4
> WCs<x,y> := LeeWeightEnumerator(Cs); WCs;
x^6 + 3*x^4*y^2 + 3*x^2*y^4 + y^6
> WCts<x,y> := LeeWeightEnumerator(Cst); WCts;
x^6 + 4*x^3*y^3 + 3*x^2*y^4
>
> #(C0 meet Ct) eq  2^((alpha+2*beta)/2 -1);
true
> #(C0 meet Cs) eq  2^((alpha+2*beta)/2 -1);
true
> #(C0 meet Cts) eq 2^((alpha+2*beta)/2 -1);
true                                                    △
```

We can now generalize the construction first described in [48], for binary singly-even self-dual codes, but greatly expanded in [75], for singly-even self-dual codes over different rings.

Theorem 4.41 ([32]). *Let* $\mathcal{C}$ *and* $\mathcal{D}$ *be* $\mathbb{Z}_2\mathbb{Z}_4$*-additive Type 0 codes in* $\mathbb{Z}_2^\alpha \times \mathbb{Z}_4^\beta$ *and* $\mathbb{Z}_2^{\alpha'} \times \mathbb{Z}_4^{\beta'}$, *respectively. Consider* $\mathcal{C}_{i,j}$ *and* $\mathcal{D}_{i,j}$ *for all* $i,j \in \{0,1\}$ *the cosets of* $\mathcal{C}_0$ *and* $\mathcal{D}_0$, *respectively. Then,*

$$(\mathcal{C}_{0,0} \times \mathcal{D}_{0,0}) \cup (\mathcal{C}_{0,1} \times \mathcal{D}_{0,1}) \cup (\mathcal{C}_{1,0} \times \mathcal{D}_{1,0}) \cup (\mathcal{C}_{1,1} \times \mathcal{D}_{1,1}),$$

and

$$(\mathcal{C}_{0,0} \times \mathcal{D}_{0,0}) \cup (\mathcal{C}_{0,1} \times \mathcal{D}_{1,1}) \cup (\mathcal{C}_{1,0} \times \mathcal{D}_{1,0}) \cup (\mathcal{C}_{1,1} \times \mathcal{D}_{0,1})$$

are $\mathbb{Z}_2\mathbb{Z}_4$-additive self-dual codes in $\mathbb{Z}_2^{\alpha+\alpha'} \times \mathbb{Z}_4^{\beta+\beta'}$.

Proof. It is easy to see that all the vectors are pairwise orthogonal and that the code is $\mathbb{Z}_2\mathbb{Z}_4$-additive.   □

4.4 $\mathbb{Z}_2\mathbb{Z}_4$-Additive Formally Self-Dual Codes

In general, a code C over any ring is said to be *formally self-dual* if its weight enumerator is the same as the weight enumerator of its dual. In this section, we consider $\mathbb{Z}_2\mathbb{Z}_4$-additive formally self-dual codes and summarise the results given in [71].

A $\mathbb{Z}_2\mathbb{Z}_4$-additive code C is *formally self-dual* if $W_{C^\perp}(x,y) = W_C(x,y)$, with respect to the weight enumerator given in (3.4). Note that any self-dual code is necessarily formally self-dual but, of course, there are formally self-dual codes that are not self-dual.

Example 4.42. Let C be the $\mathbb{Z}_2\mathbb{Z}_4$-additive code having the following generator and parity check matrices:

$$\mathcal{G} = \begin{pmatrix} 0 & 1 & 0 \\ 1 & 0 & 0 \end{pmatrix} \quad \text{and} \quad \mathcal{H} = \begin{pmatrix} 0 & 0 & 1 \end{pmatrix},$$

respectively. We have that $C \neq C^\perp$, so C is not a $\mathbb{Z}_2\mathbb{Z}_4$-additive self-dual code. The weight enumerators of C and $C^\perp$ satisfy $W_C(x,y) = W_{C^\perp}(x,y) = x^4 + 2x^3y + x^2y^2$. Hence, C is $\mathbb{Z}_2\mathbb{Z}_4$-additive formally self-dual. The code C has parameters $(2,1;2,0;2)$ whereas the code $C^\perp$ has parameters $(2,1;0,1;0)$. Note that a $\mathbb{Z}_2\mathbb{Z}_4$-additive formally self-dual code does not have necessarily the same parameters as its dual.   △

Let C be a binary (possibly non-linear) code. Then, we say that C is a *formally self-dual code* if the weight enumerator of C is held invariant by the action of the MacWilliams transform. In general, if C is a binary non-linear code, then we do not have that $W_C(x,y) = W_{C^\perp}(x,y)$ since $C^\perp = \langle C \rangle^\perp$ and $\langle C \rangle$ is larger than C. However, what we are seeking are binary codes that are images of $\mathbb{Z}_2\mathbb{Z}_4$-additive codes under the Gray map having their weight enumerator held invariant by the MacWilliams transform. The weight enumerator is held invariant by the action of the MacWilliams transform, and hence the invariant theory for binary self-dual codes described in [112, Chapter 19] also applies to C in order to study the possible weight enumerators.

The following theorem is immediate, given that the Gray map is an isometry.

Theorem 4.43 ([71]). *If $\mathcal{C}$ is a $\mathbb{Z}_2\mathbb{Z}_4$-additive formally self-dual code, then $\Phi(\mathcal{C})$ is a binary formally self-dual code.*

The results given by Lemma 4.18 and Proposition 4.4 for $\mathbb{Z}_2\mathbb{Z}_4$-additive self-dual codes can be extended to $\mathbb{Z}_2\mathbb{Z}_4$-additive formally self-dual codes as shown through the following two results.

Lemma 4.44 ([71]). *If $\mathcal{C}$ and $\mathcal{D}$ are $\mathbb{Z}_2\mathbb{Z}_4$-additive formally self-dual codes, then $\mathcal{C} \times \mathcal{D}$ is a $\mathbb{Z}_2\mathbb{Z}_4$-additive formally self-dual code.*

Corollary 4.45 ([71]). *Let $\mathcal{C}$ be a binary formally self-dual code and $\mathcal{D}$ a quaternary formally self-dual code. Then, $\mathcal{C} \times \mathcal{D}$ is a $\mathbb{Z}_2\mathbb{Z}_4$-additive formally self-dual code.*

Similarly to the results given for $\mathbb{Z}_2\mathbb{Z}_4$-additive self-dual codes in Section 4.1, we can find the possible weight enumerators, the allowable parameters α and β and the relation among weight parity, separability and antipodality for $\mathbb{Z}_2\mathbb{Z}_4$-additive formally self-dual codes. We summarise the results in Table 4.8. Even though most of the results are similar to the ones of $\mathbb{Z}_2\mathbb{Z}_4$-additive self-dual codes, there are some differences. For example, unlike for $\mathbb{Z}_2\mathbb{Z}_4$-additive self-dual codes, a $\mathbb{Z}_2\mathbb{Z}_4$-additive formally self-dual Type 0 code can be separable as shown in the following example.

Example 4.46. Consider the $\mathbb{Z}_2\mathbb{Z}_4$-additive code $\langle (1, 0\,|\,) \rangle \times \langle (|\,2) \rangle \subseteq \mathbb{Z}_2^2 \times \mathbb{Z}_4$. Its orthogonal code is $\langle (0, 1\,|\,) \rangle \times \langle (|\,2) \rangle$. These codes have vectors of weight 1 and hence are odd. Therefore, both codes are odd separable formally self-dual. Moreover, they are separable but not antipodal; i.e., $(1, 1\,|\,2)$ is not in any of these codes.   △

In [71], we can find different constructions of $\mathbb{Z}_2\mathbb{Z}_4$-additive formally self-dual codes starting from a $\mathbb{Z}_2\mathbb{Z}_4$-additive formally self-dual code $\mathcal{C}$; namely the building up and neighbour constructions. These constructions are described for self-dual codes in Section 4.3.1 and Section 4.3.2, respectively. For the building up construction, if $C = \Phi(\mathcal{C})$ is linear, then we can consider the code $\bar{\mathcal{C}}$ constructed as $\bar{\mathcal{C}} = \Phi^{-1}(\bar{C})$, where $\bar{C}$ is obtained from C by applying the binary building up construction, as in [68]. However, usually and most interestingly, $\Phi(\mathcal{C})$ is non-linear. In this case, in [71], the building up construction is generalized for $\mathbb{Z}_2\mathbb{Z}_4$-additive formally self-dual codes whose Gray map image is non-linear. Finally, for the neighbour construction, if $\mathcal{C}$ is a $\mathbb{Z}_2\mathbb{Z}_4$-additive formally self-dual Type 0 code, then a $\mathbb{Z}_2\mathbb{Z}_4$-additive even formally self-dual code is obtained with the same length as $\mathcal{C}$.

	Type 0	Type I	Type II
separability	separable or non-separable	separable or non-separable	separable or non-separable
antipodality	non-antipodal	antipodal	antipodal
allowable α, β $a, b, c \geq 0$	$\alpha = 2 + 2a$, $\beta = b$ or $\alpha = 0$, $\beta = 2 + b$	$\alpha = 2a$ $\beta = b$ $a + b > 0$	$\alpha = 8a + 4b$ $\beta = 2b + 4c$ $a + b + c > 0$
$W_{\mathcal{C}}(x, y)$	$\mathbb{C}[x^2 + y^2,$ $y(x - y)]$	$\mathbb{C}[x^2 + y^2,$ $x^8 + 14x^4y^4 + y^8]$	$\mathbb{C}[x^8 + 14x^4y^4 + y^8,$ $x^4y^4(x^4 - y^4)^4]$

Table 4.8: Separability, antipodality, weight enumerator and allowable values of α and β for $\mathbb{Z}_2\mathbb{Z}_4$-additive formally self-dual codes with different weight parity

Chapter 5

Linearity, Rank and Kernel of $\mathbb{Z}_2\mathbb{Z}_4$-Linear Codes

In previous chapters, we have seen that the binary Gray map image of a $\mathbb{Z}_2\mathbb{Z}_4$-additive code is not linear in general. Two structural properties of binary non-linear codes are the rank and the dimension of the kernel. The rank of a binary code C is simply the dimension of the linear span of the codewords of C. The kernel of a binary code C is the set of vectors that leave C invariant under translation. If C contains the all-zero vector, then the kernel is a binary linear subcode of C and we can consider its dimension.

The rank and dimension of the kernel have been studied for some families of $\mathbb{Z}_2\mathbb{Z}_4$-linear codes [34, 36, 43, 44, 102, 118, 119, 122]. These two invariants do not always give a full classification of $\mathbb{Z}_2\mathbb{Z}_4$-linear codes, since two $\mathbb{Z}_2\mathbb{Z}_4$-linear codes which are not permutation equivalent could have the same rank and dimension of the kernel. In spite of that, they can help in classification, since if two $\mathbb{Z}_2\mathbb{Z}_4$-linear codes have different ranks or dimensions of the kernel, they are not permutation equivalent. Moreover, in this case, the corresponding $\mathbb{Z}_2\mathbb{Z}_4$-additive codes are not monomially equivalent, so these two invariants can also help to distinguish between $\mathbb{Z}_2\mathbb{Z}_4$-additive codes that are not monomially equivalent.

In this chapter, we study the linearity of $\mathbb{Z}_2\mathbb{Z}_4$-linear codes and give some results on their linear span and kernel, as long as the possible values of their

There are parts of this chapter that have been previously published. Reprinted by permission from Springer Nature Customer Service Centre GmbH: C. Fernández-Córdoba, J. Pujol, and M. Villanueva. "$\mathbb{Z}_2\mathbb{Z}_4$-linear codes: rank and kernel". *Designs, Codes and Cryptography*, v. 56, 1, pp. 43–59. ©(2010). Reprinted by permission from IEEE: J. Borges, S. T. Dougherty, C. Fernández-Córdoba, and R. Ten-Valls. "$\mathbb{Z}_2\mathbb{Z}_4$-additive cyclic codes: kernel and rank". *IEEE Transactions on Information Theory*, v. 65, 4, pp. 2119–2127 ©(2019).

© Springer Nature Switzerland AG 2022
J. Borges et al., *$\mathbb{Z}_2\mathbb{Z}_4$-Linear Codes*, https://doi.org/10.1007/978-3-031-05441-9_5

dimensions. Specifically, in Section 5.1, we show different characterizations for a $\mathbb{Z}_2\mathbb{Z}_4$-linear code to be linear. In Sections 5.2 and 5.3, we determine the possible ranks and dimensions of the kernel, respectively, for $\mathbb{Z}_2\mathbb{Z}_4$-linear codes; and how to construct a $\mathbb{Z}_2\mathbb{Z}_4$-linear code having any possible rank r and dimension of the kernel k. In Section 5.4, we study the possible pairs of values (r, k) for $\mathbb{Z}_2\mathbb{Z}_4$-linear codes. In these sections, we also show that the linearity and the values of the rank and dimension of the kernel of a $\mathbb{Z}_2\mathbb{Z}_4$-linear code $C = \Phi(\mathcal{C})$ can also be given in terms of the linearity and the values of the rank and dimension of the kernel of a $\mathbb{Z}_4$-linear code that is a subcode of $\phi(\mathcal{C}_Y)$. Most of the results of this chapter are obtained from [33, 78, 79].

Unless stated otherwise, we always assume that the generator matrix of a $\mathbb{Z}_2\mathbb{Z}_4$-additive code is in the form of (2.5). In the case, we say that it is in standard form, recall that it is in the form of (2.7).

5.1 Linearity of $\mathbb{Z}_2\mathbb{Z}_4$-Linear Codes

In this section, we study the linearity of $\mathbb{Z}_2\mathbb{Z}_4$-linear codes. We give different characterizations for a $\mathbb{Z}_2\mathbb{Z}_4$-linear code to be linear. Specifically, we describe some of them in terms of the rows in a generator matrix of the corresponding $\mathbb{Z}_2\mathbb{Z}_4$-additive code $\mathcal{C}$, and another one by considering a $\mathbb{Z}_4$-linear code which is a subcode of $\phi(\mathcal{C}_Y)$.

The linearity of $\mathbb{Z}_4$-linear codes was studied in [92, 157]. The key was to consider the component-wise product $u * v$ for pairs of codewords u, v over $\mathbb{Z}_4$. Later, in [79], the results on the linearity were generalized for $\mathbb{Z}_2\mathbb{Z}_4$-linear codes as follows. Let $\mathbf{u} * \mathbf{v}$ denote the component-wise product for any $\mathbf{u}, \mathbf{v} \in \mathbb{Z}_2^\alpha \times \mathbb{Z}_4^\beta$.

Proposition 5.1 ([79]). *Let* $\mathbf{u}, \mathbf{v} \in \mathbb{Z}_2^\alpha \times \mathbb{Z}_4^\beta$. *Then,*

$$\Phi(\mathbf{u} + \mathbf{v}) = \Phi(\mathbf{u}) + \Phi(\mathbf{v}) + \Phi(2\mathbf{u} * \mathbf{v}).$$

Corollary 5.2 ([79]). *Let* $\mathbf{u} \in \mathbb{Z}_2^\alpha \times \mathbb{Z}_4^\beta$ *of order two. Then,* $\Phi(\mathbf{u} + \mathbf{v}) = \Phi(\mathbf{u}) + \Phi(\mathbf{v})$ *for all* $\mathbf{v} \in \mathbb{Z}_2^\alpha \times \mathbb{Z}_4^\beta$.

Example 5.3. Let $\mathbf{u} = (1, 0 \mid 1, 2, 2, 0, 1), \mathbf{v} = (1, 1 \mid 2, 3, 2, 1, 3), \mathbf{w} = (1, 0 \mid 2, 0, 2, 2, 0) \in \mathbb{Z}_2^2 \times \mathbb{Z}_4^5$. We have that

$$\Phi(\mathbf{u} + \mathbf{v}) = \Phi(0, 1 \mid 3, 1, 0, 1, 0) = (0, 1, 1, 0, 0, 1, 0, 0, 0, 1, 0, 0),$$
$$\Phi(\mathbf{u}) + \Phi(\mathbf{v}) = (0, 1, 1, 0, 0, 1, 0, 0, 0, 1, 1, 1),$$
$$\Phi(2\mathbf{u} * \mathbf{v}) = \Phi(0, 0 \mid 0, 0, 0, 0, 2) = (0, 0, 0, 0, 0, 0, 0, 0, 0, 0, 1, 1).$$

It is clear that $\Phi(\mathbf{u}+\mathbf{v}) = \Phi(\mathbf{u}) + \Phi(\mathbf{v}) + \Phi(2\mathbf{u} * \mathbf{v})$. Note that, since $\mathbf{w}$ is a vector of order two, $2\mathbf{u} * \mathbf{w} = 2\mathbf{v} * \mathbf{w} = \mathbf{0}$. Therefore,

$$
\begin{aligned}
\Phi(\mathbf{u}+\mathbf{w}) &= \Phi(0,0 \mid 3,2,0,2,1) \\
&= (0,0,1,0,1,1,0,0,1,1,0,1) = \Phi(\mathbf{u}) + \Phi(\mathbf{w}), \\
\Phi(\mathbf{v}+\mathbf{w}) &= \Phi(0,1 \mid 0,3,0,3,3) \\
&= (0,1,0,0,1,0,0,0,1,0,1,0) = \Phi(\mathbf{v}) + \Phi(\mathbf{w}).
\end{aligned}
$$

$\triangle$

Corollary 5.4 ([79]). *Let $\mathcal{C}$ be a $\mathbb{Z}_2\mathbb{Z}_4$-additive code. Then, $C = \Phi(\mathcal{C})$ is linear if and only if $2\mathbf{u} * \mathbf{v} \in \mathcal{C}$ for all $\mathbf{u}, \mathbf{v} \in \mathcal{C}$.*

By Corollary 5.4, if $\mathcal{C}$ is a $\mathbb{Z}_2\mathbb{Z}_4$-additive code, then the $\mathbb{Z}_2\mathbb{Z}_4$-linear code $C = \Phi(\mathcal{C})$ is linear if and only if $2\mathbf{u} * \mathbf{v} \in \mathcal{C}$ for all $\mathbf{u}, \mathbf{v} \in \mathcal{C}$. Since the component-wise product is bilinear, to check whether a $\mathbb{Z}_2\mathbb{Z}_4$-additive code with generator matrix $\mathcal{G}$ has a linear Gray map image, we do not have to check whether $2\mathbf{u} * \mathbf{v} \in \mathcal{C}$ for all $\mathbf{u}, \mathbf{v} \in \mathcal{C}$, but only for those codewords in $\mathcal{G}$. Moreover, by Corollary 5.2, it is enough to check it for those pairs of different codewords in $\mathcal{G}$ of order four. We obtain the following proposition.

Proposition 5.5 ([79]). *Let $\mathcal{C}$ be a $\mathbb{Z}_2\mathbb{Z}_4$-additive code of type $(\alpha, \beta; \gamma, \delta; \kappa)$ with generator matrix $\mathcal{G}$. Let $\{\mathbf{u}_i\}_{i=1}^{\gamma}$ and $\{\mathbf{v}_j\}_{j=1}^{\delta}$ be the row vectors of order two and four in $\mathcal{G}$, respectively. Then, $C = \Phi(\mathcal{C})$ is linear if and only if $2\mathbf{v}_j * \mathbf{v}_k \in \mathcal{C}$ for all j, k satisfying $1 \le j < k \le \delta$.*

Example 5.6. Consider the $\mathbb{Z}_2\mathbb{Z}_4$-additive code $\mathcal{C}$ of type $(3, 5; 3, 3; 3)$ generated by the following matrix:

$$
\begin{pmatrix} \mathbf{u}_1 \\ \mathbf{u}_2 \\ \mathbf{u}_3 \\ \mathbf{v}_1 \\ \mathbf{v}_2 \\ \mathbf{v}_3 \end{pmatrix} = \left(\begin{array}{ccc|ccccc} 1 & 0 & 0 & 0 & 0 & 0 & 0 & 0 \\ 0 & 1 & 0 & 2 & 0 & 0 & 0 & 0 \\ 0 & 0 & 1 & 0 & 2 & 0 & 0 & 0 \\ 0 & 0 & 1 & 1 & 0 & 1 & 0 & 0 \\ 0 & 0 & 0 & 0 & 1 & 0 & 1 & 0 \\ 0 & 0 & 0 & 0 & 1 & 0 & 0 & 1 \end{array} \right). \tag{5.1}
$$

We have that $C = \Phi(\mathcal{C})$ is not linear because $\Phi^{-1}(\Phi(\mathbf{v}_2) + \Phi(\mathbf{v}_3)) = (0, 0, 0 \mid 0, 0, 0, 1, 1) \notin \mathcal{C}$, that is, because $\Phi(\mathbf{v}_2) + \Phi(\mathbf{v}_3) \notin C$. Note that we can also see that C is not linear by Proposition 5.5 because $2\mathbf{v}_2 * \mathbf{v}_3 = (0, 0, 0 \mid 0, 2, 0, 0, 0) \notin \mathcal{C}$.

$\triangle$

Example 5.7. Consider the $\mathbb{Z}_2\mathbb{Z}_4$-additive code $\mathcal{C}$ of type $(3, 3; 3, 2; 3)$ generated by the following matrix:

$$
\begin{pmatrix} \mathbf{u}_1 \\ \mathbf{u}_2 \\ \mathbf{u}_3 \\ \mathbf{v}_1 \\ \mathbf{v}_2 \end{pmatrix} = \left(\begin{array}{ccc|ccc} 1 & 0 & 0 & 0 & 0 & 0 \\ 0 & 1 & 0 & 0 & 0 & 0 \\ 0 & 0 & 1 & 2 & 0 & 0 \\ 0 & 0 & 1 & 0 & 1 & 0 \\ 0 & 0 & 0 & 1 & 0 & 1 \end{array} \right). \tag{5.2}
$$

We have that $C = \Phi(\mathcal{C})$ is linear because for all $\mathbf{v}_i, \mathbf{v}_j \in \mathcal{C}, 1 \le j < k \le \delta = 2$, $2\mathbf{v}_i * \mathbf{v}_j = \mathbf{0} \in \mathcal{C}$; that is, because $2\mathbf{v}_1 * \mathbf{v}_2 = \mathbf{0} \in \mathcal{C}$. $\triangle$

Proposition 5.8 ([79]). *Let $\mathcal{C}$ be a $\mathbb{Z}_2\mathbb{Z}_4$-additive code of type $(\alpha, \beta; \gamma, \delta; \kappa)$. If $\delta \le 1$, then $C = \Phi(\mathcal{C})$ is linear.*

Proof. If $\delta = 0$, then C is linear by Corollary 5.2. If $\delta = 1$, then $\{\mathbf{v}_j\}_{j=1}^{\delta} = \{\mathbf{v}_1\}$ and C is linear by Proposition 5.5. $\square$

Let $\mathcal{B} = \{\mathbf{b}_1, \dots, \mathbf{b}_r\} \subseteq \mathbb{Z}_2^{\alpha} \times \mathbb{Z}_4^{\beta}$. An element $\mathbf{b} \in \mathbb{Z}_2^{\alpha} \times \mathbb{Z}_4^{\beta}$ is said to be a 2-*linear combination* of the elements of $\mathcal{B}$ if $\mathbf{b} = \sum_{i=1}^{r} \lambda_i \mathbf{b}_i$ for $\lambda_i \in \mathbb{Z}_2$. Let $\mathcal{C}$ be a $\mathbb{Z}_2\mathbb{Z}_4$-additive code. We say that $\mathcal{B}$ is a 2-*basis* of $\mathcal{C}$ if the elements in $\mathcal{B}$ are 2-linearly independent and any $\mathbf{c} \in \mathcal{C}$ is a 2-linear combination of the elements of $\mathcal{B}$.

Proposition 5.9. *Let $\mathcal{C}$ be a $\mathbb{Z}_2\mathbb{Z}_4$-additive code of type $(\alpha, \beta; \gamma, \delta; \kappa)$ with generator matrix $\mathcal{G}$. Let $\{\mathbf{u}_i\}_{i=1}^{\gamma}$ and $\{\mathbf{v}_j\}_{j=1}^{\delta}$ be the row vectors of order two and four in $\mathcal{G}$, respectively. Let $C = \Phi(\mathcal{C})$. Then,*

i) $\mathcal{B} = \{\mathbf{b}_1, \dots, \mathbf{b}_{\gamma+2\delta}\} = \{\mathbf{u}_i\}_{i=1}^{\gamma} \cup \{\mathbf{v}_j, 2\mathbf{v}_j\}_{j=1}^{\delta}$ is a 2-basis of $\mathcal{C}$.

ii) If C is linear, then $\Phi(\mathcal{B}) = \{\Phi(\mathbf{b}_1), \dots, \Phi(\mathbf{b}_{\gamma+2\delta})\}$ is a basis of C.

Proof.

i) Clearly, all elements in $\mathcal{B}$ are 2-linearly independent. Let $\mathbf{w} \in \mathcal{C}$. We have that $\mathbf{w} = \sum_{i=1}^{\gamma} \lambda_i \mathbf{u}_i + \sum_{j=1}^{\delta} \mu_j \mathbf{v}_j$, for $\lambda_i \in \mathbb{Z}_2$, $i \in \{1, \dots, \gamma\}$ and $\mu_j = \mu_j' + 2\bar{\mu}_j \in \mathbb{Z}_4$, $\mu_j', \bar{\mu}_j \in \mathbb{Z}_2$, $j \in \{1, \dots, \delta\}$. Then, $\mathbf{w} = \sum_{i=1}^{\gamma} \lambda_i \mathbf{u}_i + \sum_{j=1}^{\delta} \mu_j' \mathbf{v}_j + \sum_{j=1}^{\delta} \bar{\mu}_j 2\mathbf{v}_j$. Therefore, $\mathcal{B}$ is a 2-basis of $\mathcal{C}$.

ii) By Theorem 2.10 and the shape of the generator matrix in standard form (2.7), the $\gamma + 2\delta$ vectors in $\Phi(\mathcal{B})$ are linearly independent. Therefore, $\Phi(\mathcal{B})$ is a basis of C.

$\square$

Example 5.10. Let $\mathcal{C}$ be the $\mathbb{Z}_2\mathbb{Z}_4$-additive code given in Example 5.7. By Proposition 5.9, $\mathcal{B} = \{\mathbf{u}_1, \mathbf{u}_2, \mathbf{u}_3, \mathbf{v}_1, 2\mathbf{v}_1, \mathbf{v}_2, 2\mathbf{v}_2\}$ is a 2-basis of $\mathcal{C}$. Moreover, since $\Phi(\mathcal{C})$ is linear, we have that $\Phi(\mathcal{B})$ is a basis of $\Phi(\mathcal{C})$. Therefore, a generator matrix for $\Phi(\mathcal{C})$ is

$$
\begin{pmatrix}
\Phi(\mathbf{u}_1) \\
\Phi(\mathbf{u}_2) \\
\Phi(\mathbf{u}_3) \\
\Phi(\mathbf{v}_1) \\
\Phi(2\mathbf{v}_1) \\
\Phi(\mathbf{v}_2) \\
\Phi(2\mathbf{v}_2)
\end{pmatrix}
=
\begin{pmatrix}
1 & 0 & 0 & 0 & 0 & 0 & 0 & 0 & 0 \\
0 & 1 & 0 & 0 & 0 & 0 & 0 & 0 & 0 \\
0 & 0 & 1 & 1 & 1 & 0 & 0 & 0 & 0 \\
0 & 0 & 1 & 0 & 0 & 0 & 1 & 0 & 0 \\
0 & 0 & 0 & 0 & 0 & 1 & 1 & 0 & 0 \\
0 & 0 & 0 & 0 & 1 & 0 & 0 & 0 & 1 \\
0 & 0 & 0 & 1 & 1 & 0 & 0 & 1 & 1
\end{pmatrix}.
\tag{5.3}
$$

$\triangle$

We consider now the linearity of a $\mathbb{Z}_2\mathbb{Z}_4$-linear code $C = \Phi(\mathcal{C})$ in terms of the linearity of the Gray map image of a subcode of $\mathcal{C}_Y$. First, note that $\mathcal{C}_X$ is linear. Next lemma relates the linearity of $\Phi(\mathcal{C})$ to the linearity of $\phi(\mathcal{C}_Y)$.

Lemma 5.11. *Let $\mathcal{C}$ be a $\mathbb{Z}_2\mathbb{Z}_4$-additive code. If $\Phi(\mathcal{C})$ is linear, then $\phi(\mathcal{C}_Y)$ is linear.*

Proof. Let $u', v' \in \mathcal{C}_Y$. There exist $\mathbf{u} = (u \mid u'), \mathbf{v} = (v \mid v') \in \mathcal{C}$. Since $\Phi(\mathcal{C})$ is linear, $2\mathbf{u} * \mathbf{v} = (\mathbf{0} \mid 2u' * v') \in \mathcal{C}$ by Corollary 5.4. Hence $2u' * v' \in \mathcal{C}_Y$ and we have that $\phi(\mathcal{C}_Y)$ is linear by the same corollary or [92]. $\square$

The converse is true when $\mathcal{C}$ is separable as stated in Corollary 5.14, but it is not true in general, as it is shown in the following example.

Example 5.12. Let $\mathcal{C}$ be the $\mathbb{Z}_2\mathbb{Z}_4$-additive code given in Example 5.6. We have seen that $\Phi(\mathcal{C})$ is not linear. The code $\mathcal{C}_Y$ is generated by

$$
\begin{pmatrix}
u'_2 \\
u'_3 \\
v'_1 \\
v'_2 \\
v'_3
\end{pmatrix}
=
\begin{pmatrix}
2 & 0 & 0 & 0 & 0 \\
0 & 2 & 0 & 0 & 0 \\
1 & 0 & 1 & 0 & 0 \\
0 & 1 & 0 & 1 & 0 \\
0 & 1 & 0 & 0 & 1
\end{pmatrix}.
\tag{5.4}
$$

Note that $2v'_1 * v'_2 = 2v'_1 * v'_3 = \mathbf{0}$ and $2v'_2 * v'_3 = (0, 2, 0, 0, 0)$ are codewords in $\mathcal{C}_Y$. By Proposition 5.5, we have that $\phi(\mathcal{C}_Y)$ is linear. $\triangle$

Lemma 5.13. *Let $\mathcal{C}$ be a $\mathbb{Z}_2\mathbb{Z}_4$-additive code of type $(\alpha, \beta; \gamma, \delta; \kappa)$. Let $\kappa = \kappa_1 + \kappa_2$ and $\delta = \delta_1 + \delta_2$ given in (2.3) with $\kappa_2 = 0$. Then, $\Phi(\mathcal{C})$ is linear if and only if $\phi(\mathcal{C}_Y)$ is linear.*

Proof. By Lemma 5.11, if $\Phi(\mathcal{C})$ is linear, then $\phi(\mathcal{C}_Y)$ is also linear.

Assume that $\phi(\mathcal{C}_Y)$ is linear and $\kappa_2 = 0$. Let $\mathbf{u} = (u \mid u'), \mathbf{v} = (v \mid v') \in \mathcal{C}$. We have that $2\mathbf{u} * \mathbf{v} = (\mathbf{0} \mid 2u' * v')$. Since $\phi(\mathcal{C}_Y)$ is linear, $2u' * v' \in \mathcal{C}_Y$ by Corollary 5.4, and there exists $w \in \mathcal{C}_X$ such that $(w \mid 2u' * v') \in \mathcal{C}$. Note that $(w \mid 2u' * v') \in \mathcal{C}_b$. Since $\mathcal{C}_b$ is separable by Remark 2.5, $(\mathbf{0} \mid 2u' * v') \in \mathcal{C}_b \subseteq \mathcal{C}$. Therefore, $2\mathbf{u} * \mathbf{v} \in \mathcal{C}$ and $\Phi(\mathcal{C})$ is linear by Corollary 5.4.   □

Corollary 5.14 ([33]). *Let $\mathcal{C}$ be a separable $\mathbb{Z}_2\mathbb{Z}_4$-additive code. Then, $\Phi(\mathcal{C})$ is linear if and only if $\phi(\mathcal{C}_Y)$ is linear.*

Proof. If $\mathcal{C}$ is separable, by Remark 2.4, $\kappa_2 = 0$. Then, the statement follows by Lemma 5.13.   □

Let $\mathcal{C}$ be a $\mathbb{Z}_2\mathbb{Z}_4$-additive code with generator matrix in standard form, and let $\mathcal{C}'$ be the subcode generated by

$$\begin{pmatrix} \mathbf{0} & \mathbf{0} & \mathbf{0} & 2T_1 & 2I_{\gamma-\kappa} & \mathbf{0} \\ \mathbf{0} & \mathbf{0} & S_b & S_q & R & I_\delta \end{pmatrix}. \tag{5.5}$$

From Example 5.12, we have seen that in general it is not true that $\Phi(\mathcal{C})$ is linear if and only if $\phi(\mathcal{C}_Y)$ is linear. However, we see that $\Phi(\mathcal{C})$ is linear if and only if $\phi(\mathcal{C}'_Y)$ is linear.

Proposition 5.15 ([33]). *Let $\mathcal{C}$ be a $\mathbb{Z}_2\mathbb{Z}_4$-additive code with generator matrix in standard form, and let $\mathcal{C}'$ be the subcode generated by (5.5). Then, $\Phi(\mathcal{C})$ is linear if and only if $\phi(\mathcal{C}'_Y)$ is linear.*

Proof. Let $\mathcal{G}$ be a generator matrix of $\mathcal{C}$ in standard form. Let $\{\mathbf{u}_i = (u_i \mid u'_i)\}_{i=1}^{\gamma}$ and $\{\mathbf{v}_j = (v_j \mid v'_j)\}_{j=1}^{\delta}$ be the row vectors of order two and four in $\mathcal{G}$, respectively. Note that $\mathcal{C}'$ is generated by $\{\mathbf{u}_i : u_i = \mathbf{0}\}_{i=1}^{\gamma}$ and $\{\mathbf{v}_j\}_{j=1}^{\delta}$. By Proposition 5.5, $\Phi(\mathcal{C})$ is linear if and only if $2\mathbf{v}_j * \mathbf{v}_k \in \mathcal{C}$, for all j, k satisfying $1 \leq j < k \leq \delta$. Since $2v_j * v_k = \mathbf{0}$, $2\mathbf{v}_j * \mathbf{v}_k \in \mathcal{C}$ if and only if $2\mathbf{v}_j * \mathbf{v}_k$ is a linear combination of $\{\mathbf{u}_i : u_i = \mathbf{0}\}_{i=1}^{\gamma}$ and $\{2\mathbf{v}_j\}_{j=1}^{\delta}$; that is, if and only if $2\mathbf{v}_j * \mathbf{v}_k \in \mathcal{C}'$. By Proposition 5.5, the last condition is satisfied if and only if $\phi(\mathcal{C}'_Y)$ is linear.   □

Note that if $\mathcal{C}$ is separable, then $\mathcal{C}_Y = \mathcal{C}'_Y$ and Corollary 5.14 and Proposition 5.15 coincide.

Example 5.16. Let $\mathcal{C}$ be the $\mathbb{Z}_2\mathbb{Z}_4$-additive code defined in Example 5.6. Recall that $\Phi(\mathcal{C})$ is not linear. The code $\mathcal{C}'$ is generated by

$$\begin{pmatrix} 0 & 0 & 1 & 1 & 0 & 1 & 0 & 0 \\ 0 & 0 & 0 & 0 & 1 & 0 & 1 & 0 \\ 0 & 0 & 0 & 0 & 1 & 0 & 0 & 1 \end{pmatrix}. \tag{5.6}$$

Moreover, since $2(0,1,0,1,0) * (0,1,0,0,1) = (0,2,0,0,0) \notin \mathcal{C}'_Y$, $\phi(\mathcal{C}'_Y)$ is not linear by Proposition 5.5. △

Example 5.17. By using the MAGMA package for $\mathbb{Z}_2\mathbb{Z}_4$-additive codes [29], we construct a $\mathbb{Z}_2\mathbb{Z}_4$-additive code $\mathcal{C}$ and check that $\Phi(\mathcal{C})$ is not linear from the rows of order four of a generator matrix of $\mathcal{C}$. We check that $\phi(\mathcal{C}'_Y)$ is also non-linear.

```
> Z4 := IntegerRing(4);
> alpha := 2;
> beta := 9;
> V := RSpace(Z4, alpha + beta);
> G := Matrix([V![2,0, 0,0,2,0,0,0,0,0,0],
               V![0,0, 0,2,2,2,0,0,0,0,0],
               V![0,0, 0,0,2,0,2,0,0,0,0],
               V![0,0, 0,1,0,1,0,1,0,0,0],
               V![0,0, 1,1,0,0,0,0,1,0,0],
               V![0,2, 1,0,1,1,1,0,0,1,0],
               V![0,0, 0,0,1,0,1,0,0,0,1]]);
> C := Z2Z4AdditiveCode(G, alpha);
>
> HasLinearGrayMapImage(C);
false
> typeC := Z2Z4Type(C);
> typeC;
[ 2, 9, 3, 4, 1 ]
> gamma := typeC[3];
> delta := typeC[4];
> kappa := typeC[5];
> v1 := G[gamma+1];
> v2 := G[gamma+2];
> v3 := G[gamma+3];
> v4 := G[gamma+4];
> 2*v1*v2 in C; 2*v1*v3 in C; 2*v2*v3 in C;
false false false
> 2*v1*v4 in C; 2*v2*v4 in C; 2*v3*v4 in C;
true true true
>
> Cprime := Z2Z4AdditiveCode(G[kappa+1..gamma+delta], alpha);
> CYprime := LinearQuaternaryCode(Cprime);
> HasLinearGrayMapImage(CYprime);
false
```
△

5.2 Rank of $\mathbb{Z}_2\mathbb{Z}_4$-Linear Codes

The *rank* of a binary code C, denoted by $\mathrm{rank}(C)$, is simply the dimension of $\langle C \rangle$, which is the linear span of the codewords of C. Note that C is linear if and only if $\langle C \rangle = C$. In this section, we study the rank of $\mathbb{Z}_2\mathbb{Z}_4$-linear codes, $C = \Phi(\mathcal{C})$, and determine a basis of $\langle C \rangle$ from the rows of a generator matrix of the corresponding $\mathbb{Z}_2\mathbb{Z}_4$-additive code $\mathcal{C}$. We establish lower and upper bounds for the values of $\mathrm{rank}(C)$ and show that there exists a $\mathbb{Z}_2\mathbb{Z}_4$-linear code C of type $(\alpha, \beta; \gamma, \delta; \kappa)$ with $\mathrm{rank}(C) = r$, for any possible r. Finally, we see that $\mathrm{rank}(C)$ can also be given in terms of the dimension of $\mathcal{C}_X$ and the rank of a $\mathbb{Z}_4$-linear code which is a subcode of $\phi(\mathcal{C}_Y)$.

Proposition 5.18 ([79]). *Let $\mathcal{C}$ be a $\mathbb{Z}_2\mathbb{Z}_4$-additive code of type $(\alpha, \beta; \gamma, \delta; \kappa)$, and let $C = \Phi(\mathcal{C})$. Let $\mathcal{G}$ be a generator matrix of $\mathcal{C}$, and let $\{\mathbf{u}_i\}_{i=1}^{\gamma}$ be the rows of order two and $\{\mathbf{v}_j\}_{j=1}^{\delta}$ the rows of order four in $\mathcal{G}$. Then,*

$$\langle C \rangle = \langle \{\Phi(\mathbf{u}_i)\}_{i=1}^{\gamma}, \{\Phi(\mathbf{v}_j), \Phi(2\mathbf{v}_j)\}_{j=1}^{\delta}, \{\Phi(2\mathbf{v}_j * \mathbf{v}_k)\}_{1 \le j < k \le \delta} \rangle.$$

Proof. If $\mathbf{u} \in \mathcal{C}$ then $\mathbf{u}$ can be expressed as $\mathbf{u} = \mathbf{v}_{j_1} + \cdots + \mathbf{v}_{j_m} + \mathbf{w}$, where $\{j_1, \ldots, j_m\} \subseteq \{1, \ldots, \delta\}$ and $\mathbf{w}$ is a codeword of order two. Therefore, $\Phi(\mathbf{u}) = \Phi(\mathbf{v}_{j_1} + \cdots + \mathbf{v}_{j_m}) + \Phi(\mathbf{w})$, where $\Phi(\mathbf{w})$ is a linear combination of $\{\Phi(\mathbf{u}_i)\}_{i=1}^{\gamma}$ and $\{\Phi(2\mathbf{v}_j)\}_{j=1}^{\delta}$, and $\Phi(\mathbf{v}_{j_1} + \cdots + \mathbf{v}_{j_m}) = \Phi(\mathbf{v}_{j_1}) + \cdots + \Phi(\mathbf{v}_{j_m}) + \sum_{1 \le k < l \le m} \Phi(2\mathbf{v}_{j_k} * \mathbf{v}_{j_l})$. Therefore, $\Phi(\mathbf{u})$ is generated by $\{\Phi(\mathbf{u}_i)\}_{i=1}^{\gamma}, \{\Phi(\mathbf{v}_j), \Phi(2\mathbf{v}_j)\}_{j=1}^{\delta}$ and $\{\Phi(2\mathbf{v}_j * \mathbf{v}_k)\}_{1 \le j < k \le \delta}$. $\square$

Let $\mathcal{C}$ be a $\mathbb{Z}_2\mathbb{Z}_4$-additive code. We define the code

$$\mathcal{R}(\mathcal{C}) = \Phi^{-1}(\langle \Phi(\mathcal{C}) \rangle).$$

Note that $\Phi(\mathcal{C})$ is linear if and only if $\langle \Phi(\mathcal{C}) \rangle = \Phi(\mathcal{C})$ or, equivalently, $\mathcal{R}(\mathcal{C}) = \mathcal{C}$. By Proposition 5.18 and Corollary 5.2, we have that $\mathcal{R}(\mathcal{C})$ is in fact a $\mathbb{Z}_2\mathbb{Z}_4$-additive code, as it is stated in the following results.

Corollary 5.19 ([79]). *Let $\mathcal{C}$ be a $\mathbb{Z}_2\mathbb{Z}_4$-additive code of type $(\alpha, \beta; \gamma, \delta; \kappa)$. Let $\mathcal{G}$ be a generator matrix of $\mathcal{C}$, and let $\{\mathbf{u}_i\}_{i=1}^{\gamma}$ be the rows of order two and $\{\mathbf{v}_j\}_{j=1}^{\delta}$ the rows of order four in $\mathcal{G}$. Then,*

$$\mathcal{R}(\mathcal{C}) = \langle \mathcal{C}, \{2\mathbf{v}_j * \mathbf{v}_k\}_{1 \le j < k \le \delta} \rangle = \langle \{\mathbf{u}_i\}_{i=1}^{\gamma}, \{\mathbf{v}_j\}_{j=1}^{\delta}, \{2\mathbf{v}_j * \mathbf{v}_k\}_{1 \le j < k \le \delta} \rangle.$$

Example 5.20. Let $\mathcal{C}$ be the $\mathbb{Z}_2\mathbb{Z}_4$-additive code of type $(3, 5; 3, 3; 3)$ given in Example 5.6. Note that $2\mathbf{v}_1 * \mathbf{v}_2 = \mathbf{0}$, $2\mathbf{v}_1 * \mathbf{v}_3 = \mathbf{0}$ and $2\mathbf{v}_2 * \mathbf{v}_3 = (0, 0, 0 \mid 0, 2, 0, 0, 0) \notin \mathcal{C}$. Then, by Proposition 5.18, we have that the linear span of $C = \Phi(\mathcal{C})$ is $\langle C \rangle = \langle \Phi(\mathbf{u}_1), \Phi(\mathbf{u}_2), \Phi(\mathbf{u}_3), \Phi(\mathbf{v}_1), \Phi(2\mathbf{v}_1), \Phi(\mathbf{v}_2), \Phi(2\mathbf{v}_2), \Phi(\mathbf{v}_3), \Phi(2\mathbf{v}_3), \Phi(2\mathbf{v}_2 * \mathbf{v}_3) \rangle$; and by Corollary 5.19, $\mathcal{R}(\mathcal{C}) = \langle \mathcal{C}, 2\mathbf{v}_2 * \mathbf{v}_3 \rangle = \langle \mathbf{u}_1, \mathbf{u}_2, \mathbf{u}_3, \mathbf{v}_1, \mathbf{v}_2, \mathbf{v}_3, 2\mathbf{v}_2 * \mathbf{v}_3 \rangle.$ △

Corollary 5.21 ([79]). *Let $\mathcal{C}$ be a $\mathbb{Z}_2\mathbb{Z}_4$-additive code of type $(\alpha, \beta; \gamma, \delta; \kappa)$. Then, $\mathcal{R}(\mathcal{C})$ is a $\mathbb{Z}_2\mathbb{Z}_4$-additive code of type $(\alpha, \beta; \gamma + \bar{r}, \delta; \kappa)$, with $\bar{r} \ge 0$, and $\mathrm{rank}(\Phi(\mathcal{C})) = \log_2(|\mathcal{R}(\mathcal{C})|) = \gamma + 2\delta + \bar{r}$.*

Corollary 5.22 ([79]). *If C is a $\mathbb{Z}_2\mathbb{Z}_4$-linear code, then $\langle C \rangle$ is both linear and $\mathbb{Z}_2\mathbb{Z}_4$-linear.*

Now, we give an upper and lower bound for the rank of a $\mathbb{Z}_2\mathbb{Z}_4$-linear code in terms of the parameters of its type.

Proposition 5.23 ([79]). *Let C be a $\mathbb{Z}_2\mathbb{Z}_4$-linear code of type $(\alpha, \beta; \gamma, \delta; \kappa)$. Then, $\mathrm{rank}(C) = \gamma + 2\delta + \bar{r}$, where*

$$\bar{r} \in \left\{ 0, \ldots, \min\left\{ \beta - (\gamma - \kappa) - \delta, \ \binom{\delta}{2} \right\} \right\}.$$

Proof. Let $\mathcal{G}$ be a generator matrix of $\mathcal{C} = \Phi^{-1}(C)$ in standard form. In the generator matrix $\mathcal{G}$ there are γ rows $\{\mathbf{u}_i\}_{i=1}^{\gamma}$ of order two, and δ rows $\{\mathbf{v}_j\}_{j=1}^{\delta}$ of order four. Then, by Proposition 5.18, we can take the matrix G whose row vectors are $\{\Phi(\mathbf{u}_i)\}_{i=1}^{\gamma}$, $\{\Phi(\mathbf{v}_j), \Phi(2\mathbf{v}_j)\}_{j=1}^{\delta}$ and $\{\Phi(2\mathbf{v}_j * \mathbf{v}_k)\}_{1 \leq j < k \leq \delta}$, as a generator matrix of $\langle C \rangle$.

The binary vectors $\{\Phi(\mathbf{u}_i)\}_{i=1}^{\gamma}$ and $\{\Phi(\mathbf{v}_j), \Phi(2\mathbf{v}_j)\}_{j=1}^{\delta}$ are linearly independent over $\mathbb{Z}_2$. Thus, $\mathrm{rank}(C) = \gamma + 2\delta + \bar{r}$, where $\bar{r}$ is the number of additional independent vectors taken from $\{\Phi(2\mathbf{v}_j * \mathbf{v}_k)\}_{1 \leq j < k \leq \delta}$. Note that there are at most $\binom{\delta}{2}$ of such vectors. Using row reduction in $\Phi^{-1}(G)$, the $\binom{\delta}{2}$ vectors $\{2\mathbf{v}_j * \mathbf{v}_k\}_{1 \leq j < k \leq \delta}$ can be transformed into vectors with zeros in the last $\gamma - \kappa + \delta$ coordinates. Therefore, there are at most $\min\{\beta - (\gamma - \kappa) - \delta, \binom{\delta}{2}\}$ of such additional independent vectors, so $\bar{r} \leq \min\{\beta - (\gamma - \kappa) - \delta, \binom{\delta}{2}\}$. The lower bound of $\mathrm{rank}(C)$ follows from the case where the code C is linear. $\square$

Example 5.24. Let $\mathcal{C}$ be the $\mathbb{Z}_2\mathbb{Z}_4$-additive code given in Example 5.6, where it is shown that $C = \Phi(\mathcal{C})$ is not linear. Recall that $\mathcal{C}$ is of type $(3, 5; 3, 3; 3)$. By Proposition 5.23, $9 \leq \mathrm{rank}(C) \leq \min\{11, 12\} = 11$. From Example 5.20 and Corollary 5.21, we have that $\mathrm{rank}(C) = \log_2(|\mathcal{R}(\mathcal{C})|) = \log_2(|C|) + 1 = 10$. $\triangle$

The previous upper bound for the rank of a $\mathbb{Z}_2\mathbb{Z}_4$-linear code is tight for some families of $\mathbb{Z}_2\mathbb{Z}_4$-linear codes as, for example, perfect and extended perfect $\mathbb{Z}_2\mathbb{Z}_4$-linear codes (see Propositions 6.6, 6.13 and 6.14). We also know examples of $\mathbb{Z}_2\mathbb{Z}_4$-linear codes such that the rank is in between both bounds, e.g. the Hadamard $\mathbb{Z}_2\mathbb{Z}_4$-linear codes (see Propositions 6.19 and 6.20).

Example 5.25. Let $\overline{QRM}(r, m)$ be the class of $\mathbb{Z}_4$-linear Reed-Muller codes defined in [34]. These are $\mathbb{Z}_2\mathbb{Z}_4$-linear codes of type $(0, 2^m; 0, \delta; 0)$, where $\delta = \sum_{i=0}^{r} \binom{m}{i}$. A remarkable property is that any $\mathbb{Z}_4$-linear Kerdock-like code of length 4^m is in the class $\overline{QRM}(1, 2m - 1)$ and any extended $\mathbb{Z}_4$-linear Preparata-like code of length 4^m is in the class $\overline{QRM}(2m - 3, 2m - 1)$.

The rank of any code $C \in \overline{QRM}(r, m)$ is

$$\text{rank}(C) = \sum_{i=0}^{r} \binom{m}{i} + \sum_{i=0}^{t} \binom{m}{i},$$

where $t = \min\{2r, m\}$ [34]. Hence, if $2r \geq m$, we have that $\text{rank}(C) = \delta + \beta$, i.e. the maximum possible. A $\mathbb{Z}_4$-linear Kerdock-like code K_m of length $4^m \geq 16$ has $\text{rank}(K_m) = 2m^2 + m + 1$ and an extended $\mathbb{Z}_4$-linear Preparata-like code P_m of length $4^m \geq 64$ has $\text{rank}(P_m) = 2^{2m} - 2m$ [44], attaining the upper bound given in Proposition 5.23. $\triangle$

From the proof of Proposition 5.23, we obtain the following proposition.

Proposition 5.26. *Let C be a $\mathbb{Z}_2\mathbb{Z}_4$-additive code of type $(\alpha, \beta; \gamma, \delta; \kappa)$ with generator matrix $\mathcal{G}$. Let $\{u_i\}_{i=1}^{\gamma}$ and $\{v_j\}_{j=1}^{\delta}$ be the row vectors of order two and four in $\mathcal{G}$, respectively, and let $\mathcal{B}$ be a 2-basis of C. Then, there are $\bar{r}$ vectors $w_1, \ldots, w_{\bar{r}}$ in the set $\{2v_i * v_j\}_{1 \leq i < j \leq \delta}$ such that $\mathcal{B} \cup \{w_1, \ldots, w_{\bar{r}}\}$ is a 2-basis of $\mathcal{R}(C)$ or, equivalently, $\Phi(\mathcal{B}) \cup \{\Phi(w_1), \ldots, \Phi(w_{\bar{r}})\}$ is a basis of $\langle \Phi(C) \rangle$.*

Example 5.27. Let C be the $\mathbb{Z}_2\mathbb{Z}_4$-additive code of type $(3, 5; 3, 3; 3)$ given in Example 5.6, and let $C = \Phi(\mathcal{C})$. By Example 5.24, we have that $\text{rank}(C) = 10$ and, therefore $\bar{r} = \text{rank}(C) - (\gamma + 2\delta) = 1$. Since the vector $2v_2 * v_3 \notin C$, we have that $\{u_1, u_2, u_3, v_1, 2v_1, v_2, 2v_2, v_3, 2v_3, 2v_2 * v_3\}$ is a 2-basis of $\mathcal{R}(C)$ and $\{\Phi(u_1), \Phi(u_2), \Phi(u_3), \Phi(v_1), \Phi(2v_1), \Phi(v_2), \Phi(2v_2), \Phi(v_3), \Phi(2v_3), \Phi(2v_2 * v_3)\}$ is a basis of $\langle C \rangle$. $\triangle$

The next question is to construct $\mathbb{Z}_2\mathbb{Z}_4$-linear codes with any rank in the range of possibilities given by Proposition 5.23.

Theorem 5.28 ([79]). *Let $\alpha, \beta, \gamma, \delta, \kappa$ be integer numbers satisfying inequations (2.10). Then, there exists a $\mathbb{Z}_2\mathbb{Z}_4$-linear code C of type $(\alpha, \beta; \gamma, \delta; \kappa)$ with $\text{rank}(C) = \gamma + 2\delta + \bar{r}$, for any*

$$\bar{r} \in \left\{0, \ldots, \min\left\{\beta - (\gamma - \kappa) - \delta, \ \binom{\delta}{2}\right\}\right\}.$$

Proof. Let C be a $\mathbb{Z}_2\mathbb{Z}_4$-additive code of type $(\alpha, \beta; \gamma, \delta; \kappa)$ with generator matrix

$$\mathcal{G} = \begin{pmatrix} I_\kappa & T_b & 0 & 0 & 0 \\ 0 & 0 & 2T_1 & 2I_{\gamma-\kappa} & 0 \\ 0 & S_b & S_r & 0 & I_\delta \end{pmatrix}, \tag{5.7}$$

where S_r is a matrix over $\mathbb{Z}_4$ of size $\delta \times (\beta - (\gamma - \kappa) - \delta)$, and let $C = \Phi(\mathcal{C})$ be its corresponding $\mathbb{Z}_2\mathbb{Z}_4$-linear code. Let $\{\mathbf{u}_i\}_{i=1}^{\gamma}$ be the rows of order two and $\{\mathbf{v}_j\}_{j=1}^{\delta}$ the rows of order four in $\mathcal{G}$.

By the proof of Proposition 5.23, we have that $\{\Phi(\mathbf{u}_i)\}_{i=1}^{\gamma}$ and $\{\Phi(\mathbf{v}_j), \Phi(2\mathbf{v}_j)\}_{j=1}^{\delta}$ are $\gamma + 2\delta$ linearly independent vectors over $\mathbb{Z}_2$ and rank$(C) = \gamma + 2\delta + \bar{r}$, where $\bar{r} \in \{0, \dots, \min\{\beta - (\gamma - \kappa) - \delta, \binom{\delta}{2}\}\}$. Now, we construct a code of rank $r = \gamma + 2\delta + \bar{r}$, for each possible value of $\bar{r}$.

Let S be a set of $\bar{r} = r - (\gamma + 2\delta)$ different subsets $\{i, j\}$, where $i \neq j$ and $i, j \in \{1, \dots \delta\}$. Let e_i, $1 \leq i \leq \delta$, be the vector of length δ with 1 in the ith coordinate and zeros elsewhere. We construct S_r as a matrix over $\mathbb{Z}_4$ where in $\bar{r}$ columns there are the column vectors $e_i + e_j$ of length δ, $\{i, j\} \in S$, and in the rest of the columns, the all-zero column vector. For each one of the $\bar{r}$ column vectors $e_i + e_j$, that is, for each $\{i, j\} \in S$, the vector $2\mathbf{v}_i * \mathbf{v}_j \notin C$ and therefore the rank increases by 1. Moreover, for each $\{i, j\} \notin S$, we have that $2\mathbf{v}_i * \mathbf{v}_j = \mathbf{0} \in C$. Since the maximum number of columns of S_r is $\beta - (\gamma - \kappa) - \delta$ and the maximum number of different such columns is $\binom{\delta}{2}$, the result follows. $\square$

Note that, in the proof of the previous theorem, the matrices T_1, T_b and S_b in (5.7) can be chosen arbitrarily.

Example 5.29. By Proposition 5.23, we know that the possible ranks for $\mathbb{Z}_2\mathbb{Z}_4$-linear codes of type $(\alpha, 7; 2, 3; 2)$ are $r \in \{8, 9, 10, 11\}$. By the proof of Theorem 5.28, for each possible r, we can construct a $\mathbb{Z}_2\mathbb{Z}_4$-linear code C_r with rank$(C_r) = r$, taking the following generator matrix of $\mathcal{C}_r = \Phi^{-1}(C_r)$:

$$\mathcal{G}_r = \left(\begin{array}{cc|cc} I_2 & T_b & \mathbf{0} & \mathbf{0} \\ \mathbf{0} & S_b & S_r & I_3 \end{array} \right),$$

where T_b, S_b are matrices over $\mathbb{Z}_2$; and the matrices S_r, for each $r \in \{8, 9, 10, 11\}$, are the following: $S_8 = (\mathbf{0})$,

$$S_9 = \begin{pmatrix} 1 & 0 & 0 & 0 \\ 1 & 0 & 0 & 0 \\ 0 & 0 & 0 & 0 \end{pmatrix}, \; S_{10} = \begin{pmatrix} 1 & 0 & 0 & 0 \\ 1 & 1 & 0 & 0 \\ 0 & 1 & 0 & 0 \end{pmatrix}, \; S_{11} = \begin{pmatrix} 1 & 0 & 1 & 0 \\ 1 & 1 & 0 & 0 \\ 0 & 1 & 1 & 0 \end{pmatrix}.$$

$\triangle$

Now, we determine the rank of a $\mathbb{Z}_2\mathbb{Z}_4$-linear code $\Phi(\mathcal{C})$ in terms of the dimension of $\mathcal{C}_X$, which is linear, and the rank of the $\mathbb{Z}_4$-linear code $\phi(\mathcal{C}'_Y)$, where $\mathcal{C}'$ is the code generated by the matrix given in (5.5). Recall that if the $\mathbb{Z}_2\mathbb{Z}_4$-additive code $\mathcal{C}$ is separable, then $\mathcal{C}_Y = \mathcal{C}'_Y$.

Proposition 5.30 ([33]). *If $\mathcal{C}$ is a separable $\mathbb{Z}_2\mathbb{Z}_4$-additive code of type $(\alpha, \beta; \gamma, \delta; \kappa)$, then $\mathcal{R}(\mathcal{C}) = \mathcal{C}_X \times \mathcal{R}(\mathcal{C}_Y)$ and $\mathrm{rank}(\Phi(\mathcal{C})) = \kappa + \mathrm{rank}(\phi(\mathcal{C}_Y))$.*

Proof. If $\mathcal{C}$ is separable, then $\mathcal{C} = \mathcal{C}_X \times \mathcal{C}_Y$ and $\Phi(\mathcal{C}) = \mathcal{C}_X \times \phi(\mathcal{C}_Y)$. Therefore, since $\mathcal{C}_X$ is a linear code, $\langle \Phi(\mathcal{C}) \rangle = \mathcal{C}_X \times \langle \phi(\mathcal{C}_Y) \rangle$. Finally, since $\mathcal{R}(\mathcal{C}) = \Phi^{-1}(\langle \Phi(\mathcal{C}) \rangle)$ and $\mathcal{R}(\mathcal{C}_Y) = \Phi^{-1}(\langle \phi(\mathcal{C}_Y) \rangle)$, the statement follows. $\square$

Note that, if the $\mathbb{Z}_2\mathbb{Z}_4$-additive code $\mathcal{C}$ is not separable, $\mathrm{rank}(\Phi(\mathcal{C}))$ is not necessarily equal to $\kappa + \mathrm{rank}(\phi(\mathcal{C}_Y))$ as it is shown in the following example.

Example 5.31. Let $\mathcal{C}$ be the $\mathbb{Z}_2\mathbb{Z}_4$-additive code of type $(3, 5; 3, 3; 3)$ given in Example 5.6 whose image $\Phi(\mathcal{C})$ is not linear. By Example 5.24, we have that $\mathrm{rank}(\Phi(\mathcal{C})) = 10$. Moreover, by Example 5.12, $\phi(\mathcal{C}_Y)$ is linear and $\mathcal{C}_Y = \langle u_2', u_3', v_1', v_2', v_3' \rangle$. Therefore, we have that $\mathrm{rank}(\phi(\mathcal{C}_Y)) = 8$ and $\mathrm{rank}(\Phi(\mathcal{C})) = 10 < 3 + 8 = \kappa + \mathrm{rank}(\phi(\mathcal{C}_Y))$.

For this code, since $\phi(\mathcal{C}_Y)$ is linear, we have that $\mathcal{R}(\mathcal{C}_Y) = \mathcal{C}_Y = \langle u_2', u_3', v_1', v_2', v_3' \rangle$ and, by Example 5.20, $\mathcal{R}(\mathcal{C}) = \langle \mathbf{u}_1, \mathbf{u}_2, \mathbf{u}_3, \mathbf{v}_1, \mathbf{v}_2, \mathbf{v}_3, 2\mathbf{v}_2 * \mathbf{v}_3 \rangle$. Since $u_1' = \mathbf{0}$ and $2v_2' * v_3' = u_3'$, we have that $\mathcal{R}(\mathcal{C})_Y = \mathcal{R}(\mathcal{C}_Y)$. This result, in fact, is true for any $\mathbb{Z}_2\mathbb{Z}_4$-additive code in general, as it is proven in Lemma 5.32. $\triangle$

Lemma 5.32 ([33]). *Let $\mathcal{C}$ be a $\mathbb{Z}_2\mathbb{Z}_4$-additive code. Then, $\mathcal{R}(\mathcal{C})_Y = \mathcal{R}(\mathcal{C}_Y)$.*

Proof. Let $\mathcal{G}$ be a generator matrix of $\mathcal{C}$, and let $\{\mathbf{u}_i = (u_i \mid u_i')\}_{i=1}^{\gamma}$ be the first γ rows and $\{\mathbf{v}_j = (v_j \mid v_j')\}_{j=1}^{\delta}$ the last δ rows of $\mathcal{G}$. By Corollary 5.19, $\mathcal{R}(\mathcal{C}) = \langle \mathcal{C}, \{2\mathbf{v}_j * \mathbf{v}_k\}_{1 \leq j < k \leq \delta} \rangle$. By the same argument as in the proof of Corollary 5.19, $\mathcal{R}(\mathcal{C}_Y) = \langle \mathcal{C}_Y, \{2v_j' * v_k'\}_{1 \leq j < k \leq \delta} \rangle$. Since $2\mathbf{v}_j * \mathbf{v}_k = (\mathbf{0} \mid 2v_j' * v_k')$ for all $1 \leq j < k \leq \delta$, the statement follows. $\square$

Theorem 5.33 ([33]). *Let $\mathcal{C}$ be a $\mathbb{Z}_2\mathbb{Z}_4$-additive code of type $(\alpha, \beta; \gamma, \delta; \kappa)$ with generator matrix in standard form, and let $\mathcal{C}'$ be the subcode generated by (5.5). Then,*
$$\mathrm{rank}(\Phi(\mathcal{C})) = \kappa + \mathrm{rank}(\phi(\mathcal{C}_Y')).$$

Proof. Let $\mathcal{G}$ be a generator matrix of $\mathcal{C}$ in standard form. Let $\{\mathbf{u}_i = (u_i \mid u_i')\}_{i=1}^{\gamma}$ be the first γ rows and $\{\mathbf{v}_j = (v_j \mid v_j')\}_{j=1}^{\delta}$ the last δ rows of $\mathcal{G}$. Define the codes $\bar{\mathcal{C}} = \langle \{\mathbf{u}_i\}_{i=1}^{\kappa} \rangle$ and $\mathcal{C}' = \langle \{\mathbf{u}_i\}_{i=\kappa+1}^{\gamma}, \{\mathbf{v}_j\}_{j=1}^{\delta} \rangle$.

By Corollary 5.19, we have that
$$\begin{aligned}
\mathcal{R}(\mathcal{C}) &= \langle \{\mathbf{u}_i\}_{i=1}^{\gamma}, \{\mathbf{v}_j\}_{j=1}^{\delta}, \{2\mathbf{v}_j * \mathbf{v}_k\}_{1 \leq j < k \leq \delta} \rangle, \\
\mathcal{R}(\bar{\mathcal{C}}) &= \langle \{\mathbf{u}_i\}_{i=1}^{\kappa} \rangle, \text{ and} \\
\mathcal{R}(\mathcal{C}') &= \langle \{\mathbf{u}_i\}_{i=\kappa+1}^{\gamma}, \{\mathbf{v}_j\}_{j=1}^{\delta}, \{2\mathbf{v}_j * \mathbf{v}_k\}_{1 \leq j < k \leq \delta} \rangle.
\end{aligned}$$

Note that $\mathcal{R}(\mathcal{C}) = \mathcal{R}(\bar{\mathcal{C}}) \cup \mathcal{R}(\mathcal{C}')$. Moreover, $\mathcal{R}(\bar{\mathcal{C}}) \cap \mathcal{R}(\mathcal{C}') = \{\mathbf{0}\}$ due to the fact that for all $1 \leq j < k \leq \delta$, $2\mathbf{v}_j * \mathbf{v}_k = (\mathbf{0} \mid 2v'_j * v'_k) \notin \bar{\mathcal{C}}$. Therefore, $\mathrm{rank}(\Phi(\mathcal{C})) = \mathrm{rank}(\Phi(\bar{\mathcal{C}})) + \mathrm{rank}(\Phi(\mathcal{C}')) = \kappa + \mathrm{rank}(\Phi(\mathcal{C}'))$. Finally, by Lemma 5.32, $\mathrm{rank}(\Phi(\mathcal{C}')) = \mathrm{rank}(\phi(\mathcal{C}'_Y))$ and the statement follows. □

Example 5.34. Let $\mathcal{C}$ be the $\mathbb{Z}_2\mathbb{Z}_4$-additive code of type $(3,5;3,3,3)$ given in Example 5.6. By Example 5.24, we have that $\mathrm{rank}(\Phi(\mathcal{C})) = 10$ and, by Example 5.31, we have that $\mathrm{rank}(\Phi(\mathcal{C})) \neq \kappa + \mathrm{rank}(\phi(\mathcal{C}_Y))$. By Example 5.16, the code $\phi(\mathcal{C}'_Y)$, where $\mathcal{C}'_Y$ is generated by

$$\begin{pmatrix} 1 & 0 & 1 & 0 & 0 \\ 0 & 1 & 0 & 1 & 0 \\ 0 & 1 & 0 & 0 & 1 \end{pmatrix},$$

is not linear. In fact, it is easy to see that $\mathrm{rank}(\phi(\mathcal{C}'_Y)) = 7$ and hence $\mathrm{rank}(\Phi(\mathcal{C})) = 10 = 3 + 7 = \kappa + \mathrm{rank}(\phi(\mathcal{C}'_Y))$ satisfying Theorem 5.33. △

Example 5.35. By using the MAGMA package for $\mathbb{Z}_2\mathbb{Z}_4$-additive codes [29], we construct the same $\mathbb{Z}_2\mathbb{Z}_4$-additive code $\mathcal{C}$ as in Example 5.17. We show how to obtain the binary linear code $\langle\Phi(\mathcal{C})\rangle$, and the corresponding $\mathbb{Z}_2\mathbb{Z}_4$-additive code $\mathcal{R}(\mathcal{C})$ from all codewords and from the rows of a generator matrix of $\mathcal{C}$. We also compute the value of $\mathrm{rank}(\Phi(\mathcal{C}))$ and check that it is within the bounds given by Proposition 5.23. Finally, we see that the result given by Theorem 5.33 is satisfied for the code $\mathcal{C}$.

```
> Z4 := IntegerRing(4);
> alpha := 2;
> beta := 9;
> V := RSpace(Z4, alpha + beta);
> G := Matrix([V![2,0, 0,0,2,0,0,0,0,0,0],
              V![0,0, 0,2,2,2,0,0,0,0,0],
              V![0,0, 0,0,2,0,2,0,0,0,0],
              V![0,0, 0,1,0,1,0,1,0,0,0],
              V![0,0, 1,1,0,0,0,0,1,0,0],
              V![0,2, 1,0,1,1,1,0,0,1,0],
              V![0,0, 0,0,1,0,1,0,0,0,1]]);
> C := Z2Z4AdditiveCode(G, alpha);
> mapGray := GrayMap(Z2Z4AdditiveUniverseCode(alpha, beta));
> typeC := Z2Z4Type(C);
> gamma := typeC[3];
> delta := typeC[4];
> kappa := typeC[5];
>
> Cbin := GrayMapImage(C);
> SpanCbin := LinearCode<GF(2), alpha + 2*beta | Cbin>;
> SpanC := Z2Z4AdditiveCode([c@@mapGray : c in SpanCbin ], alpha);
```

```
> #Cbin eq #C;
true
> #SpanCbin eq #SpanC;
true
>
> v1 := G[gamma+1];
> v2 := G[gamma+2];
> v3 := G[gamma+3];
> v4 := G[gamma+4];
> 2*v1*v2 in C; 2*v1*v3 in C; 2*v2*v3 in C;
false false false
> 2*v1*v4 in C; 2*v2*v4 in C; 2*v3*v4 in C;
true true true
>
> newSpanCbin := LinearCode<GF(2), alpha + 2*beta |
                           [mapGray(u) : u in Rows(G)] cat
                           [mapGray(2*v) : v in [v1, v2, v3, v4]] cat
                           [mapGray(w) : w in [2*v1*v2, 2*v1*v3, 2*v2*v3]]>;
> newSpanCbin eq SpanCbin;
true
> newSpanC := Z2Z4AdditiveCode(Rows(G) cat
                           [2*v1*v2, 2*v1*v3, 2*v2*v3], alpha);
> newSpanC eq SpanC;
true
> Z2Z4Type(SpanC)[4] eq delta;
true
> Z2Z4Type(SpanC)[3] gt gamma;
true
>
> RankZ2(C);
14
> RankZ2(C) eq Dimension(SpanCbin);
true
> lowerBound := gamma + 2*delta;
> upperBound := Min(beta+delta+kappa, gamma+2*delta+Binomial(delta, 2));
> (lowerBound le RankZ2(C)) and (RankZ2(C) le upperBound);
true
>
> Cprime := Z2Z4AdditiveCode(G[kappa+1..gamma+delta], alpha);
> CYprime := LinearQuaternaryCode(Cprime);
> RankZ2(C) eq kappa + RankZ2(CYprime);
true                                                                    △
```

5.3   Kernel of $\mathbb{Z}_2\mathbb{Z}_4$-Linear Codes

In this section, we study the kernel of $\mathbb{Z}_2\mathbb{Z}_4$-linear codes and its dimension. We see how to compute the kernel efficiently, and how to determine a basis from the rows of a generator matrix of the corresponding $\mathbb{Z}_2\mathbb{Z}_4$-additive code. We also show that, for fixed values $\alpha, \beta, \gamma, \delta, \kappa$ satisfying (2.10), there exists a $\mathbb{Z}_2\mathbb{Z}_4$-linear code of type $(\alpha, \beta; \gamma, \delta; \kappa)$ with a kernel of dimension k, for any possible value of k. Finally, as for the rank, we prove that the dimension of the kernel can be given in terms of the dimension of $\mathcal{C}_X$ and the dimension of the kernel of a $\mathbb{Z}_4$-linear code which is a subcode of $\phi(\mathcal{C}_Y)$.

The *kernel* of a binary code C of length n, denoted by $K(C)$ and first defined in [21], is the set of vectors that leave the code C invariant under translation; that is,

$$K(C) = \{x \in \mathbb{Z}_2^n : C + x = C\}.$$

It is easy to prove that $K(C)$ is always a linear code. We denote its dimension by $\ker(C)$. Note that C is linear if and only if $K(C) = C$. If C contains the all-zero vector, then $K(C)$ is a binary linear subcode of C. Thus, if C is a $\mathbb{Z}_2\mathbb{Z}_4$-linear code, then $K(C)$ is always a linear subcode of C. In general, C can be written as the union of cosets of its kernel, and $K(C)$ is the largest such linear code for which this is true [21]. Moreover, $K(C)$ can also be defined as the intersection of all the maximal linear subspaces of C [107].

Proposition 5.36 ([79]). *Let $\mathcal{C}$ be a $\mathbb{Z}_2\mathbb{Z}_4$-additive code, and let $C = \Phi(\mathcal{C})$. Then, the kernel of C is*

$$K(C) = \{\Phi(\mathbf{u}) : \mathbf{u} \in \mathcal{C} \text{ and } 2\mathbf{u} * \mathbf{v} \in \mathcal{C}, \forall \mathbf{v} \in \mathcal{C}\}.$$

Proof. Since $K(C) \subseteq C$, if $\Phi(\mathbf{u}) \in K(C)$, $\mathbf{u} \in \mathcal{C}$. Let $\Phi(\mathbf{u}), \Phi(\mathbf{v}) \in C$. By the definition of $K(C)$, $\Phi(\mathbf{u}) \in K(C)$ if and only if $\Phi(\mathbf{u}) + \Phi(\mathbf{v}) \in C$ for all $\mathbf{v} \in \mathcal{C}$. By Proposition 5.1, $\Phi(\mathbf{u}) + \Phi(\mathbf{v}) = \Phi(\mathbf{u} + \mathbf{v} + 2\mathbf{u} * \mathbf{v}) \in C$. Therefore, $\Phi(\mathbf{u}) + \Phi(\mathbf{v}) \in C$ if and only if $2\mathbf{u} * \mathbf{v} \in \mathcal{C}$. Thus, the result follows. $\qquad\square$

Lemma 5.37 ([79]). *Let $\mathcal{C}$ be a $\mathbb{Z}_2\mathbb{Z}_4$-additive code, and let $C = \Phi(\mathcal{C})$. Given $\mathbf{u}, \mathbf{v} \in \mathcal{C}$, $\Phi(\mathbf{u}) + \Phi(\mathbf{v}) \in K(C)$ if and only if $\Phi(\mathbf{u} + \mathbf{v}) \in K(C)$.*

Proof. By Proposition 5.1 and Corollary 5.2, $\Phi(\mathbf{u} + \mathbf{v} + 2\mathbf{u} * \mathbf{v}) = \Phi(\mathbf{u}) + \Phi(\mathbf{v})$. On the one hand, if $\Phi(\mathbf{u}) + \Phi(\mathbf{v}) \in K(C)$, by Proposition 5.36, $2(\mathbf{u} + \mathbf{v} + 2\mathbf{u} * \mathbf{v}) * \mathbf{w} = 2(\mathbf{u} + \mathbf{v}) * \mathbf{w} \in \mathcal{C}$ for all $\mathbf{w} \in \mathcal{C}$. Since $\mathbf{u} + \mathbf{v} \in \mathcal{C}$, $\Phi(\mathbf{u} + \mathbf{v}) \in K(C)$. On the other hand, if $\Phi(\mathbf{u} + \mathbf{v}) \in K(C)$, again $2(\mathbf{u} + \mathbf{v}) * \mathbf{w} = 2(\mathbf{u} + \mathbf{v} + 2\mathbf{u} * \mathbf{v}) * \mathbf{w} \in \mathcal{C}$ for all $\mathbf{w} \in \mathcal{C}$. In particular, $2(\mathbf{u} + \mathbf{v}) * \mathbf{v} = 2\mathbf{u} * \mathbf{v} + 2\mathbf{v} \in \mathcal{C}$. Thus, $2\mathbf{u} * \mathbf{v} \in \mathcal{C}$ and $\mathbf{u} + \mathbf{v} + 2\mathbf{u} * \mathbf{v} \in \mathcal{C}$. By Proposition 5.36, $\Phi(\mathbf{u} + \mathbf{v} + 2\mathbf{u} * \mathbf{v}) \in K(C)$, that is, $\Phi(\mathbf{u}) + \Phi(\mathbf{v}) \in K(C)$. $\qquad\square$

Corollary 5.38 ([79]). *If C is a $\mathbb{Z}_2\mathbb{Z}_4$-linear code, then $K(C)$ is both linear and $\mathbb{Z}_2\mathbb{Z}_4$-linear.*

Let $\mathcal{C}$ be a $\mathbb{Z}_2\mathbb{Z}_4$-additive code. We define the kernel of $\mathcal{C}$, denoted by $\mathcal{K}(\mathcal{C})$, as $\mathcal{K}(\mathcal{C}) = \Phi^{-1}(K(C))$, where $C = \Phi(\mathcal{C})$ is the corresponding $\mathbb{Z}_2\mathbb{Z}_4$-linear code. By Corollary 5.38, the kernel $\mathcal{K}(\mathcal{C})$ is a $\mathbb{Z}_2\mathbb{Z}_4$-additive code, and by Proposition 5.36,

$$\mathcal{K}(\mathcal{C}) = \{\mathbf{u} \in \mathcal{C} : 2\mathbf{u} * \mathbf{v} \in \mathcal{C}, \forall \mathbf{v} \in \mathcal{C}\}. \tag{5.8}$$

Note that C is linear if and only if $\mathcal{K}(\mathcal{C}) = \mathcal{C}$. Moreover, for a generator matrix $\mathcal{G}$ of $\mathcal{C}$, it is easy to check that $\Phi(\mathbf{u}) \in K(C)$ or, equivalently, $\mathbf{u} \in \mathcal{K}(\mathcal{C})$, if and only if $\mathbf{u} \in \mathcal{C}$ and $2\mathbf{u} * \mathbf{v} \in \mathcal{C}$ for all row vector $\mathbf{v}$ in $\mathcal{G}$. We also have that $\mathcal{C}_b \subseteq \mathcal{K}(\mathcal{C})$, where $\mathcal{C}_b$ is the subcode of $\mathcal{C}$ containing all codewords of order at most two, by Proposition 5.36. Therefore, we have the following result, which can be used to compute the kernel more efficiently.

Proposition 5.39 ([79]). *Let $\mathcal{C}$ be a $\mathbb{Z}_2\mathbb{Z}_4$-additive code of type $(\alpha, \beta; \gamma, \delta; \kappa)$ with generator matrix $\mathcal{G}$. Let $\{\mathbf{u}_i\}_{i=1}^{\gamma}$ and $\{\mathbf{v}_j\}_{j=1}^{\delta}$ be the row vectors of order two and four in $\mathcal{G}$, respectively. Then,*

$$\mathcal{K}(\mathcal{C}) = \{\mathbf{u} \in \mathcal{C} \mid 2\mathbf{u} * \mathbf{v}_j \in \mathcal{C}, \forall j \in \{1, \ldots, \delta\}\}.$$

Example 5.40. Let $\mathcal{C} = \langle \mathbf{u}_1, \mathbf{u}_2, \mathbf{u}_3, \mathbf{v}_1, \mathbf{v}_2, \mathbf{v}_3 \rangle$ be the $\mathbb{Z}_2\mathbb{Z}_4$-additive code of type $(3, 5; 3, 3; 3)$ defined in Example 5.6. We have seen that $\mathcal{C}_b = \langle \mathbf{u}_1, \mathbf{u}_2, \mathbf{u}_3, 2\mathbf{v}_1, 2\mathbf{v}_2, 2\mathbf{v}_3 \rangle \subseteq \mathcal{K}(\mathcal{C})$. Thus, we only have to check which codewords of order four belong to the kernel. In fact, by the definition of the kernel, we only have to check the codewords of order four that are a 2-linear combination of $\mathbf{v}_1, \mathbf{v}_2$ and $\mathbf{v}_3$. First, by Proposition 5.39, $\mathbf{v}_1 \in \mathcal{K}(\mathcal{C})$ since $2\mathbf{v}_1 * \mathbf{v}_j \in \mathcal{C}$ for all $j \in \{1, 2, 3\}$, so $\langle \mathcal{C}_b, \mathbf{v}_1 \rangle = \langle \mathbf{u}_1, \mathbf{u}_2, \mathbf{u}_3, \mathbf{v}_1, 2\mathbf{v}_2, 2\mathbf{v}_3 \rangle \subseteq \mathcal{K}(\mathcal{C})$. By Example 5.20, $2\mathbf{v}_2 * \mathbf{v}_3 \notin \mathcal{C}$. Therefore, by Proposition 5.39, we have $\mathbf{v}_2, \mathbf{v}_3 \notin \mathcal{K}(\mathcal{C})$. We also have that $2(\mathbf{v}_2 + \mathbf{v}_3) * \mathbf{v}_2 = (0, 0, 0 \mid 0, 0, 0, 2, 0) \notin \mathcal{C}$, so $\mathbf{v}_2 + \mathbf{v}_3 \notin \mathcal{K}(\mathcal{C})$ again by Proposition 5.39. Therefore, $\mathcal{K}(\mathcal{C}) = \langle \mathbf{u}_1, \mathbf{u}_2, \mathbf{u}_3, \mathbf{v}_1, 2\mathbf{v}_2, 2\mathbf{v}_3 \rangle$. Note that $\mathcal{K}(\mathcal{C})$ is of type $(\alpha, \beta; \gamma + \bar{k}, \delta - \bar{k}; \kappa) = (3, 5; 5, 1; 3)$ for $\bar{k} = 2$. Finally, by Proposition 5.9, we have that $\{\Phi(\mathbf{u}_1), \Phi(\mathbf{u}_2), \Phi(\mathbf{u}_3), \Phi(\mathbf{v}_1), \Phi(2\mathbf{v}_1), \Phi(2\mathbf{v}_2), \Phi(2\mathbf{v}_3)\}$ is a basis of $K(\Phi(\mathcal{C}))$ and we have that $\ker(\Phi(\mathcal{C})) = \gamma + 2\delta - \bar{k} = 7$. $\triangle$

In general, as in Example 5.40, the kernel of a $\mathbb{Z}_2\mathbb{Z}_4$-additive code can be computed by checking (in the worse case) whether each one of the 2-linear combinations of the codewords of order four in a generator matrix of the code belong to the kernel. Nevertheless, now we see that we can also compute the kernel more efficiently by considering generator and parity check matrices of some related quaternary codes.

First, note that replacing ones with twos in the first α coordinates, we can see $\mathbb{Z}_2\mathbb{Z}_4$-additive codes as quaternary linear codes. Recall that χ is the map from $\mathbb{Z}_2$ to $\mathbb{Z}_4$, which is the usual inclusion from the additive structure in $\mathbb{Z}_2$ to $\mathbb{Z}_4$: $\chi(0) = 0$, $\chi(1) = 2$. This map can be extended to the map (χ, Id) : $\mathbb{Z}_2^\alpha \times \mathbb{Z}_4^\beta \rightarrow \mathbb{Z}_4^{\alpha+\beta}$, which is also denoted by χ and defined in Section 2.1. If $\mathcal{C}$ is a $\mathbb{Z}_2\mathbb{Z}_4$-additive code of type $(\alpha, \beta; \gamma, \delta; \kappa)$ with generator matrix $\mathcal{G}$, then $\chi(\mathcal{C})$ is a quaternary linear code of length $\alpha + \beta$ and type $2^\gamma 4^\delta$ with generator matrix $\mathcal{G}_{\chi(\mathcal{C})} = \chi(\mathcal{G})$. Note that $\mathcal{K}(\mathcal{C}) = \chi^{-1}\mathcal{K}(\chi(\mathcal{C}))$ and it can be proved that $\mathcal{K}(\chi(\mathcal{C}))^\perp$ is the quaternary linear code generated by the matrix

$$\begin{pmatrix} \mathcal{H}_{\chi(\mathcal{C})} \\ 2\mathcal{G}_{\chi(\mathcal{C})} * \mathcal{H}_{\chi(\mathcal{C})} \end{pmatrix}, \tag{5.9}$$

where $\mathcal{H}_{\chi(\mathcal{C})}$ is the parity check matrix of $\chi(\mathcal{C})$ and $2\mathcal{G}_{\chi(\mathcal{C})} * \mathcal{H}_{\chi(\mathcal{C})}$ is the matrix obtained computing the component-wise product $2\mathbf{u} * \mathbf{v}$ for all row vectors $\mathbf{u} \in \mathcal{G}_{\chi(\mathcal{C})}$, $\mathbf{v} \in \mathcal{H}_{\chi(\mathcal{C})}$. Note that matrix (5.9) may not be a generator matrix in the form of (2.5) since it may contain some linearly dependent rows.

Example 5.41. Let $\mathcal{C} = \langle \mathbf{u}_1, \mathbf{u}_2, \mathbf{u}_3, \mathbf{v}_1, \mathbf{v}_2, \mathbf{v}_3 \rangle$ be the $\mathbb{Z}_2\mathbb{Z}_4$-additive code defined in Example 5.6. We have that

$$\mathcal{G}_{\chi(\mathcal{C})} = \begin{pmatrix} \chi(\mathbf{u}_1) \\ \chi(\mathbf{u}_2) \\ \chi(\mathbf{u}_3) \\ \chi(\mathbf{v}_1) \\ \chi(\mathbf{v}_2) \\ \chi(\mathbf{v}_3) \end{pmatrix} = \begin{pmatrix} 2 & 0 & 0 & 0 & 0 & 0 & 0 & 0 \\ 0 & 2 & 0 & 2 & 0 & 0 & 0 & 0 \\ 0 & 0 & 2 & 0 & 2 & 0 & 0 & 0 \\ 0 & 0 & 2 & 1 & 0 & 1 & 0 & 0 \\ 0 & 0 & 0 & 0 & 1 & 0 & 1 & 0 \\ 0 & 0 & 0 & 0 & 1 & 0 & 0 & 1 \end{pmatrix},$$

and

$$\mathcal{H}_{\chi(\mathcal{C})} = \begin{pmatrix} 2 & 0 & 0 & 0 & 0 & 0 & 0 & 0 \\ 0 & 0 & 0 & 2 & 0 & 2 & 0 & 0 \\ 0 & 0 & 0 & 0 & 2 & 0 & 2 & 2 \\ 0 & 0 & 1 & 0 & 1 & 2 & 3 & 3 \\ 0 & 1 & 0 & 1 & 0 & 3 & 0 & 0 \end{pmatrix}.$$

Therefore, we have that a generator matrix of $(\mathcal{K}(\chi(\mathcal{C}))^\perp$ is

$$\begin{pmatrix} \mathcal{H}_{\chi(\mathcal{C})} \\ 2\mathcal{G}_{\chi(\mathcal{C})} * \mathcal{H}_{\chi(\mathcal{C})} \end{pmatrix} = \begin{pmatrix} 2 & 0 & 0 & 0 & 0 & 0 & 0 & 0 \\ 0 & 0 & 0 & 2 & 0 & 2 & 0 & 0 \\ 0 & 0 & 0 & 0 & 2 & 0 & 2 & 2 \\ 0 & 0 & 1 & 0 & 1 & 2 & 3 & 3 \\ 0 & 1 & 0 & 1 & 0 & 3 & 0 & 0 \\ 0 & 0 & 0 & 0 & 2 & 0 & 2 & 0 \\ 0 & 0 & 0 & 0 & 2 & 0 & 0 & 2 \end{pmatrix},$$

and hence a generator matrix of $\mathcal{K}(\chi(\mathcal{C}))$ is

$$\begin{pmatrix} 2 & 0 & 0 & 0 & 0 & 0 & 0 & 0 \\ 0 & 2 & 0 & 2 & 0 & 0 & 0 & 0 \\ 0 & 0 & 2 & 0 & 2 & 0 & 0 & 0 \\ 0 & 0 & 0 & 0 & 2 & 0 & 2 & 0 \\ 0 & 0 & 0 & 0 & 2 & 0 & 0 & 2 \\ 0 & 0 & 2 & 1 & 0 & 1 & 0 & 0 \end{pmatrix}.$$

Finally, since $\mathcal{K}(\mathcal{C}) = \chi^{-1}(\mathcal{K}(\chi(\mathcal{C})))$, a generator matrix of $\mathcal{K}(\mathcal{C})$ is

$$\left(\begin{array}{ccc|ccccc} 1 & 0 & 0 & 0 & 0 & 0 & 0 & 0 \\ 0 & 1 & 0 & 2 & 0 & 0 & 0 & 0 \\ 0 & 0 & 1 & 0 & 2 & 0 & 0 & 0 \\ 0 & 0 & 0 & 0 & 2 & 0 & 2 & 0 \\ 0 & 0 & 0 & 0 & 2 & 0 & 0 & 2 \\ 0 & 0 & 1 & 1 & 0 & 1 & 0 & 0 \end{array}\right).$$

Therefore, we have that $\mathcal{K}(\mathcal{C}) = \langle \mathbf{u}_1, \mathbf{u}_2, \mathbf{u}_3, 2\mathbf{v}_2, 2\mathbf{v}_3, \mathbf{v}_1 \rangle$ as shown in Example 5.40.   △

Now, in the next proposition, we give an upper and lower bound for the dimension of the kernel of a $\mathbb{Z}_2\mathbb{Z}_4$-linear code in terms of the parameters of its type.

Proposition 5.42 ([79]). *Let C be a $\mathbb{Z}_2\mathbb{Z}_4$-linear code of type $(\alpha, \beta; \gamma, \delta; \kappa)$. Then, $K(C)$ is a $\mathbb{Z}_2\mathbb{Z}_4$-linear subcode of C of type $(\alpha, \beta; \gamma + \bar{k}, \delta - \bar{k}; \kappa)$ and $\ker(C) = \gamma + 2\delta - \bar{k}$, where $\bar{k} \in \{0\} \cup \{2, \ldots, \delta\}$.*

Proof. Let $\mathcal{C} = \Phi^{-1}(C)$. It is easy to see that $\mathcal{K}(\mathcal{C})$ is a $\mathbb{Z}_2\mathbb{Z}_4$-additive subcode of $\mathcal{C}$ of type $(\alpha, \beta; \gamma + \bar{k}, \delta - \bar{k}; \kappa)$.

The upper bound for $\ker(C)$ is $\gamma + 2\delta$ and comes from the linear case. The lower bound $\gamma + \delta$ is straightforward, since there are $2^{\gamma+\delta}$ codewords in $\mathcal{C}_b$ (the set of codewords of order at most two) and $\mathcal{C}_b \subseteq \mathcal{K}(\mathcal{C})$. Also note that if the $\mathbb{Z}_2\mathbb{Z}_4$-linear code C is not linear, then the dimension of the kernel is less than or equal to $\gamma + 2\delta - 2$ [121]. Therefore, $\ker(C) = \gamma + 2\delta - \bar{k}$, where $\bar{k} \in \{0\} \cup \{2, \ldots, \delta\}$.   □

Given an integer $m > 0$, a set of vectors $\{\mathbf{v}_1, \mathbf{v}_2, \ldots, \mathbf{v}_m\}$ in $\mathbb{Z}_2^\alpha \times \mathbb{Z}_4^\beta$ and a subset $I = \{i_1, \ldots, i_l\} \subseteq \{1, \ldots, m\}$, we denote by $\mathbf{v}_I$ the vector $\mathbf{v}_{i_1} + \cdots + \mathbf{v}_{i_l}$. If $I = \emptyset$, then $\mathbf{v}_I = \mathbf{0}$.

Proposition 5.43 ([79])**.** *Let* $\mathcal{C}$ *be a* $\mathbb{Z}_2\mathbb{Z}_4$*-additive code of type* $(\alpha, \beta; \gamma, \delta; \kappa)$ *with generator matrix* $\mathcal{G}$, *and let* $C = \Phi(\mathcal{C})$ *be the corresponding* $\mathbb{Z}_2\mathbb{Z}_4$*-linear code with* $\ker(C) = \gamma + 2\delta - \bar{k}$, *where* $\bar{k} \in \{2, \dots, \delta\}$. *Let* $\{\mathbf{v}_j\}_{j=1}^{\delta}$ *be the rows of order four in* $\mathcal{G}$. *Then, there exists a set* $\{j_1, \dots, j_{\bar{k}}\} \subseteq \{1, \dots, \delta\}$ *such that*

$$C = \bigcup_{I \subseteq \{j_1, \dots, j_{\bar{k}}\}} (K(C) + \Phi(\mathbf{v}_I)).$$

Proof. Let $\{\mathbf{u}_i\}_{i=1}^{\gamma}$ be the rows of order two and $\{\mathbf{v}_j\}_{j=1}^{\delta}$ the rows of order four in $\mathcal{G}$. By Proposition 5.36, the Gray map image of all codewords of order at most two are in $K(C)$. Recall that there are $2^{\gamma+\delta}$ such codewords, which are generated by $\gamma + \delta$ codewords.

Moreover, there are $\delta - \bar{k}$ codewords $\mathbf{w}_j$ of order four such that $\Phi(\mathbf{w}_j) \in K(C)$ for all $j \in \{1, \dots, \delta - \bar{k}\}$, and $\Phi(\mathbf{u}_1), \dots, \Phi(\mathbf{u}_\gamma)$, $\Phi(2\mathbf{v}_1), \dots, \Phi(2\mathbf{v}_\delta)$, $\Phi(\mathbf{w}_1), \dots, \Phi(\mathbf{w}_{\delta-\bar{k}})$ are linearly independent vectors over $\mathbb{Z}_2$. The code $\mathcal{C}$ can also be generated by $\mathbf{u}_1, \dots, \mathbf{u}_\gamma, \mathbf{w}_1, \dots, \mathbf{w}_{\delta-\bar{k}}, \mathbf{v}_{j_1}, \dots, \mathbf{v}_{j_{\bar{k}}}$, where $\{j_1, j_2, \dots, j_{\bar{k}}\} \subseteq \{1, \dots, \delta\}$. Without loss of generality, we can assume that $\mathbf{v}_{j_1}, \dots, \mathbf{v}_{j_{\bar{k}}}$ are the $\bar{k}$ row vectors $\mathbf{v}_1, \dots, \mathbf{v}_{\bar{k}}$ in $\mathcal{G}$. Note that $\Phi(\mathbf{v}_I) \notin K(C)$, for any $I \subseteq \{1, \dots, \bar{k}\}$ such that $I \neq \emptyset$. In fact, if $\Phi(\mathbf{v}_I) \in K(C)$, then the set of vectors $\Phi(\mathbf{u}_1), \dots, \Phi(\mathbf{u}_\gamma), \Phi(2\mathbf{v}_1), \dots, \Phi(2\mathbf{v}_\delta), \Phi(\mathbf{w}_1), \dots, \Phi(\mathbf{w}_{\delta-\bar{k}}), \Phi(\mathbf{v}_I)$ would be linearly independent and $\ker(C) > \gamma + 2\delta - \bar{k}$.

We know that C can be written as the union of cosets of $K(C)$ [21]. Since $|K(C)| = 2^{\gamma+2\delta-\bar{k}}$ and $|C| = 2^{\gamma+2\delta}$, there are exactly $2^{\bar{k}}$ cosets. Now, we see that these cosets are the $2^{\bar{k}}$ cosets $(K(C) + \Phi(\mathbf{v}_I))$, for $I \subseteq \{1, \dots, \bar{k}\}$. Specifically, we show that the $2^{\bar{k}} - 1$ binary vectors $\Phi(\mathbf{v}_I)$, $I \subseteq \{1, \dots, \bar{k}\}$ and $I \neq \emptyset$, are in different cosets. Let $\Phi(\mathbf{v}_{I_1})$ and $\Phi(\mathbf{v}_{I_2})$ be any two of these binary vectors such that $I_1 \neq I_2$. If $\Phi(\mathbf{v}_{I_1}) \in K(C) + \Phi(\mathbf{v}_{I_2})$, then $\Phi(\mathbf{v}_{I_1}) + \Phi(\mathbf{v}_{I_2}) \in K(C)$ and, by Lemma 5.37, $\Phi(\mathbf{v}_{I_1} + \mathbf{v}_{I_2}) \in K(C)$. We also have that $\mathbf{v}_{I_1} + \mathbf{v}_{I_2} = \mathbf{v}_I + 2\mathbf{v}_{I_1 \cap I_2}$, for $I = (I_1 \cup I_2) \setminus (I_1 \cap I_2) \subseteq \{1, \dots, \bar{k}\}$. Hence, $\Phi(\mathbf{v}_{I_1} + \mathbf{v}_{I_2}) = \Phi(\mathbf{v}_I) + \Phi(2\mathbf{v}_{I_1 \cap I_2}) \in K(C)$ and $\Phi(\mathbf{v}_I) \in K(C)$, which is a contradiction. $\square$

Corollary 5.44. *Let* $\mathcal{C}$ *be a* $\mathbb{Z}_2\mathbb{Z}_4$*-additive code of type* $(\alpha, \beta; \gamma, \delta; \kappa)$ *with generator matrix* $\mathcal{G}$ *and* $\ker(\Phi(\mathcal{C})) = \gamma + 2\delta - \bar{k}$, *where* $\bar{k} \in \{2, \dots, \delta\}$. *Let* $\{\mathbf{v}_j\}_{j=1}^{\delta}$ *be the rows of order four in* $\mathcal{G}$. *Then, there exists a set* $\{j_1, \dots, j_{\bar{k}}\} \subseteq \{1, \dots, \delta\}$ *such that*

$$\mathcal{C} = \bigcup_{I \subseteq \{j_1, \dots, j_{\bar{k}}\}} (\mathcal{K}(\mathcal{C}) + \mathbf{v}_I).$$

Example 5.45. Let $\mathcal{C}$ be the $\mathbb{Z}_2\mathbb{Z}_4$-additive code given in Example 5.6. From

Examples 5.6 and 5.40, we have that

$$\mathcal{C} = \langle \mathbf{u}_1, \mathbf{u}_2, \mathbf{u}_3, \mathbf{v}_1, \mathbf{v}_2, \mathbf{v}_3 \rangle,$$
$$\mathcal{K}(\mathcal{C}) = \langle \mathbf{u}_1, \mathbf{u}_2, \mathbf{u}_3, \mathbf{v}_1, 2\mathbf{v}_2, 2\mathbf{v}_3 \rangle.$$

Then, by Proposition 5.43 and Corollary 5.44, we can write the code $\mathcal{C}$ as a union of cosets of $\mathcal{K}(\mathcal{C})$ as follows:

$$\mathcal{C} = \mathcal{K}(\mathcal{C}) \cup \left(\mathcal{K}(\mathcal{C}) + \mathbf{v}_2 \right) \cup \left(\mathcal{K}(\mathcal{C}) + \mathbf{v}_3 \right) \cup \left(\mathcal{K}(\mathcal{C}) + (\mathbf{v}_2 + \mathbf{v}_3) \right)$$

or, equivalently, we can write $C = \Phi(\mathcal{C})$ as the following union of cosets of $K(C)$:

$$C = K(C) \cup \left(K(C) + \Phi(\mathbf{v}_2) \right) \cup \left(K(C) + \Phi(\mathbf{v}_3) \right) \cup \left(K(C) + \Phi(\mathbf{v}_2 + \mathbf{v}_3) \right).$$

$$\triangle$$

From the proof of Proposition 5.43, we obtain the following proposition.

Proposition 5.46. *Let $\mathcal{C}$ be a $\mathbb{Z}_2\mathbb{Z}_4$-additive code of type $(\alpha, \beta; \gamma, \delta; \kappa)$ with generator matrix $\mathcal{G}$ and $\ker(\Phi(\mathcal{C})) = \gamma + 2\delta - \bar{k}$, where $\bar{k} \in \{2, \ldots, \delta\}$. Let $\{\mathbf{u}_i\}_{i=1}^{\gamma}$ and $\{\mathbf{v}_j\}_{j=1}^{\delta}$ be the row vectors of order two and four in $\mathcal{G}$, respectively. Let $J = \{j_1, \ldots, j_{\bar{k}}\} \subseteq \{1, \ldots, \delta\}$ such that $\mathcal{C} = \bigcup_{I \subseteq J}(\mathcal{K}(\mathcal{C}) + \mathbf{v}_I)$. Then, there exist a set of vectors $\{\mathbf{w}_\ell\}_{\ell=1}^{\delta - \bar{k}}$, for $\mathbf{w}_\ell \in \langle \{\mathbf{v}_j : j \in \{1, \ldots, \delta\} \setminus J\} \rangle$, such that*

$$\mathcal{K}(\mathcal{C}) = \langle \{\mathbf{u}_i\}_{i=1}^{\gamma}, \{2\mathbf{v}_j\}_{j \in J}, \{\mathbf{w}_\ell\}_{\ell=1}^{\delta - \bar{k}} \rangle,$$

and $\{\{\mathbf{u}_i\}_{i=1}^{\gamma}, \{2\mathbf{v}_j\}_{j=1}^{\delta}, \{\mathbf{w}_\ell\}_{\ell=1}^{\delta - \bar{k}}\}$ is a 2-basis of $\mathcal{K}(\mathcal{C})$.

Example 5.47. Let $\mathcal{C}$ be the $\mathbb{Z}_2\mathbb{Z}_4$-additive code of type $(2, 9; 3, 5; 1)$ generated by the following matrix:

$$\begin{pmatrix} \mathbf{u}_1 \\ \mathbf{u}_2 \\ \mathbf{u}_3 \\ \mathbf{v}_1 \\ \mathbf{v}_2 \\ \mathbf{v}_3 \\ \mathbf{v}_4 \\ \mathbf{v}_5 \end{pmatrix} = \left(\begin{array}{cc|ccccccccc} 1 & 0 & 0 & 0 & 0 & 0 & 0 & 0 & 0 & 0 & 0 \\ 0 & 0 & 0 & 0 & 2 & 0 & 0 & 0 & 0 & 0 & 0 \\ 0 & 0 & 0 & 2 & 0 & 2 & 0 & 0 & 0 & 0 & 0 \\ 0 & 0 & 0 & 1 & 0 & 3 & 1 & 0 & 0 & 0 & 0 \\ 0 & 1 & 0 & 3 & 0 & 3 & 0 & 1 & 0 & 0 & 0 \\ 0 & 1 & 1 & 0 & 0 & 3 & 0 & 0 & 1 & 0 & 0 \\ 0 & 1 & 0 & 1 & 0 & 3 & 0 & 0 & 0 & 1 & 0 \\ 0 & 0 & 1 & 3 & 0 & 3 & 0 & 0 & 0 & 0 & 1 \end{array} \right).$$

First, we have that $\mathcal{C}_b = \langle \mathbf{u}_1, \mathbf{u}_2, \mathbf{u}_3, 2\mathbf{v}_1, 2\mathbf{v}_2, 2\mathbf{v}_3, 2\mathbf{v}_4, 2\mathbf{v}_5 \rangle \subseteq \mathcal{K}(\mathcal{C})$. Then, in order to compute the kernel of $\mathcal{C}$, we would need to check whether $\mathbf{v}_I \in \mathcal{K}(\mathcal{C})$, for each non-empty subset $I \subseteq \{1, 2, 3, 4, 5\}$. It is easy to check that

$\mathbf{v}_1 \notin \mathcal{K}(\mathcal{C})$ since $2\mathbf{v}_1 * \mathbf{v}_3 \notin \mathcal{C}$, and $\mathbf{v}_2 \notin \mathcal{K}(\mathcal{C})$ since $2\mathbf{v}_2 * \mathbf{v}_3 \notin \mathcal{C}$. However, we have that $\mathbf{v}_1 + \mathbf{v}_2 \in \mathcal{K}(\mathcal{C})$ since $2(\mathbf{v}_1 + \mathbf{v}_2) * \mathbf{v}_i \in \mathcal{C}$ for all $i \in \{1, \ldots, 5\}$. Therefore, $\langle \mathcal{C}_b, \mathbf{v}_1 + \mathbf{v}_2 \rangle \subseteq \mathcal{K}(\mathcal{C})$ and $\mathcal{C} = \langle \mathbf{u}_1, \mathbf{u}_2, \mathbf{u}_3, \mathbf{v}_1 + \mathbf{v}_2, \mathbf{v}_2, \mathbf{v}_3, \mathbf{v}_4, \mathbf{v}_5 \rangle$.

Now, we just need to check whether $\mathbf{v}_I \in \mathcal{K}(\mathcal{C})$, for each non-empty subset I of $\{2, 3, 4, 5\}$. We have that $\mathbf{v}_3 \notin \mathcal{K}(\mathcal{C})$ since $2\mathbf{v}_2 * \mathbf{v}_3 \notin \mathcal{C}$, and $\mathbf{v}_2 + \mathbf{v}_3 \notin \mathcal{K}(\mathcal{C})$ since $2(\mathbf{v}_2 + \mathbf{v}_3) * \mathbf{v}_2 \notin \mathcal{C}$. Similarly, we obtain that $\mathbf{v}_4 \notin \mathcal{K}(\mathcal{C})$ and $\mathbf{v}_4 + \mathbf{v}_2 \in \mathcal{K}(\mathcal{C})$. Therefore, $\langle \mathcal{C}_b, \mathbf{v}_1 + \mathbf{v}_2, \mathbf{v}_4 + \mathbf{v}_2 \rangle \subseteq \mathcal{K}(\mathcal{C})$ and $\mathcal{C} = \langle \mathbf{u}_1, \mathbf{u}_2, \mathbf{u}_3, \mathbf{v}_1 + \mathbf{v}_2, \mathbf{v}_4 + \mathbf{v}_2, \mathbf{v}_2, \mathbf{v}_3, \mathbf{v}_5 \rangle$. Finally, since $\mathbf{v}_I \notin \mathcal{K}(\mathcal{C})$, for each non-empty subset $I \subseteq \{2, 3, 5\}$, we have that $\mathcal{K}(\mathcal{C})$ is the $\mathbb{Z}_2\mathbb{Z}_4$-additive code of type $(2, 9; 6, 2; 1)$ generated by the following matrix:

$$
\begin{pmatrix} \mathbf{u}_1 \\ \mathbf{u}_2 \\ \mathbf{u}_3 \\ 2\mathbf{v}_2 \\ 2\mathbf{v}_3 \\ 2\mathbf{v}_5 \\ \mathbf{v}_1 + \mathbf{v}_2 \\ \mathbf{v}_4 + \mathbf{v}_2 \end{pmatrix} = \left(\begin{array}{cc|cccccccc} 1 & 0 & 0 & 0 & 0 & 0 & 0 & 0 & 0 & 0 \\ 0 & 0 & 0 & 0 & 2 & 0 & 0 & 0 & 0 & 0 \\ 0 & 0 & 0 & 2 & 0 & 2 & 0 & 0 & 0 & 0 \\ 0 & 0 & 0 & 2 & 0 & 2 & 0 & 2 & 0 & 0 \\ 0 & 0 & 2 & 0 & 0 & 2 & 0 & 0 & 2 & 0 \\ 0 & 0 & 2 & 2 & 0 & 2 & 0 & 0 & 0 & 2 \\ 0 & 1 & 0 & 0 & 0 & 2 & 1 & 1 & 0 & 0 \\ 0 & 0 & 0 & 0 & 0 & 2 & 0 & 1 & 0 & 1 & 0 \end{array} \right).
$$

Let $C = \Phi(\mathcal{C})$. Note that $K(C) = \Phi(\mathcal{K}(\mathcal{C}))$ is a linear code of dimension 10, and C can be partitioned into $|C|/|K(C)| = 2^{13}/2^{10} = 2^3 = 8$ cosets of $K(C)$. △

Proposition 5.48 ([79]). *Let C be a $\mathbb{Z}_2\mathbb{Z}_4$-linear code of type $(\alpha, \beta; \gamma, \delta; \kappa)$. Then, $\ker(C) = \gamma + 2\delta - \bar{k}$, where*

$$
\begin{cases} \bar{k} = 0, & \text{if } s = 0, \\ \bar{k} \in \{0\} \cup \{2, \ldots, \delta\} \text{ and } \bar{k} \text{ even}, & \text{if } s = 1, \\ \bar{k} \in \{0\} \cup \{2, \ldots, \delta\}, & \text{if } s \geq 2, \end{cases}
$$

and $s = \beta - (\gamma - \kappa) - \delta$.

Proof. For $s = 0$, by Proposition 5.23, we have that $\mathrm{rank}(C) = \gamma + 2\delta$, so C is a binary linear code and $\ker(C) = \gamma + 2\delta$. For $s \geq 2$, by Proposition 5.42 we have that $\ker(C) \in \{\gamma + \delta, \ldots, \gamma + 2\delta - 2, \gamma + 2\delta\}$.

Now, we prove the result for $s = 1$. By Theorem 2.10, C is permutation equivalent to a $\mathbb{Z}_2\mathbb{Z}_4$-additive code generated by

$$
\mathcal{G}_S = \left(\begin{array}{cc|ccc} I_\kappa & T_b & 2T_2 & 0 & 0 \\ 0 & 0 & 2T_1 & 2I_{\gamma-\kappa} & 0 \\ 0 & S_b & S & R & I_\delta \end{array} \right),
$$

where S is a matrix over $\mathbb{Z}_4$ of size $\delta \times 1$. Let $\{\mathbf{u}_i\}_{i=1}^{\gamma}$ and $\{\mathbf{v}_j\}_{j=1}^{\delta}$ be the row vectors in $\mathcal{G}_S$ of order two and four, respectively.

If $\delta < 3$, then it is easy to see that $\ker(C) = \gamma + 2\delta - 2$ or $\ker(C) = \gamma + 2\delta$, by Proposition 5.42. If $\delta \geq 3$, we first show that, given four vectors $\mathbf{v}_{j_1}, \mathbf{v}_{j_2}, \mathbf{v}_{j_3}, \mathbf{v}_{j_4}$ such that $2\mathbf{v}_{j_1} * \mathbf{v}_{j_2} \notin C$ and $2\mathbf{v}_{j_3} * \mathbf{v}_{j_4} \notin C$, then $2\mathbf{v}_{j_1} * \mathbf{v}_{j_2} + 2\mathbf{v}_{j_3} * \mathbf{v}_{j_4} \in C$. Let e_k, $1 \leq k \leq \gamma - \kappa$, denote the row vector of length $\gamma - \kappa$, with a one in the kth coordinate and zeros elsewhere. Then, we can write $2\mathbf{v}_{j_1} * \mathbf{v}_{j_2} = (\mathbf{0}, \mathbf{0}, 2c, 2e_I, \mathbf{0})$, where $c \in \{0, 1\}$ and $I \subseteq \{\alpha + 2, \dots, \alpha + \gamma - \kappa + 1\}$, and $2\mathbf{v}_{j_3} * \mathbf{v}_{j_4} = (\mathbf{0}, \mathbf{0}, 2c', 2e_J, \mathbf{0})$, where $c' \in \{0, 1\}$ and $J \subseteq \{\alpha + 2, \dots, \alpha + \gamma - \kappa + 1\}$. We denote by $\mathbf{u}_I$ (resp. $\mathbf{u}_J$) the row vector obtained by adding the row vectors of order two in $\mathcal{G}_S$ with 2 in the coordinate positions given by I (resp. J). Then, $\mathbf{u}_I = (\mathbf{0}, \mathbf{0}, 2d, 2e_I, \mathbf{0}) \in C$ with $d \in \{0, 1\}$ (resp. $\mathbf{u}_J = (\mathbf{0}, \mathbf{0}, 2d', 2e_J, \mathbf{0}) \in C$ with $d' \in \{0, 1\}$). Since $2\mathbf{v}_{j_1} * \mathbf{v}_{j_2} \notin C$ (resp. $2\mathbf{v}_{j_3} * \mathbf{v}_{j_4} \notin C$) we have $2\mathbf{v}_{j_1} * \mathbf{v}_{j_2} = \mathbf{u}_I + (\mathbf{0}, \mathbf{0}, 2, \mathbf{0}, \mathbf{0})$ (resp. $2\mathbf{v}_{j_3} * \mathbf{v}_{j_4} = \mathbf{u}_J + (\mathbf{0}, \mathbf{0}, 2, \mathbf{0}, \mathbf{0})$). Therefore, $2\mathbf{v}_{j_1} * \mathbf{v}_{j_2} + 2\mathbf{v}_{j_3} * \mathbf{v}_{j_4} = \mathbf{u}_I + \mathbf{u}_J \in C$.

By Proposition 5.43, there exist $\bar{k}$ row vectors $\mathbf{v}_{j_1}, \mathbf{v}_{j_2}, \dots, \mathbf{v}_{j_{\bar{k}}}$ in $\mathcal{G}_S$, such that $\Phi(\mathbf{v}_I) \notin K(C)$ for any non-empty subset $I \subseteq \{1, \dots, \bar{k}\}$ and $\ker(C) = \gamma + 2\delta - \bar{k}$. Without loss of generality, we assume that $\{j_1, \dots, j_{\bar{k}}\} = \{1, \dots, \bar{k}\}$. Assume $\bar{k}$ is odd. We show that there exists a subset $I \subseteq \{1, \dots, \bar{k}\}$ such that $\Phi(\mathbf{v}_I) \in K(C)$. Since this is a contradiction, $\bar{k}$ can not be an odd number and the assertion is proven.

By Propositions 5.39 and 5.43, in order to prove that $\Phi(\mathbf{v}_I) \in K(C)$, it is enough to prove that $2\mathbf{v}_I * \mathbf{v}_j \in C$ for all $j \in \{1, \dots, \bar{k}\}$. That is, for each $j \in \{1, \dots, \bar{k}\}$ the number of $i \in I$ such that $2\mathbf{v}_i * \mathbf{v}_j \notin C$ is even. We define a symmetric matrix $A = (a_{ij})$, $1 \leq i, j \leq \bar{k}$, in the following way: $a_{ij} = 1$ if $2\mathbf{v}_i * \mathbf{v}_j \notin C$ and 0 otherwise. Therefore, A is a symmetric matrix of odd order and with zeros in the main diagonal. In [79, Lemma 8] it is shown that $\det(A) = 0$ and hence there exists a linear combination of some rows, $i_1, \dots, i_l$, of A equal to $\mathbf{0}$. The vector $\Phi(\mathbf{v}_I)$, where $I = \{i_1, \dots, i_l\}$, belongs to $K(C)$. This completes the proof. $\qquad\square$

Example 5.49. Let $\overline{QRM}(r, m)$ be the class of $\mathbb{Z}_4$-linear Reed-Muller codes defined in [34], as in Example 5.25. The dimension of the kernel of any code $C \in \overline{QRM}(r, m)$ is

$$\ker(C) = \sum_{i=0}^{r} \binom{m}{i} + 1 = \delta + 1,$$

except for $r = m$ (in this case, $C = \mathbb{Z}_2^{2^{m+1}}$ and $\ker(C) = 2^{m+1}$) [34].

Therefore, $\mathbb{Z}_4$-linear Kerdock-like codes and extended $\mathbb{Z}_4$-linear Preparata-like codes of length 4^m, denoted by K_m and P_m, respectively, have dimension of the kernel $\ker(K_m) = 2m + 1$ and $\ker(P_m) = 2^{2m-1} - 2m + 1$, respectively [34, 44]. △

As in Section 5.2 for the rank, the next point to be solved here is how to construct $\mathbb{Z}_2\mathbb{Z}_4$-linear codes with any dimension of the kernel in the range of possibilities given by Proposition 5.48.

Theorem 5.50 ([79]). *Let* $\alpha, \beta, \gamma, \delta, \kappa$ *be integer numbers satisfying inequations* (2.10). *Then, there exists a* $\mathbb{Z}_2\mathbb{Z}_4$*-linear code* C *of type* $(\alpha, \beta; \gamma, \delta; \kappa)$ *with* $\ker(C) = \gamma + 2\delta - \bar{k}$ *if and only if*

$$\begin{cases} \bar{k} = 0, & \text{if} \quad s = 0, \\ \bar{k} \in \{0\} \cup \{2, \dots, \delta\} \text{ and } \bar{k} \text{ even}, & \text{if} \quad s = 1, \\ \bar{k} \in \{0\} \cup \{2, \dots, \delta\}, & \text{if} \quad s \geq 2, \end{cases}$$

and $s = \beta - (\gamma - \kappa) - \delta.$

Proof. Let $\mathcal{C}$ be a $\mathbb{Z}_2\mathbb{Z}_4$-additive code of type $(\alpha, \beta; \gamma, \delta; \kappa)$ with generator matrix

$$\mathcal{G} = \begin{pmatrix} I_\kappa & T_b & 0 & 0 & 0 \\ 0 & 0 & 0 & 2I_{\gamma-\kappa} & 0 \\ 0 & S_b & S_k & 0 & I_\delta \end{pmatrix},$$

where S_k is a matrix over $\mathbb{Z}_4$ of size $\delta \times s$, and $\{\mathbf{u}_i\}_{i=1}^\gamma$ and $\{\mathbf{v}_j\}_{j=1}^\delta$ are the row vectors in $\mathcal{G}$ of order two and four, respectively. Let $C = \Phi(\mathcal{C})$ be the corresponding $\mathbb{Z}_2\mathbb{Z}_4$-linear code with $\ker(C) = k = \gamma + 2\delta - \bar{k}$. The necessary conditions for the values of k are given in Proposition 5.48.

When $s = 0$, the code C is a binary linear code, so $\ker(C) = k = \gamma + 2\delta$ and $\bar{k} = 0$. If $s > 0$ then, taking S_k the all-zero matrix over $\mathbb{Z}_4$, the code C is also a binary linear code and $\bar{k} = 0$.

When $s = 1$, for each even $\bar{k} \in \{2, \dots, \delta\}$, we can construct a matrix S_k over $\mathbb{Z}_4$ of size $\delta \times 1$ with an even number of ones, $\bar{k}$, and zeros elsewhere. We can assume that $\mathbf{v}_1, \dots, \mathbf{v}_{\bar{k}}$ are the row vectors of order four with 1 in the column from the matrix S_k, and $\mathbf{v}_{\bar{k}+1}, \dots, \mathbf{v}_\delta$ the ones with 0 in this column. By Proposition 5.39, it is easy to see that $\Phi(\mathbf{v}_j) \in K(C)$ for all $j \in \{\bar{k}+1, \dots, \delta\}$. Since the Gray map image of all codewords of order two are also in $K(C)$, $\ker(C) \geq \gamma + \delta + \delta - \bar{k} = \gamma + 2\delta - \bar{k}$. Let $K'(C)$ be the binary linear subcode of $K(C)$ generated by these $\gamma + 2\delta - \bar{k}$ linearly independent vectors. By the same arguments as in Proposition 5.43, $C = \bigcup_{I \subseteq \{1, \dots, \bar{k}\}} (K'(C) + \Phi(v_I))$. In order to prove that $\ker(C) = \gamma + 2\delta - \bar{k}$, it is enough to show that a codeword from each coset does not belong to $K(C)$, that is, $\Phi(\mathbf{v}_I) \notin K(C)$ for any non-empty subset $I \subseteq \{1, \dots, \bar{k}\}$. If $|I|$ is even, $2\mathbf{v}_I * \mathbf{v}_j \notin \mathcal{C}$ for any $j \in I$. If $|I|$ is odd, $2\mathbf{v}_I * \mathbf{v}_j \notin \mathcal{C}$ for any

$j \in \{1,\ldots,\bar{k}\}\backslash I$. Note that $\{1,\ldots,\bar{k}\}\backslash I$ is a non-empty set, since $\bar{k} \geq 2$ and $\bar{k}$ is even. Therefore, $\Phi(\mathbf{v}_I) \notin K(C)$ by Proposition 5.39.

Finally, when $s \geq 2$, for each $\bar{k} \in \{2,3,\ldots,\delta\}$, we can construct a matrix S_k over $\mathbb{Z}_4$ of size $\delta \times s$, such that only in the last $\delta - \bar{k}$ row vectors all components are zero and, moreover, in the first $\bar{k}$ coordinates of each column vector there are an even number of ones and zeros elsewhere. In this case, by the same arguments as before, it is easy to prove that $\ker(C) = \gamma + 2\delta - \bar{k}$. $\square$

Example 5.51. By Proposition 5.48, we know that the possible dimensions of the kernel for $\mathbb{Z}_2\mathbb{Z}_4$-linear codes of type $(\alpha, 9; 2, 5; 1)$ are $k \in \{12, 10, 9, 8, 7\}$. By the proof of Theorem 5.50, for each possible k, we can construct a $\mathbb{Z}_2\mathbb{Z}_4$-linear code C_k with $\ker(C_k) = k$, taking the following generator matrix of $\mathcal{C}_k = \Phi^{-1}(C_k)$:

$$\mathcal{G}_k = \begin{pmatrix} 1 & T_b & \mathbf{0} & 0 & \mathbf{0} \\ 0 & \mathbf{0} & \mathbf{0} & 2 & \mathbf{0} \\ 0 & S_b & S_k & \mathbf{0} & I_5 \end{pmatrix},$$

where T_b, S_b are matrices over $\mathbb{Z}_2$; and the matrices S_k, for each $k \in \{12, 10, 9, 8, 7\}$, are the following: $S_{12} = (\mathbf{0})$,

$$S_{10} = \begin{pmatrix} 1 & 0 & 0 \\ 1 & 0 & 0 \\ 0 & 0 & 0 \\ 0 & 0 & 0 \\ 0 & 0 & 0 \end{pmatrix}, \quad S_9 = \begin{pmatrix} 1 & 0 & 0 \\ 1 & 1 & 0 \\ 0 & 1 & 0 \\ 0 & 0 & 0 \\ 0 & 0 & 0 \end{pmatrix},$$

$$S_8 = \begin{pmatrix} 1 & 0 & 0 \\ 1 & 1 & 0 \\ 1 & 1 & 0 \\ 1 & 0 & 0 \\ 0 & 0 & 0 \end{pmatrix}, \quad S_7 = \begin{pmatrix} 1 & 0 & 0 \\ 1 & 1 & 0 \\ 1 & 1 & 0 \\ 1 & 1 & 0 \\ 0 & 1 & 0 \end{pmatrix}.$$

$\triangle$

Now, we determine the dimension of the kernel of a $\mathbb{Z}_2\mathbb{Z}_4$-linear code $\Phi(C)$ as the dimension of C_X plus the dimension of the kernel of the $\mathbb{Z}_4$-linear code $\phi(C_Y')$, where C' is the code generated by (5.5). First, we obtain some bounds that relate the dimension of the kernel of the codes $\Phi(C)$, $\phi(C_Y)$ and $\phi(C_Y')$.

Proposition 5.52 ([33]). *Let C be a $\mathbb{Z}_2\mathbb{Z}_4$-additive code of type $(\alpha, \beta; \gamma, \delta; \kappa)$. Then, $\mathcal{K}(C) \subseteq C_X \times \mathcal{K}(C_Y)$ and $\ker(\Phi(C)) \leq \kappa + \ker(\phi(C_Y))$.*

Proof. Let $\mathbf{u} = (u, u') \in \mathcal{K}(\mathcal{C}) \subseteq \mathcal{C}$. Then, for all $\mathbf{v} = (v, v') \in \mathcal{C}$, $2\mathbf{u} * \mathbf{v} \in \mathcal{C}$ by Proposition 5.36. We have that $u \in \mathcal{C}_X$ since $\mathbf{u} \in \mathcal{C}$, and $u' \in \mathcal{K}(\mathcal{C}_Y)$ since $2u' * v' \in \mathcal{C}_Y$ for all $v' \in \mathcal{C}_Y$. Therefore, $\mathbf{u} \in \mathcal{C}_X \times \mathcal{K}(\mathcal{C}_Y)$ and the result follows. □

Proposition 5.53 ([33]). *If $\mathcal{C}$ is a separable $\mathbb{Z}_2\mathbb{Z}_4$-additive code of type $(\alpha, \beta; \gamma, \delta; \kappa)$, then $\mathcal{K}(\mathcal{C}) = \mathcal{C}_X \times \mathcal{K}(\mathcal{C}_Y)$ and $\ker(\Phi(\mathcal{C})) = \kappa + \ker(\phi(\mathcal{C}_Y))$.*

Proof. Let $\mathbf{v} = (v, v') \in \mathcal{C}_X \times \mathcal{K}(\mathcal{C}_Y)$. Then $2\mathbf{v} * \mathbf{w} = (\mathbf{0} \mid 2v' * w')$ for all $\mathbf{w} = (w, w') \in \mathcal{C}$. Since $v' \in \mathcal{K}(\mathcal{C}_Y)$, we have $2v' * w' \in \mathcal{C}_Y$. Moreover, since $\mathcal{C}$ is separable, $2\mathbf{v} * \mathbf{w} = (\mathbf{0} \mid 2v' * w') \in \mathcal{C}$. Therefore, $\mathbf{v} \in \mathcal{K}(\mathcal{C})$ and $\mathcal{C}_X \times \mathcal{K}(\mathcal{C}_Y) \subseteq \mathcal{K}(\mathcal{C})$. Finally, by Proposition 5.52, we have $\mathcal{K}(\mathcal{C}) \subseteq \mathcal{C}_X \times \mathcal{K}(\mathcal{C}_Y)$. □

The following example shows that if the $\mathbb{Z}_2\mathbb{Z}_4$-additive code $\mathcal{C}$ is not separable, then $\mathcal{K}(\mathcal{C})$ is not necessarily equal to $\mathcal{C}_X \times \mathcal{K}(\mathcal{C}_Y)$.

Example 5.54. Let $\mathcal{C}$ be the $\mathbb{Z}_2\mathbb{Z}_4$-additive code of type $(1, 3; 1, 2; 1)$ generated by the following matrix:

$$\begin{pmatrix} 1 & 2 & 0 & 0 \\ 0 & 3 & 1 & 0 \\ 0 & 3 & 0 & 1 \end{pmatrix}.$$

We have that $\mathcal{C}_X = \langle 1 \rangle$. Moreover, $\phi(\mathcal{C}_Y)$ is linear since $2(3, 1, 0) * (3, 0, 1) = (2, 0, 0) \in \mathcal{C}_Y$, by Proposition 5.5. Therefore, $\mathcal{K}(\mathcal{C}_Y) = \mathcal{C}_Y$. We also have that a generator matrix of $\mathcal{K}(\mathcal{C})$ is

$$\begin{pmatrix} 1 & 2 & 0 & 0 \\ 0 & 2 & 2 & 0 \\ 0 & 2 & 0 & 2 \end{pmatrix},$$

since $2(0 \mid 3, 1, 0) * (0 \mid 3, 0, 1) = (0 \mid 2, 0, 0) \notin \mathcal{C}$ by Proposition 5.39. Therefore, $\mathcal{K}(\mathcal{C}) \subsetneq \mathcal{C}_X \times \mathcal{K}(\mathcal{C}_Y)$. For example, $(1 \mid 3, 1, 0) \in \mathcal{C}_X \times \mathcal{K}(\mathcal{C}_Y)$ but it does not belong to $\mathcal{K}(\mathcal{C})$.   △

From Proposition 5.52, we obtain that $\ker(\Phi(\mathcal{C})) \le \kappa + \ker(\phi(\mathcal{C}_Y))$. However, we can give a more accurate bound. Recall that κ_1 is the integer such that the subcode $\{(u \mid \mathbf{0}) \in \mathcal{C}\}$ is of type $(\alpha, \beta; \kappa_1, 0; \kappa_1)$.

Lemma 5.55 ([33]). *Let $\mathcal{C}$ be a $\mathbb{Z}_2\mathbb{Z}_4$-additive code. Then, $\mathcal{K}(\mathcal{C})_Y \subseteq \mathcal{K}(\mathcal{C}_Y)$.*

Proof. Let $\mathbf{v} = (v \mid v') \in \mathcal{C}$. We have that $\mathbf{v} \in \mathcal{K}(\mathcal{C})$ if and only if $2\mathbf{v} * \mathbf{w} \in \mathcal{C}$, for all $\mathbf{w} = (w \mid w') \in \mathcal{C}$. Since $2\mathbf{v} * \mathbf{w} = (0 \mid 2v' * w')$, we have that if $\mathbf{v} \in \mathcal{K}(\mathcal{C})$, then $v' \in \mathcal{K}(\mathcal{C}_Y)$ and the statement follows. □

Proposition 5.56 ([33]). *Let $\mathcal{C}$ be a $\mathbb{Z}_2\mathbb{Z}_4$-additive code of type $(\alpha, \beta; \gamma, \delta; \kappa)$. Then,*

$$\ker(\Phi(\mathcal{C})) \leq \kappa_1 + \ker(\phi(\mathcal{C}_Y)).$$

Proof. Define $\mathcal{C}_0 = \{\mathbf{v} = (v \mid \mathbf{0}) \in \mathcal{C}\}$, which is of type $(\alpha, \beta; \kappa_1, 0; \kappa_1)$. We have that $\mathcal{C}_0 \subseteq \mathcal{K}(\mathcal{C})$. It is clear that $\Phi(\mathcal{C}_0)$ is linear, so $\ker(\Phi(\mathcal{C}_0)) = \kappa_1$.

Let $\mathbf{v} = (v \mid v') \in \mathcal{K}(\mathcal{C})$. If $v' = \mathbf{0}$, then $\mathbf{v} \in \mathcal{C}_0$. Otherwise, $v' \in \mathcal{K}(\mathcal{C})_Y \subseteq \mathcal{K}(\mathcal{C}_Y)$ by Lemma 5.55 and, therefore, $\ker(\Phi(\mathcal{C})) \leq \kappa_1 + \ker(\phi(\mathcal{C}_Y))$. □

Without loss of generality, we can assume that the generator matrix of $\mathcal{C}$ is in standard form. In this case, we have that the code $\mathcal{C}_Y$ has a generator matrix of the form

$$\begin{pmatrix} 2T_2 & \mathbf{0} & \mathbf{0} \\ 2T_1 & 2I_{\gamma-\kappa} & \mathbf{0} \\ S_q & R & I_\delta \end{pmatrix}. \tag{5.10}$$

By Remark 2.4, we know that the code $\mathcal{C}_Y$ is of type $4^\delta 2^{\gamma-\kappa_1}$. Therefore, by Proposition 5.42, the minimum value for the dimension of the kernel, $\ker(\phi(\mathcal{C}_Y))$, is $\delta + \gamma - \kappa_1$.

Theorem 5.57 ([33]). *Let $\mathcal{C}$ be a $\mathbb{Z}_2\mathbb{Z}_4$-additive code. If $\mathcal{K}(\mathcal{C}_Y)$ has minimum size, then $\mathcal{K}(\mathcal{C})$ has minimum size.*

Proof. By Remark 2.4 and Proposition 5.42, if $\mathcal{K}(\mathcal{C}_Y)$ has minimum size, then $\ker(\phi(\mathcal{C}_Y)) = \delta + \gamma - \kappa_1$. By Proposition 5.56, $\ker(\Phi(\mathcal{C})) \leq \kappa_1 + \delta + \gamma - \kappa_1 = \gamma + \delta$, which is the minimum size of $\mathcal{K}(\mathcal{C})$, again by Proposition 5.42. □

In the previous statements, we compute an upper bound for the dimension of the kernel of a $\mathbb{Z}_2\mathbb{Z}_4$-additive code $\mathcal{C}$ by considering the kernel of the code $\mathcal{C}_Y$ over $\mathbb{Z}_4$. Now, we shall give the exact value of the dimension of the kernel of a $\mathbb{Z}_2\mathbb{Z}_4$-additive code $\mathcal{C}$ in terms of the dimension of $\mathcal{C}_X$ and the dimension of the kernel of the linear subcode $\mathcal{C}'$ of $\mathcal{C}_Y$. As we have seen in Section 5.1, in the case of a separable code $\mathcal{C}$, the code $\mathcal{C}'$ is exactly $\mathcal{C}_Y$, so the value $\ker(\Phi(\mathcal{C}))$ is equal to $\kappa + \ker(\phi(\mathcal{C}_Y))$, where $\kappa = \dim(\mathcal{C}_X)$.

Theorem 5.58 ([33]). *Let $\mathcal{C}$ be a $\mathbb{Z}_2\mathbb{Z}_4$-additive code of type $(\alpha, \beta; \gamma, \delta; \kappa)$ with generator matrix in standard form, and let $\mathcal{C}'$ be the subcode generated by (5.5). Then,*

$$\ker(\Phi(\mathcal{C})) = \kappa + \ker(\phi(\mathcal{C}_Y')).$$

Proof. Let $\mathcal{G}$ be a generator matrix of $\mathcal{C}$ in standard form. Let $\{\mathbf{u}_i = (u_i \mid u_i')\}_{i=1}^\gamma$ be the first γ rows and $\{\mathbf{v}_j = (v_j \mid v_j')\}_{j=1}^\delta$ the last δ rows of $\mathcal{G}$. Define the codes $\bar{\mathcal{C}} = \langle\{\mathbf{u}_i\}_{i=1}^\kappa\rangle$ and $\mathcal{C}' = \langle\{\mathbf{u}_i\}_{i=\kappa+1}^\gamma, \{\mathbf{v}_j\}_{j=1}^\delta\rangle$. Since $\mathcal{C} = \bar{\mathcal{C}} \cup \mathcal{C}'$ and $\bar{\mathcal{C}} \cap \mathcal{C}' = \emptyset$, it is easy to check that $\ker(\Phi(\mathcal{C})) = \ker(\Phi(\bar{\mathcal{C}})) + \ker(\Phi(\mathcal{C}'))$.

On the one hand, since $\bar{C} \subseteq C_b$, we have that $\bar{C} \subseteq \mathcal{K}(C)$ and $\ker(\Phi(\bar{C})) = \kappa$. On the other hand, by Proposition 5.39, for $\mathbf{v} = (v \mid v') \in C'$ we have that $\mathbf{v} \in \mathcal{K}(C)$ if and only if $2\mathbf{v} * \mathbf{v}_j \in C$ for all $j \in \{1, \ldots, \delta\}$; that is, if and only if $\mathbf{v} \in \mathcal{K}(C')$. Since $2\mathbf{v} * \mathbf{w} = (\mathbf{0} \mid 2v' * w') \in C'$ if and only if $2v' * w' \in C'_Y$, $\ker(\Phi(C')) = \ker(\phi(C'_Y))$. Therefore, the result follows. $\square$

Example 5.59. By using the MAGMA package for $\mathbb{Z}_2\mathbb{Z}_4$-additive codes [29], we construct the same $\mathbb{Z}_2\mathbb{Z}_4$-additive code C as in Example 5.17. We show how to obtain the kernel $K(\Phi(C))$, and the corresponding $\mathbb{Z}_2\mathbb{Z}_4$-additive code $\mathcal{K}(C)$ from all codewords and from the rows of a generator matrix of C. We see how to represent C as the union of cosets of $\mathcal{K}(C)$. We also compute the value of $\ker(\Phi(C))$ and check that it is within the bounds given by Proposition 5.23. Finally, we see that the result given by Theorem 5.58 is satisfied for the code C.

```
> Z4 := IntegerRing(4);
> alpha := 2;
> beta := 9;
> V := RSpace(Z4, alpha + beta);
> G := Matrix([V![2,0, 0,0,2,0,0,0,0,0,0],
               V![0,0, 0,2,2,2,0,0,0,0,0],
               V![0,0, 0,0,2,0,2,0,0,0,0],
               V![0,0, 0,1,0,1,0,1,0,0,0],
               V![0,0, 1,1,0,0,0,0,1,0,0],
               V![0,2, 1,0,1,1,1,0,0,1,0],
               V![0,0, 0,0,1,0,1,0,0,0,1]]);
> C := Z2Z4AdditiveCode(G, alpha);
> mapGray := GrayMap(C);
> typeC := Z2Z4Type(C);
> gamma := typeC[3];
> delta := typeC[4];
> kappa := typeC[5];
>
> Cbin := GrayMapImage(C);
> KernelCbin := LinearCode<GF(2), alpha+2*beta |
                [c : c in Cbin | forall{d : d in Cbin | c+d in Cbin}]>;
> KernelC := Z2Z4AdditiveCode([c@@mapGray : c in KernelCbin ], alpha);
> #Cbin eq #C;
true
> #KernelCbin eq #KernelC;
true
>
> newKernelC, newKernelCbin := KernelZ2Code(C);
> newKernelCbin eq KernelCbin;
true
> newKernelC eq KernelC;
true
>
```

```
> v1 := G[gamma+1];
> v2 := G[gamma+2];
> v3 := G[gamma+3];
> v4 := G[gamma+4];
> Grows4 := [v1, v2, v3, v4];
> &and[ 2*v4*v in C : v in Grows4] and (v4 in KernelC); // v4 in kernel
true
> for I in Subsets({1,2,3}) diff {{}} do                    // not in kernel
for>     w := &+[Grows4[i] : i in I];
for>        not &and[ 2*w*v in C : v in Grows4] and not (w in KernelC);
for> end for;
true true true true true true true
> newKernelCbin := LinearCode<GF(2), alpha+2*beta |
                              [mapGray(u) : u in Rows(G)[1..3]] cat
                              [mapGray(2*v) : v in [v1,v2,v3,v4]] cat
                              [mapGray(v4)]>;
> newKernelCbin eq KernelCbin;
true
> newKernelC := Z2Z4AdditiveCode(Rows(G)[1..3] cat
                               [2*v1, 2*v2, 2*v3, v4 ], alpha);
> newKernelC eq KernelC;
true
> (Z2Z4Type(KernelC)[3] - gamma) eq (delta - Z2Z4Type(KernelC)[4]);
true
>
> cosetRep := CosetRepresentatives(C, KernelC);
> {C!0} join Set(KernelCosetRepresentatives(C)) eq cosetRep;
true
> Set(C) eq {v + ci : v in Set(KernelC), ci in cosetRep};
true
>
> DimensionOfKernelZ2(C);
8
> DimensionOfKernelZ2(C) eq Dimension(KernelCbin);
true
> lowerBound := gamma + delta;
> upperBound := gamma + 2*delta;
> (lowerBound le DimensionOfKernelZ2(C)) and
  (DimensionOfKernelZ2(C) le upperBound);
true
>
> Cprime := Z2Z4AdditiveCode(G[kappa+1..gamma+delta], alpha);
> CYprime := LinearQuaternaryCode(Cprime);
> DimensionOfKernelZ2(C) eq kappa + DimensionOfKernelZ2(CYprime);
true                                                                      △
```

5.4 Pairs of Rank and Dimension of the Kernel

In this section, once the dimension of the kernel of a $\mathbb{Z}_2\mathbb{Z}_4$-linear code is fixed, lower and upper bounds on the rank are established. We show that there exists a $\mathbb{Z}_2\mathbb{Z}_4$-linear code C of type $(\alpha, \beta; \gamma, \delta; \kappa)$ with $r = \operatorname{rank}(C)$ and $k = \ker(C)$ for any possible pair of values (r, k).

An upper bound on the rank, once the dimension of the kernel is fixed, can be established by using the same argument as in [123].

Lemma 5.60 ([79]). *Let $\mathcal{C}$ be a $\mathbb{Z}_2\mathbb{Z}_4$-additive code of type $(\alpha, \beta; \gamma, \delta; \kappa)$, and let $C = \Phi(\mathcal{C})$ be the corresponding $\mathbb{Z}_2\mathbb{Z}_4$-linear code. If $\operatorname{rank}(C) = \gamma + 2\delta + \bar{r}$ and $\ker(C) = \gamma + 2\delta - \bar{k}$, with $\bar{k} \geq 2$, then*

$$1 \leq \bar{r} \leq \binom{\bar{k}}{2}.$$

The linear span $\langle C \rangle$ is generated by

$$\{\Phi(\mathbf{u}_i)\}_{i=1}^{\gamma}, \{\Phi(\mathbf{v}_j), \Phi(2\mathbf{v}_j)\}_{j=1}^{\delta} \text{ and } \{\Phi(2\mathbf{v}_t * \mathbf{v}_s)\}_{1 \leq s < t \leq \bar{k}}.$$

Proof. There exist $\{\mathbf{u}_i\}_{i=1}^{\gamma}$ and $\{\mathbf{v}_j\}_{j=1}^{\delta}$ vectors of order two and four respectively, such that they generate the code $\mathcal{C}$ and $C = \bigcup_{I \subseteq \{1,\ldots,\bar{k}\}} (K(C) + \Phi(\mathbf{v}_I))$ by Proposition 5.43. Note that $\Phi(\mathbf{v}_j) \in K(C)$ if and only if $j \in \{\bar{k}+1, \ldots, \delta\}$.

By Proposition 5.36, for all $j \in \{\bar{k}+1, \ldots, \delta\}$ and $i \in \{1, \ldots, \delta\}$, as $\Phi(\mathbf{v}_j) \in K(C)$, $2\mathbf{v}_j * \mathbf{v}_i \in \mathcal{C}$ and, consequently, $\Phi(2\mathbf{v}_j * \mathbf{v}_i)$ is a linear combination of $\{\Phi(\mathbf{u}_i)\}_{i=1}^{\gamma}$ and $\{\Phi(2\mathbf{v}_j)\}_{j=1}^{\delta}$. As a result, $\langle C \rangle$ is generated by $\{\Phi(\mathbf{u}_i)\}_{i=1}^{\gamma}$, $\{\Phi(\mathbf{v}_j), \Phi(2\mathbf{v}_j)\}_{j=1}^{\delta}$ and $\{\Phi(2\mathbf{v}_t * \mathbf{v}_s)\}_{1 \leq s < t \leq \bar{k}}$ and hence $\bar{r} \leq \binom{\bar{k}}{2}$, by Proposition 5.18.

Finally, since $\bar{k} \geq 2$, the binary code C is not linear and, therefore, $\bar{r} \geq 1$. □

Let C be a $\mathbb{Z}_2\mathbb{Z}_4$-linear code with $\ker(C) = \gamma + 2\delta - \bar{k}$ and $\operatorname{rank}(C) = \gamma + 2\delta + \bar{r}$. Note that if $\bar{r} = 0$ then, necessarily, $\bar{k} = 0$ (and vice versa) and C is a linear code. The next theorem determine all possible pairs of rank and dimension of the kernel for $\mathbb{Z}_2\mathbb{Z}_4$-linear codes.

Proposition 5.61 ([79]). *Let C be a $\mathbb{Z}_2\mathbb{Z}_4$-linear code of type $(\alpha, \beta; \gamma, \delta; \kappa)$ with $\ker(C) = \gamma + 2\delta - \bar{k}$ and $\operatorname{rank}(C) = \gamma + 2\delta + \bar{r}$. Then, for any $\bar{k} \in \{0\} \cup \{2, \ldots, \delta\}$,*

$$\begin{cases} \bar{r} = 0, & \text{if } \bar{k} = 0, \\ \bar{r} \in \{2, \ldots, \min\{\beta - (\gamma - \kappa) - \delta, \binom{\bar{k}}{2}\}\}, & \text{if } \bar{k} \text{ is odd}, \\ \bar{r} \in \{1, \ldots, \min\{\beta - (\gamma - \kappa) - \delta, \binom{\bar{k}}{2}\}\}, & \text{if } \bar{k} > 0 \text{ is even}. \end{cases}$$

Proof. By Proposition 5.23, $\bar{r} \in \{0, \ldots, \min\{\beta - (\gamma - \kappa) - \delta, \binom{\delta}{2}\}\}$. Recall that $\bar{r} = 0$ if $\bar{k} = 0$. Moreover, by Lemma 5.60, for a fixed $\bar{k} \geq 2$, $\bar{r} \leq \binom{\bar{k}}{2}$ and, therefore, if $\bar{k} \in \{2, \ldots, \delta\}$ then $\bar{r} \in \{1, \ldots, \min\{\beta - (\gamma - \kappa) - \delta, \binom{\bar{k}}{2}\}\}$.

In the case $\bar{r} = 1$, C is not linear and, by Proposition 5.42, $\ker(C) = \gamma + 2\delta - \bar{k}$ where $\bar{k} \in \{2, \ldots, \delta\}$. Moreover, by Proposition 5.43, there exists a generator matrix $\mathcal{G}$ of C with row vectors $\{\mathbf{u}_i\}_{i=1}^{\gamma}$ and $\{\mathbf{v}_j\}_{j=1}^{\delta}$ of order two and four, respectively, such that the $\bar{k}$ row vectors $\mathbf{v}_1, \mathbf{v}_2, \ldots, \mathbf{v}_{\bar{k}}$ satisfy $C = \bigcup_{I \subseteq \{1, \ldots, \bar{k}\}} (K(C) + \Phi(\mathbf{v}_I))$. We see that if $\bar{r} = 1$, then $\bar{k}$ is necessarily even. Assume $\bar{k}$ is odd. First, we prove that there exists $I \subseteq \{1, \ldots, \bar{k}\}$ such that $\Phi(\mathbf{v}_I) \in K(C)$, that is, $2\mathbf{v}_I * \mathbf{v}_j \in C$ for all $j \in \{1, \ldots \bar{k}\}$, which is a contradiction and, therefore, $\bar{k}$ is an even number.

By Lemma 5.60, $\langle C \rangle$ is generated by $\{\Phi(\mathbf{u}_i)\}_{i=1}^{\gamma}$, $\{\Phi(\mathbf{v}_j), \Phi(2\mathbf{v}_j)\}_{j=1}^{\delta}$ and $\{\Phi(2\mathbf{v}_j * \mathbf{v}_k)\}_{1 \leq j < k \leq \bar{k}}$. Since $\operatorname{rank}(C) = \gamma + 2\delta + 1$, for all $i, j \in \{1, \ldots, \bar{k}\}$ either $2\mathbf{v}_i * \mathbf{v}_j \in C$ or $2\mathbf{v}_i * \mathbf{v}_j = \mathbf{w}$, for an order two vector $\mathbf{w} \notin C$. If there exists $I \subseteq \{1, \ldots, \bar{k}\}$ such that, for each $j \in \{1, \ldots, \bar{k}\}$ the number of $i \in I$ verifying $2\mathbf{v}_i * \mathbf{v}_j = \mathbf{w} \notin C$ is even, then $2\mathbf{v}_I * \mathbf{v}_j \in C$. In order to prove that there exists such a set I, we define the symmetric matrix $A = (a_{ij})$, $1 \leq i, j \leq \bar{k}$, as in the proof of Proposition 5.48, leading to a contradiction. $\qquad\square$

Theorem 5.62 ([79]). *Let $\alpha, \beta, \gamma, \delta, \kappa$ be integer numbers satisfying inequalities (2.10). Then, there exists a $\mathbb{Z}_2\mathbb{Z}_4$-linear code C of type $(\alpha, \beta; \gamma, \delta; \kappa)$ with $\ker(C) = \gamma + 2\delta - \bar{k}$ and $\operatorname{rank}(C) = \gamma + 2\delta + \bar{r}$ if and only if $\bar{k} \in \{0\} \cup \{2, \ldots, \delta\}$ and*

$$\begin{cases} \bar{r} = 0, & \text{if } \bar{k} = 0, \\ \bar{r} \in \{2, \ldots, \min\{\beta - (\gamma - \kappa) - \delta, \binom{\bar{k}}{2}\}\}, & \text{if } \bar{k} \text{ is odd}, \\ \bar{r} \in \{1, \ldots, \min\{\beta - (\gamma - \kappa) - \delta, \binom{\bar{k}}{2}\}\}, & \text{if } \bar{k} > 0 \text{ is even}. \end{cases}$$

Proof. Let C be a $\mathbb{Z}_2\mathbb{Z}_4$-additive code of type $(\alpha, \beta; \gamma, \delta; \kappa)$ with generator matrix

$$\mathcal{G} = \begin{pmatrix} I_\kappa & T_b & 0 & 0 & 0 \\ 0 & 0 & 0 & 2I_{\gamma - \kappa} & 0 \\ 0 & S_b & S_{r,k} & 0 & I_\delta \end{pmatrix},$$

where $S_{r,k}$ is a matrix over $\mathbb{Z}_4$ of size $\delta \times (\beta - (\gamma - \kappa) - \delta)$, and let $C = \Phi(\mathcal{C})$ be its corresponding $\mathbb{Z}_2\mathbb{Z}_4$-linear code. The necessary conditions for the values of the pairs $\bar{r}$ and $\bar{k}$ are given in Proposition 5.61. Recall that $\bar{r} = 0$ if and only if $\bar{k} = 0$, and considering $S_{r,k}$ the zero matrix, we obtain a linear code.

Let e_k, $1 \leq k \leq \delta$, denote the column vector of length δ, with a one in the kth coordinate and zeros elsewhere. For each $\bar{k} \in \{3, \ldots, \delta\}$ and

$\bar{r} \in \{2, \ldots, \min\{\beta - (\gamma - \kappa) - \delta, \binom{\bar{k}}{2}\}\}$, we can construct $S_{r,k}$ as a matrix over $\mathbb{Z}_4$ where in one column there is the vector $e_1 + \cdots + e_{\bar{k}}$, in $\bar{r} - 1$ columns there are $\bar{r} - 1$ different column vectors $e_k + e_l$ of length δ, $1 \leq k < l \leq \bar{k}$, and in the remaining columns there is the all-zero column vector. It is easy to check that $\ker(C) = \gamma + 2\delta - \bar{k}$ and $\mathrm{rank}(C) = \gamma + 2\delta + \bar{r}$.

Finally, if $\bar{r} = 1$, we can construct $S_{r,k}$ as a matrix over $\mathbb{Z}_4$ of size $\delta \times (\beta - (\gamma - \kappa) - \delta)$ with $\bar{k}$ ones in one column and zeros elsewhere, for each even $\bar{k} \in \{2, \ldots, \delta\}$. In this case, it is also easy to check that $\mathrm{rank}(C) = \gamma + 2\delta + 1$ and $\ker(C) = \gamma + 2\delta - \bar{k}$, for any even $\bar{k} \in \{2, \ldots, \delta\}$. $\qquad\square$

Example 5.63. By Proposition 5.61, we know that the possible pairs of rank and dimension of the kernel, for $\mathbb{Z}_2\mathbb{Z}_4$-linear codes of type $(\alpha, 9; 2, 5; 1)$, are the ones given in the following table:

$k \setminus r$	12	13	14	15
12	*			
10		*		
9			*	*
8		*	*	*
7			*	*

By the proof of Theorem 5.62, for each possible pair (r, k), we can construct a $\mathbb{Z}_2\mathbb{Z}_4$-linear code $C_{r,k}$ with $\mathrm{rank}(C_{r,k}) = r$ and $\ker(C_{r,k}) = k$, taking the following generator matrix of $C_{r,k} = \Phi^{-1}(\mathcal{C}_{r,k})$:

$$\mathcal{G}_{r,k} = \begin{pmatrix} 1 & T_b & 0 & 0 & 0 \\ 0 & 0 & 0 & 2 & 0 \\ 0 & S_b & S_{r,k} & 0 & I_5 \end{pmatrix},$$

where T_b, S_b are matrices over $\mathbb{Z}_2$; and the matrices $S_{r,k}$, for each $(r, k) \in \{(12, 12), (13, 10), (13, 8), (14, 9), (14, 8), (14, 7), (15, 9), (15, 8), (15, 7)\}$, are:

$$S_{12,12} = (0), \quad S_{13,10} = \begin{pmatrix} 1 & 0 & 0 \\ 1 & 0 & 0 \\ 0 & 0 & 0 \\ 0 & 0 & 0 \\ 0 & 0 & 0 \end{pmatrix}, \quad S_{13,8} = \begin{pmatrix} 1 & 0 & 0 \\ 1 & 0 & 0 \\ 1 & 0 & 0 \\ 1 & 0 & 0 \\ 0 & 0 & 0 \end{pmatrix},$$

$$S_{14,9} = \begin{pmatrix} 1 & 1 & 0 \\ 1 & 1 & 0 \\ 1 & 0 & 0 \\ 0 & 0 & 0 \\ 0 & 0 & 0 \end{pmatrix}, \quad S_{14,8} = \begin{pmatrix} 1 & 1 & 0 \\ 1 & 1 & 0 \\ 1 & 0 & 0 \\ 1 & 0 & 0 \\ 0 & 0 & 0 \end{pmatrix}, \quad S_{14,7} = \begin{pmatrix} 1 & 1 & 0 \\ 1 & 1 & 0 \\ 1 & 0 & 0 \\ 1 & 0 & 0 \\ 1 & 0 & 0 \end{pmatrix},$$

$$S_{15,9} = \begin{pmatrix} 1 & 1 & 0 \\ 1 & 1 & 1 \\ 1 & 0 & 1 \\ 0 & 0 & 0 \\ 0 & 0 & 0 \end{pmatrix}, \quad S_{15,8} = \begin{pmatrix} 1 & 1 & 0 \\ 1 & 1 & 1 \\ 1 & 0 & 1 \\ 1 & 0 & 0 \\ 0 & 0 & 0 \end{pmatrix}, \quad S_{15,7} = \begin{pmatrix} 1 & 1 & 0 \\ 1 & 1 & 1 \\ 1 & 0 & 1 \\ 1 & 0 & 0 \\ 1 & 0 & 0 \end{pmatrix}.$$

$\triangle$

Example 5.64. By using the MAGMA package for $\mathbb{Z}_2\mathbb{Z}_4$-additive codes [29], we construct the same $\mathbb{Z}_2\mathbb{Z}_4$-additive code $\mathcal{C}$ as in Example 5.17. We show some relations between the $\mathbb{Z}_2\mathbb{Z}_4$-additive codes $\mathcal{R}(\mathcal{C})$ and $\mathcal{K}(\mathcal{C})$. We compute the dimensions of the corresponding $\mathbb{Z}_2\mathbb{Z}_4$-linear codes $\langle C \rangle$ and $K(C)$, respectively; and check that they are within the bounds given by Proposition 5.61.

```
> Z4 := IntegerRing(4);
> alpha := 2;
> beta := 9;
> V := RSpace(Z4, alpha + beta);
> G := Matrix([V![2,0, 0,0,2,0,0,0,0,0,0],
               V![0,0, 0,2,2,2,0,0,0,0,0],
               V![0,0, 0,0,2,0,2,0,0,0,0],
               V![0,0, 0,1,0,1,0,1,0,0,0],
               V![0,0, 1,1,0,0,0,0,1,0,0],
               V![0,2, 1,0,1,1,1,0,0,1,0],
               V![0,0, 0,0,1,0,1,0,0,0,1]]);
> C := Z2Z4AdditiveCode(G, alpha);
> typeC := Z2Z4Type(C);
> gamma := typeC[3];
> delta := typeC[4];
> kappa := typeC[5];
>
> SpanC, SpanCbin := SpanZ2Code(C);
> KernelC, KernelCbin := KernelZ2Code(C);
> (KernelC subset C) and (C subset SpanC);
true
> KernelCbin subset SpanCbin;
true
>
> rbar := DimensionOfSpanZ2(C) - gamma - 2*delta;
> kbar := gamma + 2*delta - DimensionOfKernelZ2(C);
> (1 le rbar) and (Binomial(kbar, 2) le rbar);
true
> upperBound := Min(Binomial(kbar, 2), beta-(gamma-kappa)-delta);
> lowerBound := 1 + (kbar mod 2);
> (lowerBound le rbar) and (rbar le upperBound);
true
```

$\triangle$

Chapter 6

Families of $\mathbb{Z}_2\mathbb{Z}_4$-Additive Codes

In this chapter, some families of $\mathbb{Z}_2\mathbb{Z}_4$-additive codes, such that the corresponding $\mathbb{Z}_2\mathbb{Z}_4$-linear codes share some common properties are described. We construct them by giving their generator or parity check matrices as $\mathbb{Z}_2\mathbb{Z}_4$-additive codes. The associated $\mathbb{Z}_2\mathbb{Z}_4$-linear codes are classified, by showing that they are all non-equivalent as binary codes. In Section 2.4, several basic and well-known families of binary linear codes that can be viewed as $\mathbb{Z}_2\mathbb{Z}_4$-linear codes are shown. In Sections 6.1–6.4, the families of $\mathbb{Z}_2\mathbb{Z}_4$-additive (extended) perfect codes, $\mathbb{Z}_2\mathbb{Z}_4$-additive Hadamard codes, $\mathbb{Z}_2\mathbb{Z}_4$-additive Reed-Muller codes, and maximum distance separable $\mathbb{Z}_2\mathbb{Z}_4$-additive codes are described, respectively. There are other well-known families of $\mathbb{Z}_2\mathbb{Z}_4$-additive codes (with $\alpha = 0$) as Preparata-like and Kerdock-like codes [44, 92], Goethals and Delsarte-Goethals codes [92], or extended dualized Kerdock codes [99].

6.1 $\mathbb{Z}_2\mathbb{Z}_4$-Additive (Extended) Perfect Codes

A *binary perfect 1-error correcting code* (briefly, *binary perfect code*) C of length n is a binary code, with minimum Hamming distance $d(C) = 3$, such that all the vectors in $\mathbb{Z}_2^n$ are within distance one from a codeword. For any $t \geq 2$ there exists exactly one binary linear perfect code of length $2^t - 1$, up to equivalence, which is the well known *Hamming code*. However, for $t \geq 4$, there exists more than $2^{2^{2^{t-1}-t}}$ non-linear codes with the same parameters (see [101, 106] for improvements on this lower bound). The *extended code*

There are parts of this chapter that have been previously published. Reprinted by permission from Springer Nature Customer Service Centre GmbH: M. Bilal, J. Borges, S. T. Dougherty, and C. Fernández-Córdoba. "Maximum distance separable codes over $\mathbb{Z}_4$ and $\mathbb{Z}_2 \times \mathbb{Z}_4$". *Designs, Codes and Cryptography*, v. 61, 1, pp. 31–40. ©(2011).

of a binary code C is the code resulting from adding an overall parity check digit to each codeword of C. The binary perfect and extended binary perfect codes are $(2^t - 1, 2^{2^t-t-1}, 3)$ and $(2^t, 2^{2^t-t-1}, 4)$ codes, respectively.

The $\mathbb{Z}_2\mathbb{Z}_4$-additive codes such that, under the Gray map, give a binary perfect code are called $\mathbb{Z}_2\mathbb{Z}_4$-*additive perfect codes*. Equivalently, the $\mathbb{Z}_2\mathbb{Z}_4$-additive codes such that, under the Gray map, give a code with the same parameters as an extended binary perfect code are called $\mathbb{Z}_2\mathbb{Z}_4$-*additive extended perfect codes*. Given a $\mathbb{Z}_2\mathbb{Z}_4$-additive (extended) perfect code, after applying the Gray map, the corresponding $\mathbb{Z}_2\mathbb{Z}_4$-linear code is called *(extended) perfect $\mathbb{Z}_2\mathbb{Z}_4$-linear code*.

Apart from the binary linear case (when $\beta = 0$), there are two different kinds of $\mathbb{Z}_2\mathbb{Z}_4$-additive extended perfect codes, those with $\alpha = 0$ and those with $\alpha \neq 0$. We distinguish between these two cases because the construction of the codes is different. In Section 6.1.1, we focus on $\mathbb{Z}_2\mathbb{Z}_4$-additive extended perfect codes with $\alpha = 0$, and in Section 6.1.2, on $\mathbb{Z}_2\mathbb{Z}_4$-additive (extended) perfect codes with $\alpha \neq 0$. In each one of these sections, we give the number of non-equivalent such codes, and a recursive construction of the parity check matrices of the corresponding $\mathbb{Z}_2\mathbb{Z}_4$-additive codes.

The (extended) perfect $\mathbb{Z}_2\mathbb{Z}_4$-linear codes can also be classified using either the rank or the dimension of the kernel, as it is proven in [43, 100, 102, 124], where these parameters are computed (see also Propositions 6.6, 6.13 and 6.14). The intersection problem for these codes, i.e., which are the possibilities for the number of codewords in the intersection of two $\mathbb{Z}_2\mathbb{Z}_4$-additive extended perfect codes of the same length is investigated in [137]. Finally, also mention that the permutation automorphism group of the corresponding extended perfect $\mathbb{Z}_2\mathbb{Z}_4$-linear codes, which include the extended perfect $\mathbb{Z}_4$-linear codes, has been studied in [104, 120, 124].

6.1.1 $\mathbb{Z}_2\mathbb{Z}_4$-Additive Extended Perfect Codes with $\alpha = 0$

The $\mathbb{Z}_2\mathbb{Z}_4$-additive extended perfect codes with $\alpha = 0$ are also called *quaternary linear extended perfect codes*, since they are also quaternary linear codes. For the same reason, after applying the Gray map to these codes, their corresponding $\mathbb{Z}_2\mathbb{Z}_4$-linear codes are extended perfect $\mathbb{Z}_2\mathbb{Z}_4$-linear codes (with $\alpha = 0$), which are also called *extended perfect $\mathbb{Z}_4$-linear codes*.

First of all, a construction of $\mathbb{Z}_4$-additive extended perfect codes $\mathcal{C}^*$ of length $\beta = 2^{t-1}$, $t \geq 2$ is given, by defining their parity check matrices. From [43, 102] a parity check matrix $\mathcal{H}^*$ of $\mathcal{C}^*$ may be represented, up to row permutation, as follows. First, consider the matrix over $\mathbb{Z}_4$ consisting of all column vectors of the form $\{0,1\}^\gamma \times \{1\} \times \mathbb{Z}_4^{\delta-1}$,

$$\begin{pmatrix} B \\ Q \end{pmatrix},$$

where B is a matrix over $\mathbb{Z}_4$ of size $\gamma \times \beta$ with all entries in $\{0,1\} \subset \mathbb{Z}_4$, and Q is a matrix over $\mathbb{Z}_4$ of size $\delta \times \beta$. Then, the matrix

$$\mathcal{H}^* = \begin{pmatrix} 2B \\ Q \end{pmatrix}.$$

is a parity check matrix of C^*. Note that the $\mathbb{Z}_4$-dual code of $C^* = \Phi(C^*)$ is of type $(0, \beta; \gamma, \delta)$. Moreover, we have that $\gamma = t + 1 - 2\delta$ since $\beta = 2^\gamma 4^{\delta-1}$ and $\beta = 2^{t-1}$. Recall that for $\mathbb{Z}_4$-additive and $\mathbb{Z}_4$-linear codes, we can omit the parameter κ since it has no sense, as seen in Section 2.1.

Example 6.1. For $t = 3$, by using the above construction, it is possible to construct two different quaternary linear extended perfect codes, C_1 and C_2 (taking $\delta = 1$ and $\delta = 2$), with parity check matrices

$$\mathcal{H}_1^* = \begin{pmatrix} 0 & 0 & 2 & 2 \\ 0 & 2 & 0 & 2 \\ 1 & 1 & 1 & 1 \end{pmatrix} \text{ and } \mathcal{H}_2^* = \begin{pmatrix} 1 & 1 & 1 & 1 \\ 0 & 1 & 2 & 3 \end{pmatrix},$$

respectively. It is well known that there is a unique extended binary perfect code of length 8 containing the all-zero vector [112]. Therefore, there is a unique (up to equivalence) extended perfect $\mathbb{Z}_4$-linear code C^* of length 8, and both codes $C_1 = \Phi(C_1)$ and $C_2 = \Phi(C_2)$ are equivalent to C^*. $\triangle$

Example 6.2. For $t = 4$, by using the above construction, we can obtain two different quaternary linear extended perfect codes (taking $\delta = 1$ and $\delta = 2$), which have the following parity check matrices:

$$\mathcal{H}_1^* = \begin{pmatrix} 0 & 0 & 0 & 0 & 2 & 2 & 2 & 2 \\ 0 & 0 & 2 & 2 & 0 & 0 & 2 & 2 \\ 0 & 2 & 0 & 2 & 0 & 2 & 0 & 2 \\ 1 & 1 & 1 & 1 & 1 & 1 & 1 & 1 \end{pmatrix} \text{ and } \mathcal{H}_2^* = \begin{pmatrix} 0 & 0 & 0 & 0 & 2 & 2 & 2 & 2 \\ 1 & 1 & 1 & 1 & 1 & 1 & 1 & 1 \\ 0 & 1 & 2 & 3 & 0 & 1 & 2 & 3 \end{pmatrix}.$$

In this case, the corresponding extended perfect $\mathbb{Z}_4$-linear codes of length 16 are non-equivalent, by Theorem 6.3. $\triangle$

In general, it is possible to construct extended perfect $\mathbb{Z}_4$-linear codes such that their $\mathbb{Z}_4$-dual codes have the parameters given in the following table, for $2 \le t \le 7$:

t	δ	$(\alpha,\beta;\gamma,\delta)$
2	1	$(0,2;1,1)$
3	1,2	$(0,4;2,1)$, $(0,4;0,2)$
4	1,2	$(0,8;3,1)$, $(0,8;1,2)$
5	1,2,3	$(0,16;4,1)$, $(0,16;2,2)$, $(0,16;0,3)$
6	1,2,3	$(0,32;5,1)$, $(0,32;3,2)$, $(0,32;1,3)$
7	1,2,3,4	$(0,64;6,1)$, $(0,64;4,2)$, $(0,64;2,3)$, $(0,64;0,4)$
...	...	...

Note that for $t = 3$ (length 8), both extended perfect $\mathbb{Z}_4$-linear codes are equivalent as shown in Example 6.1. For $t \geq 4$, all such codes are unique and pairwise non-equivalent as proved in the following result.

Theorem 6.3 ([100, 102]). *For any integer $t \geq 4$ and each $\delta \in \{1,\ldots,\lfloor(t+1)/2\rfloor\}$, there exists a unique (up to equivalence) extended perfect $\mathbb{Z}_4$-linear code C^* of length $2^t \geq 16$, such that the $\mathbb{Z}_4$-dual code of C^* is of type $(0,\beta;\gamma,\delta)$, where $\beta = 2^{t-1}$ and $\gamma = t + 1 - 2\delta$.*

The non-existence of extended perfect $\mathbb{Z}_4$-linear codes that are non-equivalent to the above constructed codes is also proved in [100]. Therefore, by Theorem 6.3, the number of non-equivalent extended perfect $\mathbb{Z}_4$-linear codes of length 2^t is $\lfloor(t+1)/2\rfloor$ for all $t \geq 4$, and it is 1 for $t = 2$ and $t = 3$. Note that the corresponding $\mathbb{Z}_2\mathbb{Z}_4$-additive codes $\mathcal{C}^* = \Phi^{-1}(C^*)$ of these extended perfect $\mathbb{Z}_4$-linear codes C^* are $\mathbb{Z}_2\mathbb{Z}_4$-additive extended perfect codes with $\alpha = 0$, or equivalently quaternary linear extended perfect codes.

We remark that if C^* is an extended perfect $\mathbb{Z}_4$-linear code of length $2^t \geq 16$, then the punctured code can not be a perfect $\mathbb{Z}_2\mathbb{Z}_4$-linear code, up to the extended Hamming code of length 16 [77].

A parity check matrix for all quaternary extended perfect codes, or equivalently for all $\mathbb{Z}_2\mathbb{Z}_4$-additive extended perfect codes with $\alpha = 0$, can also be constructed in a recursive way [100, 102], as it is described below.

Let $\mathcal{H}^*$ be a parity check matrix of a $\mathbb{Z}_2\mathbb{Z}_4$-additive extended perfect code, such that its additive dual code is of type $(0,\beta;\gamma,\delta)$, where $\beta = 2^{t-1}$, $\gamma = t + 1 - 2\delta$ and $\delta \in \{1,\ldots,\lfloor(t+1)/2\rfloor\}$. A parity check matrix for the $\mathbb{Z}_2\mathbb{Z}_4$-additive extended perfect code, such that its additive dual code is of type $(0,\beta';\gamma+1,\delta)$, where $\beta' = 2\beta = 2^t$, can be constructed as follows:

$$\begin{pmatrix} 0 & 2 \\ \mathcal{H}^* & \mathcal{H}^* \end{pmatrix}. \tag{6.1}$$

Therefore, from codes of length $2\beta = 2^t$, we obtain codes of length $2\beta' = 2^{t+1}$. Specifically, using this construction, from all the non-equivalent equivalent codes of length 2^t, we obtain all the non-equivalent codes of length 2^{t+1}, except when t is odd and $\gamma = 0$. Note also that the matrix (6.1) has the same number of row vectors of order four, δ, and one more row vector of order two, $\gamma + 1$.

When t is odd and $\gamma = 0$, we need to use another recursive construction, which gives us a matrix with the same number of row vector of order two, γ, and one more row vector of order four, $\delta + 1$. A parity check matrix for the $\mathbb{Z}_2\mathbb{Z}_4$-additive extended perfect code, such that its additive dual code is of type $(0, \beta'; \gamma, \delta + 1)$, where $\beta' = 4\beta$, can be constructed as follows:

$$\begin{pmatrix} \mathcal{H}^* & \mathcal{H}^* & \mathcal{H}^* & \mathcal{H}^* \\ 0 & 1 & 2 & 3 \end{pmatrix}. \tag{6.2}$$

Therefore, starting with a parity check matrix of the $\mathbb{Z}_2\mathbb{Z}_4$-additive extended perfect code, such that its additive dual code is of type $(0, 4; 0, 2)$, that is, with $t = 3$ and $\gamma = 0$, it is possible to construct all the cases for t odd and $\gamma = 0$.

Example 6.4. Applying matrix (6.1) to the matrices $\mathcal{H}_1^*$ and $\mathcal{H}_2^*$ of types $(0, 4; 2, 1)$ and $(0, 4; 0, 2)$, respectively, given in Example 6.1, we can obtain the parity check matrices $\mathcal{H}_1^*$ and $\mathcal{H}_2^*$, given in Example 6.2, for the quaternary linear extended perfect codes with quaternary dual of types $(0, 8; 3, 1)$ and $(0, 8; 1, 2)$, respectively. $\triangle$

Example 6.5. For $t = 5$, by Theorem 6.3, there are three non-equivalent extended perfect $\mathbb{Z}_4$-linear codes of length 32, since we have three possible parameters: $\delta = 1$, $\delta = 2$ and $\delta = 3$.

We can construct parity check matrices of the quaternary linear extended perfect codes with $\delta = 1$ and $\delta = 2$, by using (6.1) and matrices $\mathcal{H}_1^*$ and $\mathcal{H}_2^*$ given in Example 6.2. On the other hand, by using (6.2) and matrix $\mathcal{H}_2^*$ of type $(0, 4; 0, 2)$ given in Example 6.1, we can obtain the parity check matrix

$$\mathcal{H}^* = \begin{pmatrix} 1 & 1 & 1 & 1 & 1 & 1 & 1 & 1 & 1 & 1 & 1 & 1 & 1 & 1 & 1 & 1 \\ 0 & 1 & 2 & 3 & 0 & 1 & 2 & 3 & 0 & 1 & 2 & 3 & 0 & 1 & 2 & 3 \\ 0 & 0 & 0 & 0 & 1 & 1 & 1 & 1 & 2 & 2 & 2 & 2 & 3 & 3 & 3 & 3 \end{pmatrix}$$

for the quaternary linear extended perfect code with $\delta = 3$. $\triangle$

Proposition 6.6 ([43, 100, 102]). *Let $t \geq 4$ and $\delta \in \{1, \ldots, \lfloor (t+1)/2 \rfloor\}$. Let C^* be an extended perfect $\mathbb{Z}_4$-linear code of type $(0, \beta; \bar{\gamma}, \bar{\delta})$, such that its $\mathbb{Z}_4$-dual code is of type $(0, \beta; \gamma, \delta)$, where $\beta = 2^{t-1}$, $\bar{\gamma} = \gamma = t + 1 - 2\delta$ and $\bar{\delta} = 2^{t-1} - \gamma - \delta$. Then,*

$$\mathrm{rank}(C^*) = \bar{\gamma} + 2\bar{\delta} + \delta = 2^{t-1} + \bar{\delta},$$

except when $t = 4$ and $\delta = 1$, since in this case rank$(C^*) = 11$ *(the code C^* is linear); and*

$$\ker(C^*) = \begin{cases} \bar{\gamma} + \bar{\delta} + 1 & \text{if} \quad \delta \geq 3, \\ \bar{\gamma} + \bar{\delta} + 2 & \text{if} \quad \delta = 2, \\ \bar{\gamma} + \bar{\delta} + t & \text{if} \quad \delta = 1. \end{cases}$$

Note that the rank of the extended perfect $\mathbb{Z}_4$-linear codes satisfies the upper bound given in Proposition 5.23.

Example 6.7. By using the MAGMA package for $\mathbb{Z}_2\mathbb{Z}_4$-additive codes [29], we construct the three quaternary extended perfect codes which give the three non-equivalent extended perfect $\mathbb{Z}_4$-linear codes of length $2^5 = 32$. We show that the codes give extended binary perfect codes by checking that their Gray map images are $(2^5, 2^{2^5-5-1}, 4) = (32, 67108864, 4)$ codes. We also check that they are pairwise non-equivalent by computing the rank and dimension of the kernel. Since all of them have a different rank and dimension of the kernel, any of these two invariants can be used to distinguish between non-equivalent such codes.

```
> delta := 1; t := 5;
> C1 := Z2Z4ExtendedPerfectCode(delta, t: OverZ4 := true);
> Dual(C1);
(16, 64) Z2Z4-additive code of type (0, 16; 4, 1; 0)
Generator matrix:
[1 1 1 1 1 1 1 1 1 1 1 1 1 1 1 1]
[0 2 0 2 0 2 0 2 0 2 0 2 0 2 0 2]
[0 0 2 2 0 0 2 2 0 0 2 2 0 0 2 2]
[0 0 0 0 2 2 2 2 0 0 0 0 2 2 2 2]
[0 0 0 0 0 0 0 0 2 2 2 2 2 2 2 2]
> BinaryLength(C1);
32
> Z2Z4MinimumLeeDistance(C1);
4
> #C1 eq 2^(2^t-t-1);
true
> RankZ2(C1);
27
> DimensionOfKernelZ2(C1);
20
>
> delta := 2; t := 5;
> C2 := Z2Z4ExtendedPerfectCode(delta, t: OverZ4 := true);
> Dual(C2);
(16, 64) Z2Z4-additive code of type (0, 16; 2, 2; 0)
Generator matrix:
[1 0 3 2 1 0 3 2 1 0 3 2 1 0 3 2]
[0 1 2 3 0 1 2 3 0 1 2 3 0 1 2 3]
```

```
[0 0 0 0 2 2 2 2 0 0 0 2 2 2 2 2]
[0 0 0 0 0 0 0 2 2 2 2 2 2 2 2 2]
> Z2Z4MinimumLeeDistance(C2);
4
> #C2 eq 2^(2^t-t-1);
true
> RankZ2(C2);
28
> DimensionOfKernelZ2(C2);
16
>
> delta := 3; t := 5;
> C3 := Z2Z4ExtendedPerfectCode(delta, t: OverZ4 := true);
> Dual(C3);
(16, 64) Z2Z4-additive code of type (0, 16; 0, 3; 0)
Generator matrix:
[1 0 3 2 0 3 2 1 3 2 1 0 2 1 0 3]
[0 1 2 3 0 1 2 3 0 1 2 3 0 1 2 3]
[0 0 0 0 1 1 1 1 2 2 2 2 3 3 3 3]
> Z2Z4MinimumLeeDistance(C3);
4
> #C3 eq 2^(2^t-t-1);
true
> RankZ2(C3);
29
> DimensionOfKernelZ2(C3);
14                                                                           △
```

6.1.2 $\mathbb{Z}_2\mathbb{Z}_4$-Additive (Extended) Perfect Codes with $\alpha \neq 0$

Again, we start by giving a construction of perfect and extended perfect $\mathbb{Z}_2\mathbb{Z}_4$-linear codes of length $2^t - 1$ and 2^t, $t \geq 2$, respectively (denoted by C and C^*, respectively). We define them through the parity check matrices of the corresponding $\mathbb{Z}_2\mathbb{Z}_4$-additive perfect and extended perfect codes (denoted by $\mathcal{C}$ and $\mathcal{C}^*$, respectively, where $C = \Phi(\mathcal{C})$ and $C^* = \Phi(\mathcal{C}^*)$). We consider that their $\mathbb{Z}_2\mathbb{Z}_4$-dual codes are of type $(\alpha - 1, \beta; \gamma - 1, \delta)$ and $(\alpha, \beta; \gamma, \delta)$, respectively, with $\alpha \neq 0$. Note that we again omit the parameter κ in the type of these codes. This is due to the fact that for any $\mathbb{Z}_2\mathbb{Z}_4$-additive (extended) perfect code of type $(\alpha, \beta; \bar{\gamma}, \bar{\delta}; \bar{\kappa})$, we always have $\bar{\kappa} = \bar{\gamma}$, and therefore $\kappa = \gamma$. This result is proved in [137] and is easily obtained by Theorems 3.16, 6.10 and 6.11.

From [45, 124], a parity check matrix $\mathcal{H}^*$ of $\mathcal{C}^*$ may be represented, up to row permutation as follows. First, consider the matrix over $\mathbb{Z}_4$ consisting of all column vectors of the form $\{1\} \times \{0,1\}^{\gamma-1} \times \mathbb{Z}_4^{\delta}$ (up to change of sign in the last δ coordinates). Let α be the number of non-zero column vectors such that the last δ coordinates are in $\{0,2\} \subset \mathbb{Z}_4$. Note that $\alpha = 2^{t-\delta}$ for $t \geq 2$. Since we can place these column vectors in the first α columns, the matrix can be expressed as

$$\begin{pmatrix} A' & B' \\ 2D' & Q \end{pmatrix},$$

where A', B', D' are matrices over $\mathbb{Z}_4$ with all entries in $\{0,1\} \subset \mathbb{Z}_4$ of size $\gamma \times \alpha$, $\gamma \times \beta$ and $\delta \times \alpha$, respectively; and Q is a matrix over $\mathbb{Z}_4$ of size $\delta \times \beta$ with all column vectors of order four. Then, the matrix

$$\mathcal{H}^* = \begin{pmatrix} \xi(A') & 2B' \\ \xi(D') & Q \end{pmatrix}, \tag{6.3}$$

is a parity check matrix of $\mathcal{C}^*$, where ξ is the modulo two map from $\mathbb{Z}_4$ to $\mathbb{Z}_2$. It is easy to see that the $\mathbb{Z}_2\mathbb{Z}_4$-dual code of $\mathcal{C}^* = \Phi(\mathcal{C}^*)$ is of type $(\alpha, \beta; \gamma, \delta)$, where $\alpha = 2^{t-\delta}$ and $\beta = 2^{t-1} - 2^{t-\delta-1}$ since $2^t = \alpha + 2\beta$.

A parity check matrix $\mathcal{H}$ of the corresponding $\mathbb{Z}_2\mathbb{Z}_4$-additive perfect code $\mathcal{C}$ can be constructed from $\mathcal{H}^*$ deleting the first row (that is, the row vector $(\mathbf{1} \mid \mathbf{2})$) and the all-zero column vector (which is located in one of the first α columns). Since we can write the matrix $\mathcal{H}^*$ in (6.3) as a matrix of the form

$$\mathcal{H}^* = \begin{pmatrix} 1 & 1 & 2 \\ 0 & A & 2B \\ 0 & D & Q \end{pmatrix},$$

where $\xi(A') = \begin{pmatrix} 1 & 1 \\ 0 & A \end{pmatrix}$, $B' = \begin{pmatrix} 1 \\ B \end{pmatrix}$, and $\xi(D') = (\,0 \quad D\,)$, we have that

$$\mathcal{H} = \begin{pmatrix} A & 2B \\ D & Q \end{pmatrix}, \tag{6.4}$$

where A and D are matrices over $\mathbb{Z}_2$ of size $(\gamma - 1) \times (\alpha - 1)$ and $\delta \times (\alpha - 1)$, respectively; B is a matrix with all entries in $\{0,1\} \subset \mathbb{Z}_4$ of size $(\gamma - 1) \times \beta$; and Q is a matrix over $\mathbb{Z}_4$ of size $\delta \times \beta$.

Example 6.8. For $t = 2$, the parity check matrices of the $\mathbb{Z}_2\mathbb{Z}_4$-additive extended perfect codes, with $\delta = 0$ and $\delta = 1$, are

$$\mathcal{H}'_0 = \begin{pmatrix} 1 & 1 & 1 & 1 \\ 0 & 0 & 1 & 1 \\ 0 & 1 & 0 & 1 \end{pmatrix} \text{ and } \mathcal{H}'_1 = \begin{pmatrix} 1 & 1 & 2 \\ 0 & 1 & 1 \end{pmatrix},$$

respectively. And for $t = 3$, the matrices for $\delta = 0$ and $\delta = 1$ are

$$\mathcal{H}_0'' = \begin{pmatrix} 1 & 1 & 1 & 1 & 1 & 1 & 1 & 1 \\ 0 & 0 & 0 & 0 & 1 & 1 & 1 & 1 \\ 0 & 0 & 1 & 1 & 0 & 0 & 1 & 1 \\ 0 & 1 & 0 & 1 & 0 & 1 & 0 & 1 \end{pmatrix} \quad \text{and } \mathcal{H}_1'' = \left(\begin{array}{cccc|cc} 0 & 0 & 1 & 1 & 0 & 2 \\ 1 & 1 & 1 & 1 & 2 & 2 \\ 0 & 1 & 0 & 1 & 1 & 1 \end{array} \right),$$

respectively. Since the extended perfect codes of length 4 (length 8) are unique up to equivalence, both extended perfect $\mathbb{Z}_2\mathbb{Z}_4$-linear codes with $\alpha \neq 0$ for $t = 2$ ($t = 3$) are equivalent. Similarly, both perfect $\mathbb{Z}_2\mathbb{Z}_4$-linear codes with $\alpha \neq 0$ for $t = 2$ and $t = 3$ are equivalent. △

Example 6.9. For $t = 4$, by using the above construction, we can obtain three extended perfect $\mathbb{Z}_2\mathbb{Z}_4$-linear codes of length 16, which are not $\mathbb{Z}_4$-linear codes, so with $\alpha \neq 0$ (taking $\delta = 0$, $\delta = 1$ and $\delta = 2$).

For the case $\delta = 1$, we first construct the matrix over $\mathbb{Z}_4$ consisting of all column vectors of the form $\{1\} \times \{0,1\}^2 \times \mathbb{Z}_4^1$, up to change of sign in the last coordinate. Note that there are exactly 8 column vectors such that the last coordinate contains an element from $\{0,2\} \subset \mathbb{Z}_4$. Rearranging columns in order to have these column vectors in the first columns, we obtain the matrix

$$\left(\begin{array}{cccccccc|cccc} 1 & 1 & 1 & 1 & 1 & 1 & 1 & 1 & 1 & 1 & 1 & 1 \\ 0 & 0 & 1 & 1 & 0 & 0 & 1 & 1 & 0 & 0 & 1 & 1 \\ 0 & 1 & 0 & 1 & 0 & 1 & 0 & 1 & 0 & 1 & 0 & 1 \\ 0 & 0 & 0 & 0 & 2 & 2 & 2 & 2 & 1 & 1 & 1 & 1 \end{array} \right).$$

Therefore, the following matrix $\mathcal{H}^*$ is a parity check matrix of the corresponding $\mathbb{Z}_2\mathbb{Z}_4$-additive extended perfect code C^* with $\delta = 1$, so that its additive dual code $C^{*\perp}$ is of type $(8, 4; 3, 1)$:

$$\mathcal{H}^* = \left(\begin{array}{cccccccc|cccc} 1 & 1 & 1 & 1 & 1 & 1 & 1 & 1 & 2 & 2 & 2 & 2 \\ 0 & 0 & 1 & 1 & 0 & 0 & 1 & 1 & 0 & 0 & 2 & 2 \\ 0 & 1 & 0 & 1 & 0 & 1 & 0 & 1 & 0 & 2 & 0 & 2 \\ 0 & 0 & 0 & 0 & 1 & 1 & 1 & 1 & 1 & 1 & 1 & 1 \end{array} \right).$$

Note that deleting the first row and column of $\mathcal{H}^*$, we obtain the following matrix $\mathcal{H}$ which is a parity check matrix of the $\mathbb{Z}_2\mathbb{Z}_4$-additive perfect code C with $\delta = 1$, so that its additive dual code $C^{\perp}$ is of type $(7, 4; 2, 1)$:

$$\mathcal{H} = \left(\begin{array}{ccccccc|cccc} 0 & 1 & 1 & 0 & 0 & 1 & 1 & 0 & 2 & 0 & 2 \\ 0 & 0 & 0 & 1 & 1 & 1 & 1 & 0 & 0 & 2 & 2 \\ 1 & 1 & 0 & 1 & 0 & 1 & 0 & 1 & 1 & 1 & 1 \end{array} \right).$$

△

By using the above construction, it is possible to construct extended perfect $\mathbb{Z}_2\mathbb{Z}_4$-linear codes with $\alpha \neq 0$ such that their $\mathbb{Z}_2\mathbb{Z}_4$-dual codes have the parameters given in the following table, for $2 \leq t \leq 7$:

t	δ	$(\alpha, \beta; \gamma, \delta)$
2	0,1	$(4,0;3,0), (2,1;1,1)$
3	0,1	$(8,0;4,0), (4,2;2,1)$
4	0,1,2	$(16,0;5,0), (8,4;3,1), (4,6;1,2)$
5	0,1,2	$(32,0;6,0), (16,8;4,1), (8,12;2,2)$
6	0,1,2,3	$(64,0;7,0), (32,16;5,1), (16,24;3,2), (8,28;1,3)$
7	0,1,2,3	$(128,0;8,0), (64,32;6,1), (32,48;4,2), (16,56;2,3)$
...	...	...

From these codes, it is also possible to construct perfect $\mathbb{Z}_2\mathbb{Z}_4$-linear codes with $\alpha \neq 0$ having a $\mathbb{Z}_2\mathbb{Z}_4$-dual code with the parameters given in the following table, for $2 \leq t \leq 7$:

t	δ	$(\alpha, \beta; \gamma, \delta)$
2	0,1	$(3,0;2,0), (1,1;0,1)$
3	0,1	$(7,0;3,0), (3,2;1,1)$
4	0,1,2	$(15,0;4,0), (7,4;2,1), (3,6;0,2)$
5	0,1,2	$(31,0;5,0), (15,8;3,1), (7,12;1,2)$
6	0,1,2,3	$(63,0;6,0), (31,16;4,1), (15,24;2,2), (7,28;0,3)$
7	0,1,2,3	$(127,0;7,0), (63,32;5,1), (31,48;3,2), (15,56;1,3)$
...	...	...

Note that for $t = 2$ and $t = 3$ (length 4 and 8, respectively), both perfect and extended perfect $\mathbb{Z}_2\mathbb{Z}_4$-linear codes are equivalent as shown in Example 6.8, and for $t \geq 4$, all such codes are pairwise non-equivalent and unique as proved in the following results.

Theorem 6.10 ([43]). *For any integer $t \geq 4$ and each $\delta \in \{0, \ldots, \lfloor t/2 \rfloor\}$, there exists a unique (up to equivalence) extended perfect $\mathbb{Z}_2\mathbb{Z}_4$-linear code C^* of length $2^t \geq 16$, such that the $\mathbb{Z}_2\mathbb{Z}_4$-dual code of C^* is of type $(\alpha, \beta; \gamma, \delta)$, where $\alpha = 2^{t-\delta}$, $\beta = 2^{t-1} - 2^{t-\delta-1}$, and $\gamma = t + 1 - 2\delta$.*

Theorem 6.11 ([45]). *For any integer $t \geq 4$ and each $\delta \in \{0, \ldots, \lfloor t/2 \rfloor\}$, there exists a unique (up to equivalence) perfect $\mathbb{Z}_2\mathbb{Z}_4$-linear code C of length $n = 2^t - 1 \geq 15$, such that the $\mathbb{Z}_2\mathbb{Z}_4$-dual code of C is of type $(\alpha, \beta; \gamma, \delta)$, where $\alpha = 2^{t-\delta} - 1$, $\beta = 2^{t-1} - 2^{t-\delta-1}$, and $\gamma = t - 2\delta$.*

By Theorems 6.10 and 6.11, the number of non-equivalent (extended) perfect $\mathbb{Z}_2\mathbb{Z}_4$-linear codes of length 2^t with $\alpha \neq 0$ is $\lfloor (t+2)/2 \rfloor$ for all $t \geq 4$, and it is 1 for $t = 2$ and $t = 3$. Note that the corresponding $\mathbb{Z}_2\mathbb{Z}_4$-additive codes of these (extended) perfect $\mathbb{Z}_2\mathbb{Z}_4$-linear codes are $\mathbb{Z}_2\mathbb{Z}_4$-additive extended perfect codes with $\alpha \neq 0$.

The study of $\mathbb{Z}_2\mathbb{Z}_4$-additive extended perfect codes is absolutely different if they come from the extended code of a perfect $\mathbb{Z}_4$-linear code (with $\alpha = 0$) as in Theorem 6.3 or from an extended perfect $\mathbb{Z}_2\mathbb{Z}_4$-linear codes (with $\alpha \neq 0$) as in Theorem 6.10. Note that, in the first case, the quaternary all-one vector is always in both codes, the $\mathbb{Z}_2\mathbb{Z}_4$-additive extended perfect code and its additive dual code. However, in the second case, the vector with binary ones in the binary part and quaternary twos in the quaternary part is the one which is always in these two codes, the $\mathbb{Z}_2\mathbb{Z}_4$-additive extended perfect code and its additive dual code.

As for quaternary extended perfect codes, a parity check matrix for all $\mathbb{Z}_2\mathbb{Z}_4$-additive extended perfect codes with $\alpha \neq 0$ can also be constructed in a recursive way [137].

Let $(\mathcal{H}_\alpha \mid \mathcal{H}_\beta)$ be a parity check matrix of a $\mathbb{Z}_2\mathbb{Z}_4$-additive extended perfect code, such that its additive dual code is of type $(\alpha, \beta; \gamma, \delta)$, where $\alpha = 2^{t-\delta}$, $\beta = 2^{t-1} - 2^{t-\delta-1}$, $\gamma = t + 1 - 2\delta$ and $\delta \in \{0, \ldots, \lfloor t/2 \rfloor\}$. A parity check matrix for the $\mathbb{Z}_2\mathbb{Z}_4$-additive extended perfect code, such that its additive dual code is of type $(\alpha', \beta'; \gamma + 1, \delta)$, where $\alpha' = 2\alpha = 2^{t-\delta+1}$ and $\beta' = 2\beta = 2^t - 2^{t-\delta}$, can be constructed as follows:

$$\begin{pmatrix} 0 & 1 & 0 & 2 \\ \mathcal{H}_\alpha & \mathcal{H}_\alpha & \mathcal{H}_\beta & \mathcal{H}_\beta \end{pmatrix}. \tag{6.5}$$

Therefore, from codes of length $\alpha + 2\beta = 2^t$ and $\alpha = 2^{t-\delta}$, we obtain codes of length $\alpha' + 2\beta' = 2^{t+1}$ and $\alpha' = 2^{t-\delta+1}$. This means that using this construction, from all the non-equivalent codes of length 2^t, we obtain all the codes of length 2^{t+1}, except when t is even and $\gamma = 1$.

Again, when t is even and $\gamma = 1$, we need to use another recursive construction. A parity check matrix for the $\mathbb{Z}_2\mathbb{Z}_4$-additive extended perfect code, such that its additive dual code is of type $(\alpha', \beta'; \gamma, \delta + 1)$, where $\alpha' = 2\alpha = 2^{t-\delta+1}$ and $\beta' = \alpha + 4\beta = 2^{t-\delta} + 2^{t+1} - 2^{t-\delta+1} = 2^{t+1} - 2^{t-\delta}$, can be constructed as follows:

$$\begin{pmatrix} \mathcal{H}_\alpha & \mathcal{H}_\alpha & 2\mathcal{H}_\alpha & \mathcal{H}_\beta & \mathcal{H}_\beta & \mathcal{H}_\beta & \mathcal{H}_\beta \\ 0 & 1 & 1 & 0 & 1 & 2 & 3 \end{pmatrix}. \tag{6.6}$$

Therefore, starting with a parity check matrix of the $\mathbb{Z}_2\mathbb{Z}_4$-additive extended perfect code, such that its additive code is of type $(2, 1; 1, 1)$, that is, with $t = 2$ and $\gamma = 1$, it is possible to construct all the cases for t even and $\gamma = 1$.

Example 6.12. For $t = 4$, by Theorem 6.10, there are three non-equivalent extended perfect $\mathbb{Z}_2\mathbb{Z}_4$-linear codes of length 16, since we have three possible parameters: $\delta = 0$, $\delta = 1$ and $\delta = 2$.

We can construct parity check matrices of the $\mathbb{Z}_2\mathbb{Z}_4$-additive extended perfect codes with $\delta = 0$ and $\delta = 1$, by using (6.5) and the matrices $\mathcal{H}_0''$ and $\mathcal{H}_1''$ given in Example 6.8. On the other hand, by using (6.6) and the matrix $\mathcal{H}_1'$ of type $(2, 1; 1, 1)$, given in Example 6.8, we can obtain a parity check matrix for the $\mathbb{Z}_2\mathbb{Z}_4$-additive extended perfect code with $\delta = 2$. $\triangle$

Proposition 6.13 ([43]). *Let $t \geq 4$ and $\delta \in \{0, \ldots, \lfloor t/2 \rfloor\}$. Let C^* be an extended perfect $\mathbb{Z}_2\mathbb{Z}_4$-linear code of type $(\alpha, \beta; \bar{\gamma}, \bar{\delta})$, such that its $\mathbb{Z}_2\mathbb{Z}_4$-dual code is of type $(\alpha, \beta; \gamma, \delta)$, where $\alpha = 2^{t-\delta}, \beta = 2^{t-1} - 2^{t-\delta-1}, \gamma = t + 1 - 2\delta$, $\bar{\gamma} = \alpha - \gamma$ and $\bar{\delta} = \beta - \delta$. Then,*

$$\mathrm{rank}(C^*) = \bar{\gamma} + 2\bar{\delta} + \delta = \beta + \bar{\delta} + \bar{\gamma},$$

and

$$\mathrm{ker}(C^*) = \begin{cases} \bar{\gamma} + \bar{\delta} + 1 & \text{if } \delta \geq 1, \\ \bar{\gamma} + 2\bar{\delta} & \text{if } \delta = 0. \end{cases}$$

Note that the rank of the extended perfect $\mathbb{Z}_2\mathbb{Z}_4$-linear codes with $\alpha \neq 0$ satisfies the upper bound given in Proposition 5.23.

Proposition 6.14 ([124]). *Let $t \geq 4$ and $\delta \in \{0, \ldots, \lfloor t/2 \rfloor\}$. Let C be a perfect $\mathbb{Z}_2\mathbb{Z}_4$-linear code of type $(\alpha, \beta; \bar{\gamma}, \bar{\delta})$, such that its $\mathbb{Z}_2\mathbb{Z}_4$-dual code is of type $(\alpha, \beta; \gamma, \delta)$, where $\alpha = 2^{t-\delta} - 1$, $\beta = 2^{t-1} - 2^{t-\delta-1}$, $\gamma = t - 2\delta$, $\bar{\gamma} = \alpha - \gamma$ and $\bar{\delta} = \beta - \delta$. Then,*

$$\mathrm{rank}(C) = \bar{\gamma} + 2\bar{\delta} + \delta,$$

and

$$\mathrm{ker}(C) = \begin{cases} \bar{\gamma} + \bar{\delta} + 1 & \text{if } \delta \geq 1, \\ \bar{\gamma} & \text{if } \delta = 0. \end{cases}$$

Example 6.15. By using the MAGMA package for $\mathbb{Z}_2\mathbb{Z}_4$-additive codes [29], we construct the three $\mathbb{Z}_2\mathbb{Z}_4$-additive extended perfect codes which give the three non-equivalent extended perfect $\mathbb{Z}_2\mathbb{Z}_4$-linear codes of length $2^5 = 32$ with $\alpha \neq 0$. We show that the codes give extended binary perfect codes by checking that their Gray map images are $(2^5, 2^{2^5 - 5 - 1}, 4) = (32, 67108864, 4)$ codes. We also check that they are pairwise non-equivalent by computing the rank and dimension of the kernel. Since all of them have a different rank and dimension of the kernel, any of these two invariants can be used to distinguish between non-equivalent such codes.

```
> delta := 0; t := 5;
> C1 := Z2Z4ExtendedPerfectCode(delta, t);
> Dual(C1);
(32, 64) Z2Z4-additive code of type (32, 0; 6, 0; 6)
Generator matrix:
[2 0 0 2 0 2 2 0 0 2 2 0 2 0 0 2 0 2 2 0 2 0 0 2 2 0 0 2 0 2 2 0]
[0 2 0 2 0 2 0 2 0 2 0 2 0 2 0 2 0 2 0 2 0 2 0 2 0 2 0 2 0 2 0 2]
[0 0 2 2 0 0 2 2 0 0 2 2 0 0 2 2 0 0 2 2 0 0 2 2 0 0 2 2 0 0 2 2]
[0 0 0 0 2 2 2 2 0 0 0 0 2 2 2 2 0 0 0 0 2 2 2 2 0 0 0 0 2 2 2 2]
[0 0 0 0 0 0 0 0 2 2 2 2 2 2 2 2 0 0 0 0 0 0 0 0 2 2 2 2 2 2 2 2]
[0 0 0 0 0 0 0 0 0 0 0 0 0 0 0 0 2 2 2 2 2 2 2 2 2 2 2 2 2 2 2 2]
> BinaryLength(C1);
32
> Z2Z4MinimumLeeDistance(C1);
4
> #C1 eq 2^(2^t-t-1);
true
> RankZ2(C1);
26
> DimensionOfKernelZ2(C1);
26
>
> delta := 1; t := 5;
> C2 := Z2Z4ExtendedPerfectCode(delta, t);
> Dual(C2);
(24, 64) Z2Z4-additive code of type (16, 8; 4, 1; 4)
Generator matrix:
[2 0 0 2 0 2 2 0 0 2 2 0 2 0 0 2 1 3 3 1 3 1 1 3]
[0 2 0 2 0 2 0 2 0 2 0 2 0 2 0 2 1 1 1 1 1 1 1 1]
[0 0 2 2 0 0 2 2 0 0 2 2 0 0 2 2 0 2 0 2 0 2 0 2]
[0 0 0 0 2 2 2 2 0 0 0 0 2 2 2 2 0 0 2 2 0 0 2 2]
[0 0 0 0 0 0 0 0 2 2 2 2 2 2 2 2 0 0 0 2 2 2 2]
[0 0 0 0 0 0 0 0 0 0 0 0 0 0 0 0 2 2 2 2 2 2 2 2]
> BinaryLength(C2);
32
> Z2Z4MinimumLeeDistance(C2);
4
> #C2 eq 2^(2^t-t-1);
true
> RankZ2(C2);
27
> DimensionOfKernelZ2(C2);
20
>
> delta := 2; t := 5;
> C3 := Z2Z4ExtendedPerfectCode(delta, t);
> Dual(C3);
(20, 64) Z2Z4-additive code of type (8, 12; 2, 2; 2)
Generator matrix:
```

```
[2 0 0 2 0 2 2 0 1 3 1 0 3 2 3 1 3 2 1 0]
[0 2 0 2 0 2 0 2 0 2 1 1 1 1 0 2 1 1 1 1]
[0 0 2 2 0 0 2 2 1 1 0 1 2 3 1 1 0 1 2 3]
[0 0 0 0 2 2 2 2 0 0 0 0 0 2 2 2 2 2 2 2]
[0 0 0 0 0 0 0 0 2 2 0 2 0 2 2 2 0 2 0 2]
[0 0 0 0 0 0 0 0 0 0 2 2 2 0 0 2 2 2 2 2]
> BinaryLength(C3);
32
> Z2Z4MinimumLeeDistance(C3);
4
> #C3 eq 2^(2^t-t-1);
true
> RankZ2(C3);
28
> DimensionOfKernelZ2(C3);
17                                                                    △
```

6.2 $\mathbb{Z}_2\mathbb{Z}_4$-Additive Hadamard Codes

A *Hadamard matrix* H_n of order n is an $n \times n$ matrix of $+1$'s and -1's such that $H_n H_n^T = nI_n$, where I_n is the identity matrix of size $n \times n$. It is well known that if a Hadamard matrix H_n of order n exists, then n is 1, 2 or a multiple of 4 [5, 112]. Two Hadamard matrices are *equivalent* if one matrix can be obtained from the other by permuting rows and (or) columns and multiplying rows and (or) columns by -1. We can change the first row and column of H_n into $+1$'s, and we obtain an equivalent Hadamard matrix H_n', which is called *normalized*.

If $+1$'s are replaced by 0's and -1's by 1's, H_n' is changed into a *binary Hadamard matrix* $c(H_n')$. Since the rows of H_n' are orthogonal, any two rows of $c(H_n')$ agree in $n/2$ places and differ in $n/2$ places, and so are at Hamming distance $n/2$ apart. The binary code consisting of the rows of $c(H_n')$ and their complements is called a *(binary) Hadamard code* [5, 112] and we use H to denote it. A Hadamard code of length n has $2n$ codewords and minimum distance $n/2$, so it is a $(n, 2n, n/2)$ binary code.

Example 6.16. We consider the following normalized Hadamard matrix of order 4:

$$H_4 = \begin{pmatrix} 1 & 1 & 1 & 1 \\ 1 & 1 & -1 & -1 \\ 1 & -1 & 1 & -1 \\ 1 & -1 & -1 & 1 \end{pmatrix}$$

Then, the corresponding binary Hadamard matrix is

$$
c(H_4') = \begin{pmatrix} 0 & 0 & 0 & 0 \\ 0 & 0 & 1 & 1 \\ 0 & 1 & 0 & 1 \\ 0 & 1 & 1 & 0 \end{pmatrix},
$$

and the Hadamard code H is given by the set $H = \{(0,0,0,0), (0,0,1,1), (0,1,0,1), (0,1,1,0), (1,1,1,1), (1,1,0,0), (1,0,1,0), (1,0,0,1)\}$. In general, a Hadamard code is not necessarily linear. However, in this case, it is easy to check that H is a linear code with the following generator matrix:

$$
\begin{pmatrix} 1 & 1 & 1 & 1 \\ 0 & 0 & 1 & 1 \\ 0 & 1 & 0 & 1 \end{pmatrix}.
$$

$\triangle$

The $\mathbb{Z}_2\mathbb{Z}_4$-additive codes such that, under the Gray map, give a Hadamard code are called $\mathbb{Z}_2\mathbb{Z}_4$-additive Hadamard codes. Given a $\mathbb{Z}_2\mathbb{Z}_4$-additive Hadamard code, after applying the Gray map, the corresponding $\mathbb{Z}_2\mathbb{Z}_4$-linear code is called Hadamard $\mathbb{Z}_2\mathbb{Z}_4$-linear code. The $\mathbb{Z}_2\mathbb{Z}_4$-additive Hadamard codes with $\alpha = 0$ are also called quaternary linear Hadamard codes, and their corresponding $\mathbb{Z}_2\mathbb{Z}_4$-linear codes are also called Hadamard $\mathbb{Z}_4$-linear codes.

The $\mathbb{Z}_2\mathbb{Z}_4$-additive Hadamard codes can also be seen as the additive dual codes of the $\mathbb{Z}_2\mathbb{Z}_4$-additive extended perfect codes [43, 102]. In other words, the Hadamard $\mathbb{Z}_2\mathbb{Z}_4$-linear codes H are the $\mathbb{Z}_2\mathbb{Z}_4$-dual of the extended perfect $\mathbb{Z}_2\mathbb{Z}_4$-linear codes C^*, that is, $H = C_\perp^*$. Therefore, the parity check matrices of the $\mathbb{Z}_2\mathbb{Z}_4$-additive extended perfect codes, described in Sections 6.1.1 and 6.1.2, can be taken as generator matrices for the $\mathbb{Z}_2\mathbb{Z}_4$-additive Hadamard codes.

On the one hand, when $\alpha = 0$, Proposition 6.17 gives the result on the classification of Hadamard $\mathbb{Z}_4$-linear codes, up to equivalence. Note that the number of non-equivalent Hadamard $\mathbb{Z}_4$-linear codes of length 2^t is $\lfloor (t-1)/2 \rfloor$ for all $t \geq 3$, and it is 1 for $t = 2$.

Proposition 6.17 ([102]). *For any integer $t \geq 4$ and each $\delta \in \{1, \ldots, \lfloor (t+1)/2 \rfloor\}$, there exists a unique Hadamard $\mathbb{Z}_4$-linear code H of type $(0, \beta; \gamma, \delta)$, where $\beta = 2^{t-1}$ and $\gamma = t+1-2\delta$. Moreover, all these codes H are pairwise non-equivalent, except for $\delta = 1$ and $\delta = 2$, where the codes H are equivalent to the binary dual of the extended Hamming code.*

On the other hand, Theorem 6.18 gives an analogous result for the Hadamard $\mathbb{Z}_2\mathbb{Z}_4$-linear codes with $\alpha \neq 0$. Note that, in this case, the number of non-equivalent Hadamard $\mathbb{Z}_2\mathbb{Z}_4$-linear codes with $\alpha \neq 0$ of length 2^t is $\lfloor t/2 \rfloor$ for all $t \geq 2$.

Theorem 6.18 ([43]). *For any integer $t \geq 4$ and each $\delta \in \{0, \ldots, \lfloor t/2 \rfloor\}$, there exists a unique Hadamard $\mathbb{Z}_2\mathbb{Z}_4$-linear code H of type $(\alpha, \beta; \gamma, \delta)$, where $\alpha = 2^{t-\delta}$, $\beta = 2^{t-1} - 2^{t-\delta-1}$, and $\gamma = t + 1 - 2\delta$. Moreover, all these codes H are pairwise non-equivalent, except for $\delta = 0$ and $\delta = 1$, where the codes H are equivalent to the binary dual of the extended Hamming code.*

In [105], it is shown that each Hadamard $\mathbb{Z}_2\mathbb{Z}_4$-linear code with $\alpha = 0$ is equivalent to a Hadamard $\mathbb{Z}_2\mathbb{Z}_4$-linear code with $\alpha \neq 0$, so indeed there are $\lfloor t/2 \rfloor$ non-equivalent Hadamard $\mathbb{Z}_2\mathbb{Z}_4$-linear codes of length 2^t.

Proposition 6.19 ([102, 122]). *Let $t \geq 3$ and $\delta \in \{1, \ldots, \lfloor (t+1)/2 \rfloor\}$. Let H be a Hadamard $\mathbb{Z}_4$-linear code of type $(0, \beta; \gamma, \delta)$, where $\beta = 2^{t-1}$ and $\gamma = t + 1 - 2\delta$. Then,*

$$\text{rank}(H) = \begin{cases} \gamma + 2\delta + \binom{\delta-1}{2} & \text{if} \quad \delta \geq 3, \\ \gamma + 2\delta & \text{if} \quad \delta = 1, 2, \end{cases}$$

and

$$\ker(H) = \begin{cases} \gamma + \delta + 1 & \text{if} \quad \delta \geq 3, \\ \gamma + 2\delta & \text{if} \quad \delta = 1, 2. \end{cases}$$

Proposition 6.20 ([122]). *Let $t \geq 3$ and $\delta \in \{0, \ldots, \lfloor t/2 \rfloor\}$. Let H be a Hadamard $\mathbb{Z}_2\mathbb{Z}_4$-linear code of type $(\alpha, \beta; \gamma, \delta)$, where $\alpha = 2^{t-\delta}$, $\beta = 2^{t-1} - 2^{t-\delta-1}$, and $\gamma = t + 1 - 2\delta$. Then,*

$$\text{rank}(H) = \begin{cases} \gamma + 2\delta + \binom{\delta}{2} & \text{if} \quad \delta \geq 2, \\ \gamma + 2\delta & \text{if} \quad \delta = 0, 1, \end{cases}$$

and

$$\ker(H) = \begin{cases} \gamma + \delta & \text{if} \quad \delta \geq 2, \\ \gamma + 2\delta & \text{if} \quad \delta = 0, 1. \end{cases}$$

Note that, for the Hadamard $\mathbb{Z}_2\mathbb{Z}_4$-linear codes with $\alpha \neq 0$, the rank satisfies the upper bound given in Proposition 5.23, ant the dimension of the kernel satisfies the lower bound given by Proposition 5.42.

By Propositions 6.19 and 6.20, as for extended perfect $\mathbb{Z}_2\mathbb{Z}_4$-linear codes, the Hadamard $\mathbb{Z}_2\mathbb{Z}_4$-linear codes can also be classified using either the rank or the dimension of the kernel, as it is proven in [102, 122]. The intersection problem for these codes, i.e., which are the possibilities for the number of codewords in the intersection of two $\mathbb{Z}_2\mathbb{Z}_4$-additive codes of the same length is investigated in [136]. The permutation automorphism group of these $\mathbb{Z}_2\mathbb{Z}_4$-additive Hadamard codes and the corresponding $\mathbb{Z}_2\mathbb{Z}_4$-linear codes has been studied in [104, 105, 120]. Furthermore, a general class of these codes such as the translation invariant Hadamard propelinear class and the Hadamard full propelinear class has been studied in [15, 56, 113, 138].

Example 6.21. Continuing with Examples 6.7 and 6.15, we construct the Hadamard $\mathbb{Z}_2\mathbb{Z}_4$-linear codes of length 32 with $\alpha = 0$ and $\alpha \neq 0$. We see that they are the dual of the extended perfect $\mathbb{Z}_2\mathbb{Z}_4$-linear codes. Moreover, we check that the $\mathbb{Z}_4$-linear codes with $\delta = 1$ and $\delta = 2$ are equivalent; and the $\mathbb{Z}_2\mathbb{Z}_4$-linear codes such that $\alpha \neq 0$ are equivalent when $\delta = 0$ and $\delta = 1$. Finally, we also see that each Hadamard $\mathbb{Z}_4$-linear code is equivalent to a Hadamard $\mathbb{Z}_2\mathbb{Z}_4$-linear code.

```
> t := 5;
> HZ4delta1 := Z2Z4HadamardCode(1, t : OverZ4 := true);
> Dual(HZ4delta1) eq Z2Z4ExtendedPerfectCode(1, t : OverZ4 := true);
true
> HZ4delta2 := Z2Z4HadamardCode(2, t : OverZ4 := true);
> Dual(HZ4delta2) eq Z2Z4ExtendedPerfectCode(2, t : OverZ4 := true);
true
> HZ4delta3 := Z2Z4HadamardCode(3, t : OverZ4 := true);
> Dual(HZ4delta3) eq Z2Z4ExtendedPerfectCode(3, t : OverZ4 := true);
true
> HZ4delta1 eq HZ4delta2;
false
> DZ4delta1 := Design<2, 32 | [ c : c in GrayMapImage(HZ4delta1)
                                     | Weight(c) eq 16 ]>;
> DZ4delta2 := Design<2, 32 | [ c : c in GrayMapImage(HZ4delta2)
                                     | Weight(c) eq 16 ]>;
> DZ4delta3 := Design<2, 32 | [ c : c in GrayMapImage(HZ4delta3)
                                     | Weight(c) eq 16 ]>;
> IsIsomorphic(DZ4delta1, DZ4delta2);
true Mapping from: Dsgn: DZ4delta1 to Dsgn: DZ4delta2

> HZ2Z4delta0 := Z2Z4HadamardCode(0, t);
> Dual(HZ2Z4delta0) eq Z2Z4ExtendedPerfectCode(0, t);
true
> HZ2Z4delta1 := Z2Z4HadamardCode(1, t);
> Dual(HZ2Z4delta1) eq Z2Z4ExtendedPerfectCode(1, t);
true
> HZ2Z4delta2 := Z2Z4HadamardCode(2, t);
> Dual(HZ2Z4delta2) eq Z2Z4ExtendedPerfectCode(2, t);
true
> DZ2Z4delta0 := Design<2, 32 | [ c : c in GrayMapImage(HZ2Z4delta0)
                                      | Weight(c) eq 16 ]>;
> DZ2Z4delta1 := Design<2, 32 | [ c : c in GrayMapImage(HZ2Z4delta1)
```

```
                                          | Weight(c) eq 16 ]>;
> DZ2Z4delta2 := Design<2, 32 | [ c : c in GrayMapImage(HZ2Z4delta2)
                                          | Weight(c) eq 16 ]>;
> IsIsomorphic(DZ2Z4delta0, DZ2Z4delta1);
true Mapping from: Dsgn: DZ2Z4delta0 to Dsgn: DZ2Z4delta1
>
> IsIsomorphic(DZ4delta1, DZ2Z4delta0);
true Mapping from: Dsgn: DZ4delta1 to Dsgn: DZ2Z4delta0
> IsIsomorphic(DZ4delta2, DZ2Z4delta1);
true Mapping from: Dsgn: DZ4delta2 to Dsgn: DZ2Z4delta1
> IsIsomorphic(DZ4delta3, DZ2Z4delta2);
true Mapping from: Dsgn: DZ4delta3 to Dsgn: DZ2Z4delta2                  △
```

6.3 $\mathbb{Z}_2\mathbb{Z}_4$-Additive Reed-Muller Codes

Binary linear Reed-Muller codes (RM codes) have been extensively studied
and have good combinatorics properties [112]. Several families of quaternary linear codes have been proposed and studied trying to generalize RM
codes [34, 36, 150]. However, when the corresponding $\mathbb{Z}_4$-linear codes are
considered, they do not satisfy all the properties of RM codes. In this section, we follow [129, 130], constructing families of quaternary linear codes
and, more general, $\mathbb{Z}_2\mathbb{Z}_4$-additive codes such that, after the Gray map, the
corresponding $\mathbb{Z}_2\mathbb{Z}_4$-linear codes have the same parameters and properties as
RM codes. In contrast to the binary linear case, where there is just one RM
family for each length 2^m, in the $\mathbb{Z}_2\mathbb{Z}_4$-additive case, we have several families
of codes. We refer to them as $\mathcal{RM}_s$ and $\mathcal{ARM}_s$ families, when the codes
are quaternary linear and $\mathbb{Z}_2\mathbb{Z}_4$-additive with $\alpha \neq 0$, respectively.

First, we give a summary of the parameters and properties of RM codes.
For $m \geq 1$ and $0 \leq r \leq m$, there is a binary linear Reed-Muller code
$\mathrm{RM}(r, m)$ such that,

i) the length is $n = 2^m$, the minimum distance is $d = 2^{m-r}$, the dimension
   is $k = \sum_{i=0}^{r} \binom{m}{i}$;

ii) $\mathrm{RM}(r - 1, m)$ is a subcode of $\mathrm{RM}(r, m)$ for $r > 0$; and

iii) $\mathrm{RM}(r, m)$ is the dual code of $\mathrm{RM}(m - 1 - r, m)$ for $r < m$.

Moreover, $\mathrm{RM}(1, m)$ is equivalent to the linear Hadamard code and $\mathrm{RM}(m -
2, m)$ is equivalent to the extended Hamming code of the same length. Fig-
ure 6.1 shows an scheme of the inclusion and duality properties.

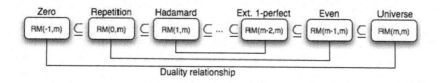

Figure 6.1: Sequence of RM(r,m) codes for $r \in \{-1,\ldots,m\}$

The $\text{RM}(r,m)$ code with $m \geq 2$ and $0 \leq r \leq m$ can be described using the Plotkin construction as follows [112]:

$$\text{RM}(r,m) = \{(u, u+v) : u \in \text{RM}(r, m-1),$$
$$v \in \text{RM}(r-1, m-1)\},$$

where $\text{RM}(0,m)$ is the repetition code $\{\mathbf{0},\mathbf{1}\}$ and $\text{RM}(m,m)$ is the universe code. For $m = 1$, there are only two codes: the repetition code $\text{RM}(0,1)$ and the universe code $\text{RM}(1,1)$.

In [92], it is proved that $\text{RM}(r,m)$ is $\mathbb{Z}_4$-linear for $r \in \{0,1,2,m-1\}$ and not $\mathbb{Z}_4$-linear for $r = m-2$ ($m \geq 5$). In a subsequent work [94], it is proved that $\text{RM}(r,m)$ is not $\mathbb{Z}_4$-linear for $3 \leq r \leq m-2$.

We start by describing quaternary linear Reed-Muller families, denoted by $\mathcal{RM}_s$, such that the corresponding $\mathbb{Z}_4$-linear codes have the same parameters and properties as the binary linear Reed-Muller family. Moreover, for $r = 1$, they include the quaternary linear Hadamard codes and, for $r = m-2$, the $\mathbb{Z}_2\mathbb{Z}_4$-additive extended perfect codes, constructed Sections 6.1 and 6.2, respectively. After that, we will see similar results for $\mathbb{Z}_2\mathbb{Z}_4$-additive Reed-Muller families with $\alpha \neq 0$, denoted by $\mathcal{ARM}_s$.

The construction of these families of codes, $\mathcal{RM}_s$ and $\mathcal{ARM}_s$, is based on the Plotkin construction and some variations.

Definition 6.22 (Plotkin Construction). *Let $\mathcal{X}$ and $\mathcal{Y}$ be any two $\mathbb{Z}_2\mathbb{Z}_4$-additive codes of types $(\alpha,\beta;\gamma_{\mathcal{X}},\delta_{\mathcal{X}})$, $(\alpha,\beta;\gamma_{\mathcal{Y}},\delta_{\mathcal{Y}})$ and minimum distances $d_{\mathcal{X}}$, $d_{\mathcal{Y}}$, respectively. If $\mathcal{G}_{\mathcal{X}}$ and $\mathcal{G}_{\mathcal{Y}}$ are the generator matrices of $\mathcal{X}$ and $\mathcal{Y}$, then the matrix*

$$\mathcal{G}_P = \begin{pmatrix} \mathcal{G}_{\mathcal{X}} & \mathcal{G}_{\mathcal{X}} \\ 0 & \mathcal{G}_{\mathcal{Y}} \end{pmatrix}$$

is the generator matrix of a $\mathbb{Z}_2\mathbb{Z}_4$-additive code, denoted by $\mathcal{P}(\mathcal{X},\mathcal{Y})$.

Proposition 6.23 ([130]). *The code $\mathcal{P}(\mathcal{X},\mathcal{Y})$ is a $\mathbb{Z}_2\mathbb{Z}_4$-additive code of type $(2\alpha, 2\beta; \gamma, \delta)$, where $\gamma = \gamma_{\mathcal{X}} + \gamma_{\mathcal{Y}}$, $\delta = \delta_{\mathcal{X}} + \delta_{\mathcal{Y}}$, and minimum distance $d = \min\{2d_{\mathcal{X}}, d_{\mathcal{Y}}\}$.*

Applying twice the Plotkin construction, one after another, but slightly changing the submatrices in the generator matrix, we obtain a new construction with interesting properties with regard to the minimum distance of the generated code. We call this new construction *BQ-Plotkin* when $\alpha = 0$ and *BA-Plotkin* when $\alpha \neq 0$.

Definition 6.24 (BQ-Plotkin Construction). *Let* $\mathcal{A}$, $\mathcal{B}$, *and* $\mathcal{C}$ *be quaternary linear codes of length* β *with generators matrices* $\mathcal{G}_A$, $\mathcal{G}_B$, *and* $\mathcal{G}_C$, *respectively. We define the code* $\mathcal{BQ}(\mathcal{A}, \mathcal{B}, \mathcal{C})$ *as the quaternary linear code of length* 4β *generated by*

$$\mathcal{G}_{BQ} = \begin{pmatrix} \mathcal{G}_A & \mathcal{G}_A & \mathcal{G}_A & \mathcal{G}_A \\ 0 & \mathcal{G}_B' & 2\mathcal{G}_B' & 3\mathcal{G}_B' \\ 0 & 0 & \hat{\mathcal{G}}_B & \hat{\mathcal{G}}_B \\ 0 & 0 & 0 & \mathcal{G}_C \end{pmatrix},$$

where $\mathcal{G}_B'$ *is the matrix obtained from* $\mathcal{G}_B$ *after replacing twos by ones in their* γ_B *rows of order two and* $\hat{\mathcal{G}}_B$ *is the matrix obtained from* $\mathcal{G}_B$ *after removing their* γ_B *rows of order two.*

Recall that any generator matrix $\mathcal{G}$ of a $\mathbb{Z}_2\mathbb{Z}_4$-additive code of type $(\alpha, \beta; \gamma, \delta; \kappa)$ can be written as in (2.5)

$$\mathcal{G} = \left(\begin{array}{c|c} B_1 & 2B_3 \\ B_2 & Q \end{array} \right),$$

where B_1, B_2 are matrices over $\mathbb{Z}_2$ of size $\gamma \times \alpha$ and $\delta \times \alpha$, respectively; B_3 is a matrix over $\mathbb{Z}_4$ of size $\gamma \times \beta$ with all entries in $\{0, 1\} \subset \mathbb{Z}_4$; and Q is a matrix over $\mathbb{Z}_4$ of size $\delta \times \beta$ with quaternary row vectors of order four. Now, we denote by $\mathcal{G}[b_1]$, $\mathcal{G}[b_2]$, $\mathcal{G}[b_3]$ and $\mathcal{G}[q_4]$ the four submatrices B_1, B_2, B_3, and Q of $\mathcal{G}$, respectively; and by $\mathcal{G}[b]$ and $\mathcal{G}[q]$ the submatrices $\begin{pmatrix} B_1 \\ B_2 \end{pmatrix}$ and $\begin{pmatrix} 2B_3 \\ Q \end{pmatrix}$, respectively.

Definition 6.25 (BA-Plotkin Construction). *Let* $\mathcal{X}$, $\mathcal{Y}$, *and* $\mathcal{Z}$ *be* $\mathbb{Z}_2\mathbb{Z}_4$*-additive codes in* $\mathbb{Z}_2^\alpha \times \mathbb{Z}_4^\beta$ *with generators matrices* $\mathcal{G}_X$, $\mathcal{G}_Y$, *and* $\mathcal{G}_Z$, *respectively. We define the code* $\mathcal{BA}(\mathcal{X}, \mathcal{Y}, \mathcal{Z})$ *as the* $\mathbb{Z}_2\mathbb{Z}_4$*-additive code in* $\mathbb{Z}_2^{2\alpha} \times \mathbb{Z}_4^{\alpha+4\beta}$ *generated by*

$$\mathcal{G}_{BA} = \left(\begin{array}{cc|cc cccc} \mathcal{G}_X[b] & \mathcal{G}_X[b] & 2\iota(\mathcal{G}_X[b]) & \mathcal{G}_X[q] & \mathcal{G}_X[q] & \mathcal{G}_X[q] & \mathcal{G}_X[q] \\ 0 & \mathcal{G}_Y[b_1] & \iota(\mathcal{G}_Y[b_1]) & 0 & 2\mathcal{G}_Y[b_3] & \mathcal{G}_Y[b_3] & 3\mathcal{G}_Y[b_3] \\ 0 & \mathcal{G}_Y[b_2] & \iota(\mathcal{G}_Y[b_2]) & 0 & \mathcal{G}_Y[q_4] & 2\mathcal{G}_Y[q_4] & 3\mathcal{G}_Y[q_4] \\ \mathcal{G}_Y[b_2] & \mathcal{G}_Y[b_2] & 0 & 0 & 0 & \mathcal{G}_Y[q_4] & \mathcal{G}_Y[q_4] \\ 0 & \mathcal{G}_Z[b] & 0 & 0 & 0 & 0 & \mathcal{G}_{Z[q]} \end{array} \right),$$

where $\iota(\mathcal{G})$ is the matrix obtained from $\mathcal{G}$ after considering its entries as elements in $\mathbb{Z}_4$.

Note that it is easy to check that if $\mathcal{X}$, $\mathcal{Y}$ and $\mathcal{Z}$ are quaternary linear codes, the above BA-Plotkin construction give a code $\mathcal{BA}(\mathcal{X}, \mathcal{Y}, \mathcal{Z})$ which is permutation equivalent to the code $\mathcal{BQ}(\mathcal{X}, \mathcal{Y}, \mathcal{Z})$ constructed by using the BQ-Plotkin construction defined in Definition 6.24.

Both Plotkin constructions give us the following two propositions.

Proposition 6.26 ([130]). *Let $\mathcal{A}$, $\mathcal{B}$, and $\mathcal{C}$ be quaternary linear codes of types $(0, \beta; \gamma_A, \delta_A)$, $(0, \beta; \gamma_B, \delta_B)$, and $(0, \beta; \gamma_C, \delta_C)$, respectively. The code $\mathcal{BQ}(\mathcal{A}, \mathcal{B}, \mathcal{C})$ is a quaternary linear code of type $(0, 4\beta; \gamma, \delta)$, where $\gamma = \gamma_A + \gamma_C$ and $\delta = \delta_A + \gamma_B + 2\delta_B + \delta_C$. Moreover, if d_A, d_B, and d_C is the minimum distance of $\mathcal{A}$, $\mathcal{B}$, and $\mathcal{C}$, respectively, then the minimum distance of $\mathcal{BQ}(\mathcal{A}, \mathcal{B}, \mathcal{C})$ is $d = \min\{4d_A, 2d_B, d_C\}$.*

Proposition 6.27 ([129]). *Let $\mathcal{X}$, $\mathcal{Y}$, and $\mathcal{Z}$ be $\mathbb{Z}_2\mathbb{Z}_4$-additive codes of types $(\alpha, \beta; \gamma_X, \delta_X)$, $(\alpha, \beta; \gamma_Y, \delta_Y)$, and $(\alpha, \beta; \gamma_Z, \delta_Z)$, respectively. The code $\mathcal{BA}(\mathcal{X}, \mathcal{Y}, \mathcal{Z})$ is a $\mathbb{Z}_2\mathbb{Z}_4$-additive code of type $(2\alpha, \alpha + 4\beta; \gamma, \delta)$, where $\gamma = \gamma_X + \gamma_Z$ and $\delta = \delta_X + \gamma_Y + 2\delta_Y + \delta_Z$. Moreover, if d_X, d_Y, and d_Z is the minimum distance of $\mathcal{X}$, $\mathcal{Y}$, and $\mathcal{Z}$, respectively, then the minimum distance of $\mathcal{BA}(\mathcal{X}, \mathcal{Y}, \mathcal{Z})$ is $d = \min\{4d_X, 2d_Y, d_Z\}$.*

Now, we have the tools to define the families of quaternary linear Reed-Muller codes $\mathcal{RM}_s$ and $\mathbb{Z}_2\mathbb{Z}_4$-additive Reed-Muller codes $\mathcal{ARM}_s$ with $\alpha \neq 0$.

We begin by showing the construction of the families of quaternary linear Reed-Muller codes $\mathcal{RM}_s$. In this case, for each integer $m \geq 1$, we describe $\lfloor \frac{m+1}{2} \rfloor$ such families, having $\mathbb{Z}_4$-linear codes of length 2^m. Each one of these families is identified by a subindex $s \in \{0, \ldots, \lfloor \frac{m-1}{2} \rfloor\}$ and contains exactly $m + 1$ codes, denoted by $\mathcal{RM}_s(r, m)$, $0 \leq r \leq m$.

For the recursive construction, it is convenient to define the codes $\mathcal{RM}_s(r, m)$ also for $r < 0$ and $r > m$. The code $\mathcal{RM}_s(r, m)$ with $r < 0$ is defined as the zero code. The code $\mathcal{RM}_s(0, m)$ is defined as the repetition code $\{\mathbf{0}, \mathbf{2}\}$. The code $\mathcal{RM}_s(r, m)$ with $r \geq m$ is defined as the universe code, that is, the whole space $\mathbb{Z}_4^{2^{m-1}}$. For $m = 1$, there is only one family with $s = 0$, and in this family there are only the zero, the repetition, and the universe codes for $r < 0$, $r = 0$ and $r \geq 1$, respectively, so we have that

$$\mathcal{RM}_0(-1, 1) = \{\mathbf{0}\} \subseteq \mathcal{RM}_0(0, 1) = \{\mathbf{0}, \mathbf{2}\} \subseteq \mathcal{RM}_0(1, 1) = \mathbb{Z}_4.$$

In this case, the generator matrix of $\mathcal{RM}_0(0, 1)$ is $\mathcal{G}_0(0, 1) = (2)$, and the generator matrix of $\mathcal{RM}_0(r, 1)$ is $\mathcal{G}_0(r, 1) = (1)$, where $r \geq 1$.

For any $m \geq 2$, given two codes $\mathcal{RM}_s(r, m-1)$ and $\mathcal{RM}_s(r-1, m-1)$, where $0 \leq s \leq \lfloor \frac{m-2}{2} \rfloor$, the $\mathcal{RM}_s(r, m)$ code can be constructed in a recursive way using the Plotkin construction given by Definition 6.22 as follows:

$$\mathcal{RM}_s(r, m) = P(\mathcal{RM}_s(r, m-1), \mathcal{RM}_s(r-1, m-1)).$$

Note that when m is odd, the $\mathcal{RM}_s$ family with $s = \frac{m-1}{2}$ can not be generated using the Plotkin construction. In this case, for any $m \geq 3$, m odd and $s = \frac{m-1}{2}$, given $\mathcal{RM}_{s-1}(r, m-2)$, $\mathcal{RM}_{s-1}(r-1, m-2)$, and $\mathcal{RM}_{s-1}(r-2, m-2)$, the $\mathcal{RM}_s(r, m)$ code can be constructed using the BQ-Plotkin construction given by Definition 6.24 as follows:

$$\mathcal{RM}_s(r, m) = BQ(\,\mathcal{RM}_{s-1}(r, m-2),$$
$$\mathcal{RM}_{s-1}(r-1, m-2),$$
$$\mathcal{RM}_{s-1}(r-2, m-2)).$$

Example 6.28. For $m = 2$, there is one family, which can be generated using the Plotkin construction. The generator matrices of $\mathcal{RM}_0(r, 2)$, $0 \leq r \leq 2$, are the following: $\mathcal{G}_0(0, 2) = (\ 2\ \ 2\ )$,

$$\mathcal{G}_0(1, 2) = \begin{pmatrix} 1 & 1 \\ 0 & 2 \end{pmatrix}, \quad \text{and} \quad \mathcal{G}_0(2, 2) = \begin{pmatrix} 1 & 1 \\ 0 & 1 \end{pmatrix}.$$

$\triangle$

Example 6.29. For $m = 3$, there are two families. The $\mathcal{RM}_0$ family can be generated using the Plotkin construction. The generator matrices are the following:

$$\mathcal{G}_0(0, 3) = (\ 2\ \ 2\ \ 2\ \ 2\ ); \quad \mathcal{G}_0(1, 3) = \begin{pmatrix} 1 & 1 & 1 & 1 \\ 0 & 2 & 0 & 2 \\ 0 & 0 & 2 & 2 \end{pmatrix};$$

$$\mathcal{G}_0(2, 3) = \begin{pmatrix} 1 & 1 & 1 & 1 \\ 0 & 1 & 0 & 1 \\ 0 & 0 & 1 & 1 \\ 0 & 0 & 0 & 2 \end{pmatrix}; \quad \mathcal{G}_0(3, 3) = \begin{pmatrix} 1 & 1 & 1 & 1 \\ 0 & 1 & 0 & 1 \\ 0 & 0 & 1 & 1 \\ 0 & 0 & 0 & 1 \end{pmatrix}.$$

On the other hand, the $\mathcal{RM}_1$ family has to be generated using the BQ-Plotkin construction. The generator matrices of $\mathcal{RM}_1(r, 3)$, $0 \leq r \leq 3$, are the following:

$$\mathcal{G}_1(0, 3) = (\ 2\ \ 2\ \ 2\ \ 2\ ); \quad \mathcal{G}_1(1, 3) = \begin{pmatrix} 1 & 1 & 1 & 1 \\ 0 & 1 & 2 & 3 \end{pmatrix};$$

$$\mathcal{G}_1(2, 3) = \begin{pmatrix} 1 & 1 & 1 & 1 \\ 0 & 1 & 2 & 3 \\ 0 & 0 & 1 & 1 \\ 0 & 0 & 0 & 2 \end{pmatrix}; \quad \mathcal{G}_1(3, 3) = \begin{pmatrix} 1 & 1 & 1 & 1 \\ 0 & 1 & 2 & 3 \\ 0 & 0 & 1 & 1 \\ 0 & 0 & 0 & 1 \end{pmatrix}.$$

$\triangle$

Example 6.30. For $m = 4$ and $m = 5$, there are two and three families of quaternary linear Reed-Muller codes, respectively. Tables 6.1 and 6.2 show the parameters $(\alpha, \beta; \gamma, \delta)$ and minimum distances d for all these $\mathcal{RM}_s(r, m)$ codes with $m = 4$ and $m = 5$, respectively. $\triangle$

(r, m)	$(0, 4)$	$(1, 4)$	$(2, 4)$	$(3, 4)$	$(4, 4)$
$s = 0$	$(0, 8; 1, 0)$	$(0, 8; 3, 1)$	$(0, 8; 3, 4)$	$(0, 8; 1, 7)$	$(0, 8; 0, 8)$
$s = 1$	$(0, 8; 1, 0)$	$(0, 8; 3, 1)$	$(0, 8; 1, 5)$	$(0, 8; 1, 7)$	$(0, 8; 0, 8)$

Table 6.1: Parameters $(\alpha, \beta; \gamma, \delta)$ for all $\mathcal{RM}_s(r, 4)$ codes

(r, m)	$(0, 5)$	$(1, 5)$	$(2, 5)$	$(3, 5)$	$(4, 5)$	$(5, 5)$
$s = 0$	$(0, 16; 1, 0)$	$(0, 16; 4, 1)$	$(0, 16; 6, 5)$	$(0, 16; 4, 11)$	$(0, 16; 1, 15)$	$(0, 16; 0, 16)$
$s = 1$	$(0, 16; 1, 0)$	$(0, 16; 2, 2)$	$(0, 16; 2, 7)$	$(0, 16; 2, 12)$	$(0, 16; 1, 15)$	$(0, 16; 0, 16)$
$s = 2$	$(0, 16; 1, 0)$	$(0, 16; 0, 3)$	$(0, 16; 2, 7)$	$(0, 16; 0, 13)$	$(0, 16; 1, 15)$	$(0, 16; 0, 16)$

Table 6.2: Parameters $(\alpha, \beta; \gamma, \delta)$ for all $\mathcal{RM}_s(r, 5)$ codes

The following theorem summarizes the parameters and properties of these $\mathcal{RM}_s$ families of codes.

Theorem 6.31 ([130]). *For $m \geq 1$, the $\lfloor \frac{m+1}{2} \rfloor$ families of quaternary linear codes $\mathcal{RM}_s(r, m)$, $0 \leq s \leq \lfloor \frac{m-1}{2} \rfloor$, $0 \leq r \leq m$, have the following properties:*

i) *The length of the corresponding $\mathbb{Z}_4$-linear codes is 2^m, the minimum Lee weight is $d = 2^{m-r}$, and the size is $2^{\sum_{i=0}^{r} \binom{m}{i}}$.*

ii) *$\mathcal{RM}_s(0, m)$ is the repetition code, $\mathcal{RM}_s(m, m)$ is the whole space $\mathbb{Z}_4^{2^{m-1}}$, and $\mathcal{RM}_s(m-1, m)$ is the even code.*

iii) *$\mathcal{RM}_s(r-1, m)$ is a subcode of $\mathcal{RM}_s(r, m)$ for $r > 0$.*

iv) *$\mathcal{RM}_s(1, m)$ is a quaternary linear Hadamard code and $\mathcal{RM}_s(m-2, m)$ is a quaternary linear extended perfect code.*

v) *$\mathcal{RM}_s(r, m)$ is monomially equivalent to the dual of $\mathcal{RM}_s(m-1-r, m)$ for $-1 \leq r \leq m$.*

The monomial matrix giving the equivalence in item *(v)* of the above theorem is K_m, where $K_1 = \begin{pmatrix} 1 \end{pmatrix}$, $K_2 = \begin{pmatrix} 1 & 0 \\ 0 & 3 \end{pmatrix}$ and, in general, K_m denotes the Kronecker product of $m-1$ factors, $K_2 \otimes \cdots \otimes K_2$.

Note that each one of the $\lfloor \frac{m+1}{2} \rfloor$ families of quaternary linear Reed-Muller codes contains one of the $\lfloor \frac{m+1}{2} \rfloor$ non-equivalent extended perfect $\mathbb{Z}_4$-linear codes of length 2^m, given by Theorem 6.3 for $m \geq 4$, which is the Gray map image of the code $\mathcal{RM}_s(m-2, m)$.

The quaternary linear Reed-Muller codes, $\mathcal{RM}_s(r, m)$, $m \geq 1$, $0 \leq s \leq \lfloor \frac{m-1}{2} \rfloor$, and $0 \leq r \leq m$, were completely classified in [119], computing the dimension of the kernel of the corresponding $\mathbb{Z}_4$-linear codes and generalizing the known results about the dimension of the kernel for Hadamard $\mathbb{Z}_4$-linear and extended perfect $\mathbb{Z}_4$-linear codes given in Propositions 6.6 and 6.19.

Example 6.32. By using MAGMA [46], we construct the family of quaternary linear Reed-Muller codes $\mathcal{RM}_1$, with $\mathbb{Z}_4$-linear codes of length $2^4 = 16$. Then, we check some parameters and properties given in Theorem 6.31.

```
> RM1 := ReedMullerCodesRMZ4(1, 4);
> #RM1;
5
>
> (RM1[1] subset RM1[2]) and (RM1[2] subset RM1[3]) and
  (RM1[3] subset RM1[4]) and (RM1[4] subset RM1[5]);
true
>
> // RM1[i] is monomially equivalent to the dual of RM1[5-i]
> K4 := Matrix(Integers(4), [[1,0,0,0,0,0,0,0],
                              [0,3,0,0,0,0,0,0],
                              [0,0,3,0,0,0,0,0],
                              [0,0,0,1,0,0,0,0],
                              [0,0,0,0,3,0,0,0],
                              [0,0,0,0,0,1,0,0],
                              [0,0,0,0,0,0,1,0],
                              [0,0,0,0,0,0,0,3]]);
> LinearCode(GeneratorMatrix(RM1[1])*K4) eq Dual(RM1[4]);
true
> LinearCode(GeneratorMatrix(RM1[2])*K4) eq Dual(RM1[3]);
true
>
> // The minimum Lee distance is 2^(4-r)
> MinimumLeeDistance(RM1[1]) eq 2^4;
true
> MinimumLeeDistance(RM1[2]) eq 2^3;
true
> MinimumLeeDistance(RM1[3]) eq 2^2;
```

```
true
> MinimumLeeDistance(RM1[4]) eq 2^1;
true
> MinimumLeeDistance(RM1[5]) eq 2^0;
>
> // RM1[1] is the repetition code
> #RM1[1] eq 2;
true
> // RM1[2] is a quaternary linear Hadamard code
> #RM1[2] eq 4^2*2;
true
> // RM1[3] is a quaternary linear extended perfect code
> #RM1[3] eq 4^5*2;
true
> // RM1[4] is the even code
> #RM1[4] eq 4^7*2;
true
> // RM1[5] is the universe code
> #RM1[5] eq 4^8;
true                                                            △
```

Regarding the families of $\mathbb{Z}_2\mathbb{Z}_4$-additive Reed-Muller codes with $\alpha \neq 0$, $\mathcal{ARM}_s$, there are $\lfloor \frac{m+2}{2} \rfloor$ such families for each integer $m \geq 1$. We identify each one by a subindex $s \in \{0, \ldots, \lfloor \frac{m}{2} \rfloor\}$. Each one of these families contains exactly $m+1$ codes, denoted by $\mathcal{ARM}_s(r, m)$, $0 \leq r \leq m$. Moreover, each family contains one of the $\lfloor \frac{m+2}{2} \rfloor$ non-equivalent extended perfect $\mathbb{Z}_2\mathbb{Z}_4$-linear codes, given by Theorem 6.10 for $m \geq 4$, which is the Gray map image of the code $\mathcal{ARM}_s(m-2, m)$.

To be coherent with all notations, the code $\mathcal{ARM}_s(-1, m)$ is defined as the zero code, $\mathcal{ARM}_s(0, m)$ is defined as the repetition code $\{(\mathbf{0} \mid \mathbf{0}), (\mathbf{1} \mid \mathbf{2})\}$, $\mathcal{ARM}_s(m-1, m)$ is defined as the even code, whereas the codes $\mathcal{ARM}_s(r, m)$, with $r \geq m$, are defined as the whole space $\mathbb{Z}_2^\alpha \times \mathbb{Z}_4^\beta$, respectively.

We start by considering the case $m = 1$, that is, when the corresponding $\mathbb{Z}_2\mathbb{Z}_4$-linear codes have length $n = 2$. In this case, there is only one family with $s = 0$. The $\mathbb{Z}_2\mathbb{Z}_4$-additive Reed-Muller code $\mathcal{ARM}_0(0, 1)$ is the repetition code of type $(2, 0; 1, 0)$, which is equal to the binary code $\{(0, 0), (1, 1)\}$. The code $\mathcal{ARM}_0(1, 1)$ is the whole space $\mathbb{Z}_2^2$, thus a $\mathbb{Z}_2\mathbb{Z}_4$-additive code of type $(2, 0; 2, 0)$. Both codes $\mathcal{ARM}_0(0, 1)$ and $\mathcal{ARM}_0(1, 1)$ are binary linear codes with the same parameters and properties as the $RM(r, 1)$ codes. Generator matrices for $\mathcal{ARM}_0(0, 1)$ and $\mathcal{ARM}_0(1, 1)$ are, respectively,

$$\mathcal{G}_0(0, 1) = \left(1 \quad 1 \mid \right) \text{ and } \mathcal{G}_0(1, 1) = \begin{pmatrix} 1 & 1 \\ 0 & 1 \end{pmatrix} \Bigg| \, .$$

For $m = 2$, there are two families of $\mathbb{Z}_2\mathbb{Z}_4$-additive Reed-Muller codes, with $s = 0$ and $s = 1$. The corresponding $\mathbb{Z}_2\mathbb{Z}_4$-linear codes have length 2^2. The family $\mathcal{ARM}_0(r, 2)$ consists of binary codes obtained from applying the Plotkin construction defined in Definition 6.22 to the family $\mathcal{ARM}_0(r, 1)$. The generator matrices for $\mathcal{ARM}_0(0, 2)$, $\mathcal{ARM}_0(1, 2)$ and $\mathcal{ARM}_0(2, 2)$ are the following: $\mathcal{G}_0(0, 2) = \begin{pmatrix} 1 & 1 & 1 & 1 \end{pmatrix}$,

$$\mathcal{G}_0(1, 2) = \begin{pmatrix} 1 & 1 & 1 & 1 \\ 0 & 1 & 0 & 1 \\ 0 & 0 & 1 & 1 \end{pmatrix}, \text{ and } \mathcal{G}_0(2, 2) = \begin{pmatrix} 1 & 1 & 1 & 1 \\ 0 & 1 & 0 & 1 \\ 0 & 0 & 1 & 1 \\ 0 & 0 & 0 & 1 \end{pmatrix}.$$

For $s = 1$, we define $\mathcal{ARM}_1(0, 2)$, $\mathcal{ARM}_1(1, 2)$, and $\mathcal{ARM}_1(2, 2)$ as the codes generated by the following matrices: $\mathcal{G}_1(0, 2) = \begin{pmatrix} 1 & 1 & | & 2 \end{pmatrix}$,

$$\mathcal{G}_1(1, 2) = \begin{pmatrix} 1 & 1 & | & 2 \\ 0 & 1 & | & 1 \end{pmatrix}, \text{ and } \mathcal{G}_1(2, 2) = \begin{pmatrix} 1 & 1 & | & 2 \\ 0 & 1 & | & 0 \\ 0 & 1 & | & 1 \end{pmatrix}.$$

In general, let $\mathcal{ARM}_s(r, m - 1)$, $\mathcal{ARM}_s(r - 1, m - 1)$, and $\mathcal{ARM}_s(r - 2, m - 1)$, with $0 \le s \le \lfloor \frac{m-1}{2} \rfloor$, be three consecutive $\mathbb{Z}_2\mathbb{Z}_4$-additive Reed-Muller codes of types $(\alpha, \beta; \gamma', \delta')$, $(\alpha, \beta; \gamma'', \delta'')$, and $(\alpha, \beta; \gamma''', \delta''')$ with $\alpha = 2^{m-s}$, $\beta = 2^{m-1} - 2^{m-s-1}$; minimum distances 2^{m-r-1}, 2^{m-r}, and 2^{m-r+1}; and generator matrices $\mathcal{G}_s(r, m - 1)$, $\mathcal{G}_s(r - 1, m - 1)$, and $\mathcal{G}_s(r - 2, m - 1)$, respectively. By using Propositions 6.23 and 6.27, we have the following results:

Theorem 6.33 ([129]). *For any $m \ge 2$, $0 < r < m$ and $0 \le s \le \lfloor \frac{m-1}{2} \rfloor$, the code $\mathcal{ARM}_s(r, m)$ obtained by using the Plotkin construction given by Definition 6.22 from codes $\mathcal{ARM}_s(r, m - 1)$ and $\mathcal{ARM}_s(r - 1, m - 1)$ is a $\mathbb{Z}_2\mathbb{Z}_4$-additive code of type $(2\alpha, 2\beta; \gamma, \delta)$, where $\gamma = \gamma' + \gamma''$ and $\delta = \delta' + \delta''$; has 2^k codewords, where $k = \sum_{i=0}^{r} \binom{m}{i}$; minimum distance 2^{m-r}; and $\mathcal{ARM}_s(r - 1, m) \subset \mathcal{ARM}_s(r, m)$.*

Theorem 6.34 ([129]). *For any $m \ge 3$ odd, $0 < r < m$, and $s = \lfloor (m-1)/2 \rfloor$, the code $\mathcal{ARM}_{s+1}(r, m + 1)$ obtained by using the BA-Plotkin construction given by Definition 6.25, where generator matrices $\mathcal{G}_X$, $\mathcal{G}_Y$, $\mathcal{G}_Z$ stand for $\mathcal{G}_s(r, m - 1)$, $\mathcal{G}_s(r - 1, m - 1)$ and $\mathcal{G}_s(r - 2, m - 1)$, respectively, is a $\mathbb{Z}_2\mathbb{Z}_4$-additive code of type $(2\alpha, \alpha + 4\beta; \gamma, \delta)$, where $\gamma = \gamma' + \gamma'''$ and $\delta = \delta' + \gamma'' + 2\delta'' + \delta'''$; has 2^k codewords, where $k = \sum_{i=0}^{r} \binom{m + 1}{i}$; minimum distance 2^{m-r+1}; and $\mathcal{ARM}_{s+1}(r - 1, m + 1) \subset \mathcal{ARM}_{s+1}(r, m + 1)$.*

Using Theorems 6.33 and 6.34, we can construct all $\mathcal{ARM}_s(r,m)$ codes for $m \geq 3$. Note that when the Plotkin construction given in Definition 6.22 is applied to $\mathbb{Z}_2\mathbb{Z}_4$-additive codes with $\alpha \neq 0$, it is necessary to move all binary coordinates at the first positions.

Once applied the Gray map, all $\mathcal{ARM}_s(r,m)$ codes give rise to binary codes with the same parameters and properties as the RM codes, except the duality property. Moreover, when $m = 2$, $m = 3$, or $s = 0$, they also have the same codewords. The duality property is not assured for $m \geq 5$ and $2 \leq r \leq m-3$. Note that $\mathcal{ARM}_s(-1,m)$, $\mathcal{ARM}_s(0,m)$, and $\mathcal{ARM}_s(1,m)$ are the additive dual codes of $\mathcal{ARM}_s(m,m)$, $\mathcal{ARM}_s(m-1,m)$, and $\mathcal{ARM}_s(m-2,m)$, respectively, up to monomially equivalence.

Other families of $\mathcal{ARM}_s$ codes could be obtained by using the same construction, but placing the matrix $\mathcal{G}_{\mathcal{Z}[q]}$ in the 4th, 5th, or 6th column, instead of in the 7th column as it appears in the matrix $\mathcal{G}_{BA}$ given in Definition 6.25. For example, for $m = 6$ and $s = 2$, it is possible to obtain a family of codes with the same properties as $\mathcal{ARM}_2(r,6)$, having codes which are not equivalent to $\mathcal{ARM}_2(r,6)$ for $r \in \{2,3\}$.

Example 6.35. For $m = 3$, using Theorem 6.33, we obtain the codes $\mathcal{ARM}_s(r,m)$ having the parameters $(\alpha,\beta;\gamma,\delta)$ shown in Table 6.3.

(r,m)	$(0,3)$	$(1,3)$	$(2,3)$	$(3,3)$
$s=0$	$(8,0;1,0)$	$(8,0;4,0)$	$(8,0;7,0)$	$(8,0;8,0)$
$s=1$	$(4,2;1,0)$	$(4,2;2,1)$	$(4,2;3,2)$	$(4,2;4,2)$

Table 6.3: Parameters $(\alpha,\beta;\gamma,\delta)$ for all $\mathcal{ARM}_s(r,3)$ codes

For $m = 4$, by Theorem 6.10, we know that there exists three non-equivalent extended perfect $\mathbb{Z}_2\mathbb{Z}_4$-linear codes with $\alpha \neq 0$. Our goal is to construct three families of $\mathbb{Z}_2\mathbb{Z}_4$-additive Reed-Muller codes $\mathcal{ARM}_s(r,4)$, such that each family contains, for $r = 2$, one of those three $\mathbb{Z}_2\mathbb{Z}_4$-additive extended perfect codes. Table 6.4 shows the parameters $(\alpha,\beta;\gamma,\delta)$ of all these codes.

(r,m)	$(0,4)$	$(1,4)$	$(2,4)$	$(3,4)$	$(4,4)$
$s=0$	$(16,0;1,0)$	$(16,0;5,0)$	$(16,0;11,0)$	$(16,0;15,0)$	$(16,0;16,0)$
$s=1$	$(8,4;1,0)$	$(8,4;3,1)$	$(8,4;5,3)$	$(8,4;7,4)$	$(8,4;8,4)$
$s=2$	$(4,6;1,0)$	$(4,6;1,2)$	$(4,6;3,4)$	$(4,6;3,6)$	$(4,6;4,6)$

Table 6.4: Parameters $(\alpha,\beta;\gamma,\delta)$ for all $\mathcal{ARM}_s(r,4)$ codes

The codes $\mathcal{ARM}_s(r,4)$ with $s = 0$ and $s = 1$ of Table 6.4 can be obtained by applying the Plotkin construction given by Definition 6.22 to the codes $\mathcal{ARM}_s(r,3)$ with $s = 0$ and $s = 1$ of Table 6.3. However, the codes $\mathcal{ARM}_2(r,4)$ in Table 6.4 can not be obtained by using this Plotkin construction. It is in this point that we need to use the BA-Plotkin construction given by Definition 6.25. $\triangle$

Example 6.36. By using Magma [46], and attaching the package for $\mathbb{Z}_2\mathbb{Z}_4$-additive codes [29], we construct the family of $\mathbb{Z}_2\mathbb{Z}_4$-additive Reed-Muller codes $\mathcal{ARM}_2$, with $\mathbb{Z}_2\mathbb{Z}_4$-linear codes of length $2^5 = 32$. Then, we check some parameters and properties given above.

```
> m := 5; s := 2;
> ARM2 := Z2Z4ReedMullerCodes(s, m);
> ARM2b := [ Z2Z4ReedMullerCode(s, r, m) : r in [0..m] ];
> ARM2 eq ARM2b;
true
>
> (ARM2[1] subset ARM2[2]) and (ARM2[2] subset ARM2[3]) and
  (ARM2[3] subset ARM2[4]) and (ARM2[4] subset ARM2[5]) and
  (ARM2[5] subset ARM2[6]);
true
>
> // The minimum Lee distance is 2^(m-r)
> for r in [0..m] do
for>   Z2Z4MinimumLeeDistance(ARM2[r+1]) eq 2^(m-r);
for> end for;
true
true
true
true
true
true
>
> // ARM2[1] is the repetition code
> #ARM2[1] eq 2;
true
> // ARM2[2] is a Z2Z4-additive Hadamard code
> #ARM2[2] eq 2^(m+1);
true
> // ARM2[4] is a Z2Z4-additive extended perfect code
> #ARM2[4] eq 2^(2^m-m-1);
true
> // ARM2[5] is the even code
> #ARM2[5] eq 2^(2^m-1);
true
> // ARM2[6] is the universe code
> #ARM2[6] eq 2^(2^m);
```

true   △

6.4 MDS $\mathbb{Z}_2\mathbb{Z}_4$-Additive Codes

In this section, known upper bounds on the minimum distance of codes over rings are applied to the case of $\mathbb{Z}_2\mathbb{Z}_4$-additive codes. Two kinds of maximum distance separable codes are studied. All possible parameters of these codes are determined and, in certain cases, the codes are characterized. These results are also valid when $\alpha = 0$, namely for quaternary linear codes. The contents in this section can be found in [28].

The usual Singleton bound [149] for a code C of length n over an alphabet of size q is given by

$$d(C) \leq n - \log_q |C| + 1.$$

This is a combinatorial bound and does not rely on the algebraic structure of the code C. It is well known [112] that for the binary case, $q = 2$, the only codes achieving this bound are the repetition code (with $d_H(C) = n$), the codes with minimum distance 2 and size 2^{n-1}, and the universe code containing all 2^n vectors. We remark that sometimes codes with just one codeword are also considered in this class, but it depends on the definition of minimum distance for such codes.

Let $\mathcal{C}$ be a $\mathbb{Z}_2\mathbb{Z}_4$-additive code of type $(\alpha, \beta; \gamma, \delta; \kappa)$ and let $C = \Phi(\mathcal{C})$. Since $d(\mathcal{C}) = d_H(C)$, we immediately obtain that

$$d(\mathcal{C}) \leq \alpha + 2\beta - \gamma - 2\delta + 1. \tag{6.7}$$

This version of the Singleton bound was previously stated for quaternary linear codes ($\alpha = 0$) in [74].

From [74], we also know that if $\mathcal{C}$ is a code of length n over a ring R with minimum distance $d(\mathcal{C})$, then

$$\left\lfloor \frac{d(\mathcal{C}) - 1}{2} \right\rfloor \leqslant n - r(\mathcal{C}), \tag{6.8}$$

where $r(\mathcal{C})$ is the minimal cardinality of a generating set for $\mathcal{C}$. Recall that the rank of a binary code C is the dimension of the linear span of C, as also defined in Section 5.2. For binary codes, we have that $r(C) = \text{rank}(C)$. However, if $\mathcal{C}$ is a $\mathbb{Z}_2\mathbb{Z}_4$-additive code of type $(\alpha, \beta; \gamma, \delta; \kappa)$, then $r(\mathcal{C}) = \gamma + \delta$, and in general this value is different from $\text{rank}(\Phi(\mathcal{C}))$.

Theorem 6.37 ([28]). *Let $\mathcal{C}$ be a $\mathbb{Z}_2\mathbb{Z}_4$-additive code of type $(\alpha, \beta; \gamma, \delta; \kappa)$. Then,*

$$\frac{d(\mathcal{C}) - 1}{2} \leqslant \frac{\alpha}{2} + \beta - \frac{\gamma}{2} - \delta; \tag{6.9}$$

$$\left\lfloor \frac{d(\mathcal{C}) - 1}{2} \right\rfloor \leqslant \alpha + \beta - \gamma - \delta. \tag{6.10}$$

Let $\mathcal{C}$ be a $\mathbb{Z}_2\mathbb{Z}_4$-additive code. Recall that if $\mathcal{C} = \mathcal{C}_X \times \mathcal{C}_Y$, then $\mathcal{C}$ is called separable.

Theorem 6.38 ([28]). *If $\mathcal{C}$ is a separable $\mathbb{Z}_2\mathbb{Z}_4$-additive code of type $(\alpha, \beta; \gamma, \delta; \kappa)$, then $d(\mathcal{C}) = \min\{d(\mathcal{C}_X), d(\mathcal{C}_Y)\}$ and $d(\mathcal{C}) \leq \min\{\alpha - \kappa + 1, \overline{d}\}$, where $\overline{d}$ is the maximum value satisfying both bounds given in (6.9) and (6.10).*

We say that a $\mathbb{Z}_2\mathbb{Z}_4$-additive code $\mathcal{C}$ is *maximum distance separable* (MDS) if $d(\mathcal{C})$ meets the bound given in (6.9) or (6.10). In the first case, we say that $\mathcal{C}$ is MDS with respect to the Singleton bound, briefly MDSS. In the second case, $\mathcal{C}$ is MDS with respect to the rank bound, briefly MDSR.

Now, we give the characterization of all MDSS $\mathbb{Z}_2\mathbb{Z}_4$-additive codes. By the even code, we mean the set of all even weight vectors. By the repetition code, we mean the code such that its Gray map image is the binary repetition code with the all-zero and the all-one codewords. These codes are defined in Section 2.4.

Theorem 6.39 ([28]). *Let $\mathcal{C}$ be an MDSS $\mathbb{Z}_2\mathbb{Z}_4$-additive code of type $(\alpha, \beta; \gamma, \delta; \kappa)$ such that $1 < |\mathcal{C}| < 2^{\alpha+2\beta}$. Then, $\mathcal{C}$ is either*

   i) the repetition code of type $(\alpha, \beta; 1, 0; \kappa)$ and minimum distance $d(\mathcal{C}) = \alpha + 2\beta$, where $\kappa = 1$ if $\alpha > 0$ and $\kappa = 0$ otherwise; or

   ii) the even code with minimum distance $d(\mathcal{C}) = 2$ and type $(\alpha, \beta; \alpha - 1, \beta; \alpha - 1)$ if $\alpha > 0$, or type $(0, \beta; 1, \beta - 1; 0)$ otherwise.

Since the codes described in Theorem 6.39-(i) and Theorem 6.39-(ii) are additive dual codes of one another, it is still true that the dual of an MDSS code is again MDSS, which is a well-known property for linear codes over finite fields [112].

We can also give a strong condition for a $\mathbb{Z}_2\mathbb{Z}_4$-additive code to be MDSR.

Theorem 6.40 ([28]). *Let $\mathcal{C}$ be an MDSR $\mathbb{Z}_2\mathbb{Z}_4$-additive code of type $(\alpha, \beta; \gamma, \delta; \kappa)$ such that $1 < |\mathcal{C}| < 2^{\alpha+2\beta}$. Then, either*

   i) $\mathcal{C}$ is the repetition code as in Theorem 6.39-(i) with $\alpha \leq 1$; or *ii)* $\mathcal{C}$ *is of type* $(\alpha, \beta; \gamma, \alpha + \beta - \gamma - 1; \alpha)$, *where* $\alpha \leq 1$ *and* $d(\mathcal{C}) = 4 - \alpha \in \{3, 4\}$; *or*

iii) $\mathcal{C}$ *is of type* $(\alpha, \beta; \gamma, \alpha + \beta - \gamma; \alpha)$, *where* $\alpha \leq 1$ *and* $d(\mathcal{C}) \leq 2 - \alpha \in \{1, 2\}$.

Note that it is not true that the additive dual code of an MDSR code is again MDSR, as it is shown in Examples 6.41 and 6.42. Moreover, these examples show MDS $\mathbb{Z}_2\mathbb{Z}_4$-additive codes satisfying the bound given in (6.9), that is, MDSS $\mathbb{Z}_2\mathbb{Z}_4$-additive codes, which are not MDSR. The code given in Example 6.41 is MDS with $\gamma = 0$, and the code given in Example 6.42 is MDS with $\alpha > 1$.

Example 6.41. Let $\mathcal{C}_2$ be the $\mathbb{Z}_2\mathbb{Z}_4$-additive code of length 2 with generator matrix $\mathcal{G}_2 = (1 \mid 1)$. The code $\mathcal{C}_2$ is of type $(1, 1; 0, 1; 0)$ and $d(\mathcal{C}_2) = 2$. The code $\mathcal{C}_2$ is an MDSS code, since satisfies the bound given in (6.9). In fact, it is the even code with $\alpha = \beta = 1$. Its additive dual code $\mathcal{C}_2^{\perp}$ is the repetition code $\{(0, 0), (1, 2)\}$ of type $(1, 1; 1, 0; 1)$, which is MDSS and MDSR. However, $\mathcal{C}_2$ does not satisfy the bound given in (6.10), so it is not MDSR. $\triangle$

Example 6.42. Let $\mathcal{C}_3$ be the $\mathbb{Z}_2\mathbb{Z}_4$-additive code with generator matrix

$$\mathcal{G}_3 = \left(\begin{array}{cc|c} 1 & 1 & 0 \\ 0 & 1 & 1 \end{array} \right).$$

The code $\mathcal{C}_3$ is of type $(2, 1; 1, 1; 1)$ and $d(\mathcal{C}_3) = 2$. This is again an MDSS code, which is the even code for $\alpha = 2$ and $\beta = 1$. The code $\mathcal{C}_3$ is not an MDSR code, but $\mathcal{C}_3^{\perp} = \{(0, 0, 0), (1, 1, 2)\}$ is again MDSS and MDSR. $\triangle$

In Examples 6.41 and 6.42, we have seen $\mathbb{Z}_2\mathbb{Z}_4$-additive codes which are MDSS but not MDSR. In Example 6.43, a $\mathbb{Z}_2\mathbb{Z}_4$-additive code which is MDSR but not MDSS is given.

Example 6.43. Let $\mathcal{C}_4$ be the $\mathbb{Z}_2\mathbb{Z}_4$-additive code $\mathcal{C}_4$ with generator matrix

$$\mathcal{G}_4 = \left(\begin{array}{c|ccc} 1 & 2 & 0 & 0 \\ 1 & 0 & 2 & 0 \\ 1 & 0 & 0 & 2 \end{array} \right).$$

The code $\mathcal{C}_4$ is of type $(1, 3; 3, 0; 1)$ and $d(\mathcal{C}_4) = 3$. Thus, it is an MDSR code, but not MDSS. $\triangle$

The next example gives a general construction for MDS $\mathbb{Z}_2\mathbb{Z}_4$-additive codes meeting the bound given in (6.10) starting from binary MDS codes.

Example 6.44. Let C be a binary linear $[n, k, d]$ code which is MDS. Recall that χ is the usual inclusion map from $\mathbb{Z}_2$ to $\mathbb{Z}_4$, so $\chi(0) = 0$ and $\chi(1) = 2$. Applying χ to all but one coordinate of the codewords in C, we obtain a $\mathbb{Z}_2\mathbb{Z}_4$-additive code $\mathcal{C}$ of type $(1, n-1; k, 0; 1)$ and $d(\mathcal{C}) = 2d - 1$. Then, since C is MDS, $d = n - k + 1 = \alpha + \beta - \gamma + 1$, so $\lfloor \frac{d(\mathcal{C})-1}{2} \rfloor = d - 1 = \alpha + \beta - (\gamma + \delta)$ and $\mathcal{C}$ meets the bound given in (6.10). Of course, this construction works for the even binary code and the repetition binary code which are the possible binary linear MDS codes with more than one codeword. $\triangle$

For MDS $\mathbb{Z}_2\mathbb{Z}_4$-additive codes, we can state which are the possible values for the rank of their corresponding $\mathbb{Z}_2\mathbb{Z}_4$-linear codes.

Corollary 6.45 ([28]). *If $\mathcal{C}$ is an MDS $\mathbb{Z}_2\mathbb{Z}_4$-additive code, then $C = \Phi(\mathcal{C})$ is a linear code or its rank is equal to $\log_2 |\mathcal{C}| + 1$. In this last case, $\mathcal{C}$ is a MDSR code with minimum distance 3 or 4.*

Finally, the next example shows an MDSR code which has a non-linear Gray map image.

Example 6.46. Let $\mathcal{C}_8$ be the $\mathbb{Z}_2\mathbb{Z}_4$-additive code given by the following generator matrix

$$\mathcal{G}_8 = \begin{pmatrix} 1 & 2 & 0 & 0 & 0 & 0 & 0 & 0 \\ 0 & 2 & 2 & 0 & 0 & 0 & 0 & 0 \\ 0 & 2 & 0 & 2 & 0 & 0 & 0 & 0 \\ 0 & 2 & 0 & 0 & 2 & 0 & 0 & 0 \\ 0 & 2 & 0 & 0 & 0 & 2 & 0 & 0 \\ 0 & 1 & 1 & 1 & 0 & 0 & 1 & 0 \\ 0 & 1 & 0 & 0 & 1 & 1 & 0 & 1 \end{pmatrix}.$$

The code $\mathcal{C}_8$ is of type $(1, 7; 5, 2; 1)$ with $d(\mathcal{C}_8) = 3$. It is an MDSR code since it meets the bound given in (6.10). Moreover, we have that $2(0 \mid 1, 1, 1, 0, 0, 1, 0) * (0 \mid 1, 0, 0, 1, 1, 0, 1) = (0 \mid 2, 0, 0, 0, 0, 0, 0) \notin \mathcal{C}_8$, where $*$ denotes the component-wise product. Therefore, from [79] or Corollary 5.4, the rank of $\Phi(\mathcal{C}_8)$ is $\log_2 |\mathcal{C}_8| + 1 = 10$, so $\mathcal{C}_8$ has a non-linear Gray map image. $\triangle$

As a summary, we enumerate the possible MDS $\mathbb{Z}_2\mathbb{Z}_4$-additive codes of type $(\alpha, \beta; \gamma, \delta; \kappa)$, with $\beta > 0$ and $\gamma + 2\delta < \alpha + 2\beta$:

i) Repetition codes of type $(\alpha, \beta; 1, 0; 1)$, $\alpha > 0$; or $(0, \beta; 1, 0; 0)$ in the quaternary linear case. These are MDSS codes which are also MDSR if and only if $\alpha \leq 1$. Their minimum distance is $\alpha + 2\beta$.

ii) Even codes of type $(\alpha, \beta; \alpha - 1, \beta; \alpha - 1)$, $\alpha > 0$, which are MDSS codes but not MDSR; or $(0, \beta; 1, \beta - 1; 0)$ in the quaternary linear case, which are MDSS and MDSR codes. In any case, these codes have minimum distance 2.

iii) Codes of type $(1, \beta; \gamma, \beta - \gamma; 1)$ with minimum distance 3. These are MDSR codes but not MDSS, except for $\beta = \gamma = 1$, which is a repetition code. Note that, for $\beta > 1$ and $\gamma = 1$, it is not possible to have minimum distance 3; otherwise the binary Gray image would be an MDSS code that does not exist.

iv) Quaternary linear codes of type $(0, \beta; \gamma, \beta - \gamma - 1; 0)$ with minimum distance 4. Again, these are MDSR codes but not MDSS, except for $\gamma = 1$ and $\beta = 2$, which is a repetition code. For $\beta \neq 2$ and $\gamma = 1$, it is not possible to have minimum distance 4; otherwise the binary Gray image would be an MDSS code that does not exist.

v) Codes of type $(\alpha, \beta; \gamma, \alpha + \beta - \gamma; \alpha)$, where $\alpha \leq 1$ and minimum distance at most $2 - \alpha$. These are MDSR codes but not MDSS, except for the case $(0, \beta; 1, \beta - 1; 0)$ which is already included in (ii).

In the first two cases, the binary Gray map images are linear codes. In cases (iii) and (iv), if C is the binary Gray image of such a code, then C is linear or its linear span has size $2|C|$. In case (v), the binary Gray images are linear codes. As a conclusion, we have that all MDS $\mathbb{Z}_2\mathbb{Z}_4$-additive codes are zero or one error-correcting codes with an exception for the trivial repetition codes containing two codewords.

Chapter 7

$\mathbb{Z}_2\mathbb{Z}_4$-Additive Cyclic Codes

Cyclic codes over finite fields are one of the most studied families of codes. They are a subfamily of linear codes and the process of codification is simplified compared to general linear codes. It is also remarkable the rich algebraic structure of these codes. The family of cyclic codes includes some important codes as Hamming, Reed-Solomon and BCH codes. From [92], cyclic codes over $\mathbb{Z}_4$ were studied and, later, over other rings in general. In this chapter, cyclic codes over a product of rings are defined. Specifically, we consider cyclic codes over $\mathbb{Z}_2^\alpha \times \mathbb{Z}_4^\beta$ with odd β. The algebraic structure of these codes as submodules of $\mathbb{Z}_2[x]/\langle x^\alpha - 1 \rangle \times \mathbb{Z}_4[x]/\langle x^\beta - 1 \rangle$ is studied. We describe the generator polynomials of these codes, their duality and also their linear images under the Gray map as well as their values for the invariants, rank and dimension of the kernel.

There are parts of this chapter that have been previously published. Reprinted by permission from IEEE: J. Borges, C. Fernández-Córdoba, and R. Ten-Valls. "$\mathbb{Z}_2\mathbb{Z}_4$-additive cyclic codes, generator polynomials, and dual codes". *IEEE Transactions on Information Theory*, v. 62, 11, pp. 6348–6354. ©(2016); J. Borges, C. Fernández-Córdoba, and R. Ten-Valls. "Binary images of $\mathbb{Z}_2\mathbb{Z}_4$-additive cyclic codes". *IEEE Transactions on Information Theory*, v. 64, 12, pp. 7551–7556 ©(2018); and J. Borges, S. T. Dougherty, C. Fernández-Córdoba, and R. Ten-Valls. "$\mathbb{Z}_2\mathbb{Z}_4$-additive cyclic codes: kernel and rank". *IEEE Transactions on Information Theory*, v. 65, 4, pp. 2119–2127 ©(2019). Reprinted by permission from Springer Nature Customer Service Centre GmbH: J. Borges, and C. Fernández-Córdoba. "There is exactly one $\mathbb{Z}_2\mathbb{Z}_4$-cyclic 1-perfect code". *Designs, Codes and Cryptography*, v. 85, 3, pp. 557–566 ©(2017).

7.1   Parameters and Generator Polynomials

Let $\mathbf{u} = (u \mid u') \in \mathbb{Z}_2^\alpha \times \mathbb{Z}_4^\beta$ and i be an integer. Then, we denote by

$$\mathbf{u}^{(i)} = (u^{(i)} \mid u'^{(i)})$$
$$= (u_{1+i}, u_{2+i}, \ldots, u_{\alpha+i} \mid u'_{1+i}, u'_{2+i}, \ldots, u'_{\beta+i})$$

the cyclic ith shift of $\mathbf{u}$, where the subscripts are taken modulo α and β, respectively, and then one is added.

Definition 7.1. *A $\mathbb{Z}_2\mathbb{Z}_4$-additive code $\mathcal{C}$ is cyclic if for any codeword $\mathbf{u} \in \mathcal{C}$ we have $\mathbf{u}^{(1)} \in \mathcal{C}$.*

Example 7.2. Consider the basic families of $\mathbb{Z}_2\mathbb{Z}_4$-additive codes described in Section 2.4. It is easy to check that these codes are in fact cyclic codes. The families are the following:

  i) the zero code $\mathcal{C} = \{\mathbf{0} \mid \mathbf{0}\}$,

 ii) the universe code $\mathcal{C} = \mathbb{Z}_2^\alpha \times \mathbb{Z}_4^\beta$,

iii) the repetition code $\mathcal{C} = \{(\mathbf{0} \mid \mathbf{0}), (\mathbf{1} \mid \mathbf{2})\}$, and

 iv) the even code $\mathcal{C} = \{\mathbf{u} \in \mathbb{Z}_2^\alpha \times \mathbb{Z}_4^\beta : \operatorname{wt}(\mathbf{u}) \equiv 0 \bmod (2)\}$.

$\triangle$

Example 7.3. Consider the $\mathbb{Z}_2\mathbb{Z}_4$-additive code of type $(3, 3; 2, 1; 2)$

$$\begin{aligned}
\mathcal{C} = \{&(0,0,0 \mid 0,0,0), \\
&(0,0,0 \mid 1,1,1), (0,0,0 \mid 2,2,2), (0,0,0 \mid 3,3,3), \\
&(1,0,1 \mid 2,0,0), (1,1,0 \mid 0,2,0), (0,1,1 \mid 0,0,2), \\
&(0,1,1 \mid 2,2,0), (1,0,1 \mid 0,2,2), (1,1,0 \mid 2,0,2), \\
&(1,0,1 \mid 3,1,1), (1,1,0 \mid 1,3,1), (0,1,1 \mid 1,1,3), \\
&(1,0,1 \mid 1,3,3), (1,1,0 \mid 3,1,3), (0,1,1 \mid 3,3,1)\}.
\end{aligned}$$

It is easy to check that for all $\mathbf{u} \in \mathcal{C}$, $\mathbf{u}^{(1)} \in \mathcal{C}$ and hence $\mathcal{C}$ is cyclic. A generator matrix of $\mathcal{C}$ is

$$\mathcal{G} = \left(\begin{array}{ccc|ccc} 1 & 0 & 1 & 2 & 0 & 0 \\ 0 & 1 & 1 & 2 & 2 & 0 \\ 0 & 0 & 0 & 1 & 1 & 1 \end{array} \right).$$

$\triangle$

Proposition 7.4. *Let $\mathcal{C}$ be a $\mathbb{Z}_2\mathbb{Z}_4$-additive code with generator matrix $\mathcal{G}$. Then, $\mathcal{C}$ is cyclic if and only if $\mathbf{u}^{(1)} \in \mathcal{C}$ for all row vector $\mathbf{u}$ of $\mathcal{G}$.*

Proof. If $\mathcal{C}$ is cyclic then, by definition, $\mathbf{u}^{(1)} \in \mathcal{C}$ for all row vector $\mathbf{u}$ of $\mathcal{G}$. Let $\mathbf{u}_1, \ldots, \mathbf{u}_{\gamma+\delta}$ be the row vectors of $\mathcal{G}$ and assume that $\mathbf{u}_i^{(1)} \in \mathcal{C}$ for all $1 \le i \le \gamma + \delta$. For $\mathbf{v} \in \mathcal{C}$, we have that $\mathbf{v} = \sum_{i=1}^{\gamma+\delta} \lambda_i \mathbf{u}_i$ where $\lambda_i \in \mathbb{Z}_4$, $1 \le i \le \gamma + \delta$. Therefore, $\mathbf{v}^{(1)} = \sum_{i=1}^{\gamma+\delta} \lambda_i \mathbf{u}_i^{(1)} \in \mathcal{C}$ and hence $\mathcal{C}$ is cyclic. $\qquad\square$

Example 7.5. Consider the code $\mathcal{C}$ generated by

$$
\mathcal{G} = \left(\begin{array}{ccc|ccccc}
1 & 0 & 1 & 0 & 0 & 0 & 0 & 0 \\
0 & 1 & 1 & 0 & 0 & 0 & 0 & 0 \\
0 & 0 & 0 & 1 & 0 & 0 & 0 & 3 \\
0 & 0 & 0 & 0 & 1 & 0 & 0 & 3 \\
0 & 0 & 0 & 0 & 0 & 1 & 0 & 3 \\
0 & 0 & 0 & 0 & 0 & 0 & 1 & 3
\end{array}\right).
$$

Note that for any row vector $\mathbf{v}$ of $\mathcal{G}$ we have that $\mathbf{v}^{(1)}$ is also a codeword in $\mathcal{C}$. Therefore, the $\mathbb{Z}_2\mathbb{Z}_4$-additive code $\mathcal{C}$ is cyclic. $\qquad\triangle$

In order to study $\mathbb{Z}_2\mathbb{Z}_4$-additive cyclic codes, we consider the polynomial rings $\mathbb{Z}_2[x]$ and $\mathbb{Z}_4[x]$. We can relate polynomials over $\mathbb{Z}_2[x]$ and $\mathbb{Z}_4[x]$ by using the modulo 2 map and the Hensel lift. First, for $\lambda(x) \in \mathbb{Z}_4[x]$, denote by $\tilde{\lambda}(x)$ the polynomial $\lambda(x)$ after taking the coefficients modulo 2. Second, we consider Hensel's lemma and the definition of the Hensel lift of a polynomial in $\mathbb{Z}_2[x]$.

Lemma 7.6 (Hensel's Lemma [157]). *Let $f(x)$ be a monic polynomial in $\mathbb{Z}_4[x]$ and assume that*

$$
\tilde{f}(x) = \tilde{f}_1(x)\tilde{f}_2(x)\cdots\tilde{f}_r(x),
$$

where $\tilde{f}_1(x), \tilde{f}_2(x), \ldots, \tilde{f}_r(x)$ are pairwise coprime polynomials in $\mathbb{Z}_2[x]$. Then, there exist monic polynomials $g_1(x), g_2(x), \ldots, g_r(x) \in \mathbb{Z}_4[x]$ such that they are pairwise coprime in $\mathbb{Z}_4[x]$, $f(x) = g_1(x)g_2(x)\cdots g_r(x)$ and, for all $i = 1, \ldots, r$, $\tilde{g}_i(x) = \tilde{f}_i(x)$ and $\deg(g_i(x)) = \deg(\tilde{f}_i(x))$.

Over $\mathbb{Z}_2$, we have that $x^n - 1$ can be written as a unique factorization of irreducible polynomials over $\mathbb{Z}_2$. Moreover, if n is odd, then these polynomials are also pairwise coprime. Therefore, for n odd, we obtain the following statement as a consequence of Hensel's lemma.

Proposition 7.7 ([157]). *Let n be an odd positive integer. Then, over $\mathbb{Z}_4$,*

$$x^n - 1 = g_1(x)g_2(x)\cdots g_r(x),$$

where $g_1(x), g_2(x), \ldots, g_r(x)$ are pairwise coprime in $\mathbb{Z}_4[x]$ and they are uniquely determined up to rearrangement.

Proposition 7.8 ([157]). *Let n be an odd positive integer and $f_2(x) \in \mathbb{Z}_2[x]$ dividing $x^n - 1$. Then, there exists a unique polynomial $f(x) \in \mathbb{Z}_4[x]$ dividing $x^n - 1$ and satisfying $\tilde{f}(x) = f_2(x)$.*

With the notation used in the last proposition, the polynomial $f(x)$ is called the Hensel lift of $f_2(x)$. By [157], to get $f(x)$ from $f_2(x)$ we can use Algorithm 7.1.

Algorithm 7.1 Hensel lift

Require: A polynomial $f_2(x) \in \mathbb{Z}_2[x]$ without multiple roots in any extension field of $\mathbb{Z}_2$.

1: Write $f_2(x) = e(x) - d(x)$, where $e(x)$ contains only even power terms and $d(x)$ only odd power terms.
2: Calculate $f(x^2) = \pm(e(x)^2 - d(x)^2)$, where we take the $-$ sign only if $\deg(e(x)) < \deg(d(x))$.
3: **return** $f(x)$, the Hensel lift of $f_2(x)$.

Let $R_{\alpha,\beta} = \mathbb{Z}_2[x]/\langle x^\alpha - 1\rangle \times \mathbb{Z}_4[x]/\langle x^\beta - 1\rangle$, and define the operation $\star : \mathbb{Z}_4[x] \times R_{\alpha,\beta} \to R_{\alpha,\beta}$ as

$$\lambda(x) \star (p(x) \mid q(x)) = (\tilde{\lambda}(x)p(x) \mid \lambda(x)q(x)).$$

Let $u = (u_1, \ldots, u_\alpha) \in \mathbb{Z}_2^\alpha$, $u' = (u'_1, \ldots, u'_\beta) \in \mathbb{Z}_4^\beta$ and $\mathbf{u} = (u \mid u') \in \mathbb{Z}_2^\alpha \times \mathbb{Z}_4^\beta$. We define the polynomials $u(x) = u_1 + u_2 x + \cdots + u_\alpha x^{\alpha-1} \in \mathbb{Z}_2[x]/\langle x^\alpha - 1\rangle$, $u'(x) = u'_1 + u'_2 x + \cdots + u'_\beta x^{\beta-1} \in \mathbb{Z}_4[x]/\langle x^\beta - 1\rangle$, and $\mathbf{u}(x) = (u(x) \mid u'(x)) \in \mathbb{Z}_2[x]/\langle x^\alpha - 1\rangle \times \mathbb{Z}_4[x]/\langle x^\beta - 1\rangle$. Note that $xu(x) = u^{(1)}(x)$, $xu'(x) = u'^{(1)}(x)$ and hence $x\mathbf{u}(x) = \mathbf{u}^{(1)}(x)$. For a code $\mathcal{C} \subseteq \mathbb{Z}_2^\alpha \times \mathbb{Z}_4^\beta$, define $\mathcal{C}(x) = \{\mathbf{u}(x) \in R_{\alpha,\beta} : \mathbf{u} \in \mathcal{C}\}$. However, we usually identify $\mathcal{C}$ with $\mathcal{C}(x)$.

Proposition 7.9 ([1]). *A subset $\mathcal{C} \subseteq R_{\alpha,\beta}$ is a $\mathbb{Z}_2\mathbb{Z}_4$-additive cyclic code if and only if $\mathcal{C}$ is a $\mathbb{Z}_4[x]$-submodule of $R_{\alpha,\beta}$.*

From now on, in this chapter, unless it is stated otherwise, $\mathcal{C}$ is considered as a $\mathbb{Z}_4[x]$-submodule of $R_{\alpha,\beta}$, and any polynomial $f(x)$ from $\mathbb{Z}_2[x]$ or $\mathbb{Z}_4[x]$

is denoted simply by f. Moreover, we need to assure that $x^\alpha - 1$ and $x^\beta - 1$ factorize uniquely as a product of pairwise coprime basic irreducible polynomials. In the case of $x^\alpha - 1$ over $\mathbb{Z}_2[x]$, this property is satisfied for any value of α. However, the polynomial $x^\beta - 1$ factorizes uniquely in $\mathbb{Z}_4[x]$, by using Hensel's lemma, if β is odd. Therefore, we only consider $\mathbb{Z}_2\mathbb{Z}_4$-additive cyclic codes of type $(\alpha, \beta; \gamma, \delta; \kappa)$ with odd β and hence, unless otherwise stated, β is odd.

Example 7.10. Let $\mathcal{C}$ be the $\mathbb{Z}_2\mathbb{Z}_4$-additive cyclic code $\mathcal{C}$ defined in Example 7.3. The codewords in $\mathcal{C}$ considered as elements in $R_{3,3}$ are the following:

$$\mathcal{C} = \{(0 \mid 0),$$
$$(0 \mid 1 + x + x^2), (0 \mid 2 + 2x + 2x^2), (0 \mid 3 + 3x + 3x^2),$$
$$(1 + x^2 \mid 2), (1 + x \mid 2x), (x + x^2 \mid 2x^2),$$
$$(x + x^2 \mid 2 + 2x), (1 + x^2 \mid 2x + 2x^2), (1 + x \mid 2 + 2x^2),$$
$$(1 + x^2 \mid 3 + x + x^2), (1 + x \mid 1 + 3x + x^2), (x + x^2 \mid 1 + x + 3x^2),$$
$$(1 + x^2 \mid 1 + 3x + 3x^2), (1 + x \mid 3 + x + 3x^2), (x + x^2 \mid 3 + 3x + x^2)\}.$$

Note that the code $\mathcal{C}$ is generated by the polynomials $(1 + x^2 \mid 2), (x + x^2 \mid 2 + 2x)$, and $(0 \mid 1 + x + x^2)$, which correspond to the rows of the generator matrix $\mathcal{G}$ given in Example 7.3. △

Any ideal in $\mathbb{Z}_2[x]/\langle x^\alpha - 1 \rangle$ is a principal ideal of the form $\langle b \rangle$, where $b \mid (x^\alpha - 1)$ [112]. Moreover, from [49], we have that every ideal in $\mathbb{Z}_4[x]/\langle x^\beta - 1 \rangle$ is a principal ideal of the form $\langle fh + 2f \rangle$, where $fgh = x^\beta - 1$, and f, g and h are pairwise coprime. The following theorem gives generator polynomials of a $\mathbb{Z}_2\mathbb{Z}_4$-additive cyclic code seen as a $\mathbb{Z}_4[x]$-submodule of $R_{\alpha,\beta}$.

Theorem 7.11 ([1]). *Let $\mathcal{C}$ be a $\mathbb{Z}_2\mathbb{Z}_4$-additive cyclic code of type $(\alpha, \beta; \gamma, \delta; \kappa)$. Then, we can define $\mathcal{C}$ as*

$$\mathcal{C} = \langle (b \mid 0), (\ell \mid fh + 2f) \rangle, \tag{7.1}$$

where the polynomials b, ℓ, f and h satisfy the following conditions:

i) $x^\beta - 1 = fhg$, where $f, h, g \in \mathbb{Z}_4[x]$ are pairwise coprime,

ii) $b \in \mathbb{Z}_2[x]$ divides $x^\alpha - 1$,

iii) $\ell \in \mathbb{Z}_2[x]$ fulfils $\deg(\ell) < \deg(b)$,

iv) b divides $\ell(x^\beta - 1)/\tilde{f}$.

Proof. Let $\mathcal{C}$ be a $\mathbb{Z}_4[x]$-submodule of $R_{\alpha,\beta}$. Note that $\mathcal{C}_X$ and $\mathcal{C}_Y$ are submodules of $\mathbb{Z}_2[x]/\langle x^\alpha - 1 \rangle$ and $\mathbb{Z}_4[x]/\langle x^\beta - 1 \rangle$, respectively.

From [49], $\mathcal{C}_Y = \langle fh + 2f \rangle$, where $fgh = x^\beta - 1$, and f, g and h are pairwise coprime. Therefore, f and h satisfy Condition (*i*).

Define $\mathcal{C}_0 = \{(p \mid 0) \in \mathcal{C}\}$. Note that $(\mathcal{C}_0)_X$ is a submodule of $\mathbb{Z}_2[x]/\langle x^\alpha - 1 \rangle$. It is easy to check that $\mathcal{C}_0 \cong (\mathcal{C}_0)_X$ and hence $\mathcal{C}_0$ is finitely generated. Let b be a generator of $(\mathcal{C}_0)_X$, then $b \mid (x^\alpha - 1)$ and $(b \mid 0)$ is a generator of $\mathcal{C}_0$ satisfying Condition (*ii*).

We have that $fh + 2f \in \mathcal{C}_Y$. Let $\ell' \in \mathbb{Z}_2[x]/\langle x^\alpha - 1 \rangle$ such that $(\ell' \mid fh + 2f) \in \mathcal{C}$. We consider $\ell = \ell' \pmod{b}$. Note that, since $(b \mid 0) \in \mathcal{C}, (\ell \mid fh + 2f) \in \mathcal{C}$, and ℓ satisfies Condition (*iii*).

We prove that $\mathcal{C} = \langle (b \mid 0), (\ell \mid fh + 2f) \rangle$. Let $(p \mid q) \in \mathcal{C}$. Since $q \in \mathcal{C}_Y = \langle fh + 2f \rangle$, there exists $\lambda \in \mathbb{Z}_4[x]$ such that $q = \lambda(fh + 2f)$. Now,

$$(p \mid q) - \lambda \star (\ell \mid fh + 2f) = (p - \tilde{\lambda}\ell \mid 0) \in \mathcal{C}_0.$$

Then, there exists $\mu \in \mathbb{Z}_4[x]$ such that $(p - \tilde{\lambda}\ell \mid 0) = \mu \star (b \mid 0)$. Thus,

$$(p \mid q) = \mu \star (b \mid 0) + \lambda \star (\ell \mid fh + 2f).$$

Finally, $\frac{x^\beta - 1}{f} \star (\ell \mid fh + 2f) = (\frac{x^\beta - 1}{f}\ell \mid 0) \in \mathcal{C}$ and, therefore $\frac{x^\beta - 1}{f}\ell$ is a multiple of b, satisfying Condition (*iv*). □

When the generator polynomials of a $\mathbb{Z}_2\mathbb{Z}_4$-additive code satisfy the conditions of Theorem 7.11, we say that they are in *standard form*.

Corollary 7.12. *Let* $\mathcal{C} = \langle (b \mid 0), (\ell \mid fh+2f) \rangle$ *be a* $\mathbb{Z}_2\mathbb{Z}_4$*-additive cyclic code of type* $(\alpha, \beta; \gamma, \delta; \kappa)$ *with generator polynomials in standard form. Then, b divides* $\gcd(b, \ell)(x^\beta - 1)/\tilde{f}$ *and $\tilde{h}$* $\gcd(b, \ell\tilde{g})$*, where* $fgh = x^\beta - 1$.

Proof. First, since b divides $\ell(x^\beta - 1)/\tilde{f}$ by Theorem 7.11, we have that b divides $\gcd(b, \ell)(x^\beta - 1)/\tilde{f}$.

Let $d = \gcd(b, \ell\tilde{g})$. By Bezout's identity, there exists $\lambda_2, \mu_2 \in \mathbb{Z}_2[x]$ such that $\lambda_2 b + \mu_2 \ell\tilde{g} = d$. Let λ and μ be the Hensel lift of λ_2 and μ_2, respectively. We have that $\lambda \star (b \mid 0) + \mu \star (\ell\tilde{g} \mid 2f) = (d \mid 2f\mu) \in \mathcal{C}$. Therefore, $h \star (d \mid 2f\mu) = (\tilde{h}d \mid 0) \in \mathcal{C}$, and b divides $\tilde{h}d$. □

Let $\mathcal{D}$ be a quaternary linear code of length β. Recall that $\mathrm{Tor}(\mathcal{D})$ is the torsion code of $\mathcal{D}$, and $\mathrm{Res}(\mathcal{D})$ is the residue code of $\mathcal{D}$, both defined in Section 2.3. The following lemma is easily proven.

Lemma 7.13. *Let* $\mathcal{D} = \langle fh+2f \rangle$ *be a quaternary cyclic code. Then,* $\mathrm{Tor}(\mathcal{D})$ *is a binary cyclic code generated by $\tilde{f}$.*

Algorithm 7.2 computes the generator polynomials in standard form for a $\mathbb{Z}_2\mathbb{Z}_4$-additive cyclic code. When in the context it is not clear, in order to distinguish between elements in $\mathbb{Z}_2^\alpha \times \mathbb{Z}_4^\beta$ and $R_{\alpha,\beta}$, we define the bijective map $\theta : \mathbb{Z}_2^\alpha \times \mathbb{Z}_4^\beta \longrightarrow R_{\alpha,\beta}$ such that

$$\theta(v_1,\ldots,v_\alpha \mid v_1',\ldots,v_\beta') = (v_1 + v_2 x + \cdots + v_\alpha x^{\alpha-1} \mid v_1' + v_2' x + \cdots + v_\beta' x^{\beta-1}).$$

Algorithm 7.2 Generator polynomials in standard form

Require: A $\mathbb{Z}_2\mathbb{Z}_4$-additive cyclic code, $\mathcal{C}$, of type $(\alpha, \beta; \gamma, \delta; \kappa)$ with odd β and generator matrix $\mathcal{G} = (\mathcal{G}_X \mid \mathcal{G}_Y)$.

1: Compute the generator polynomial f_2 of the binary code $\mathrm{Tor}(\mathcal{C}_Y)$.
2: Compute the generator polynomial $f_2 h_2$ of the binary code $\mathrm{Res}(\mathcal{C}_Y)$.
3: Compute f and h, the Hensel lift of f_2 and h_2, respectively.
4: Compute the generator polynomial b of the code $(\mathcal{C}_0)_X = \{w \in \mathbb{Z}_2^\alpha : (w \mid \mathbf{0}) \in \mathcal{C}\}$. If $(\mathcal{C}_0)_X = \{\mathbf{0}\}$, then $b = x^\alpha - 1$.
5: Find $\mathbf{v} \in \mathbb{Z}_2^\gamma \times \mathbb{Z}_4^\delta$ such that $\theta(\mathbf{v}\mathcal{G}_Y)$ is $fh + 2f$.
6: Compute the polynomial $\ell = \theta(\mathbf{v}\mathcal{G}_X) \pmod{b}$.
7: **return** The generator polynomials b, ℓ, f and h.

Example 7.14. Let $\mathcal{C}$ be the $\mathbb{Z}_2\mathbb{Z}_4$-additive cyclic code given in Example 7.10. A generator matrix of $\mathcal{C}$, given in Example 7.3, is

$$\mathcal{G} = (\mathcal{G}_X \mid \mathcal{G}_Y) = \begin{pmatrix} 1 & 0 & 1 & 2 & 0 & 0 \\ 0 & 1 & 1 & 2 & 2 & 0 \\ 0 & 0 & 0 & 1 & 1 & 1 \end{pmatrix}.$$

Applying Algorithm 7.2, we obtain the following results:

i) $\mathrm{Tor}(\mathcal{C}_Y) = \{0, 1 + x + x^2, 1, x, x^2, 1 + x, x + x^2, 1 + x^2\}$. Therefore, $\mathrm{Tor}(\mathcal{C}_Y) = \langle 1 \rangle$ and $f_2 = 1$.

ii) $\mathrm{Res}(\mathcal{C}_Y) = \{0, 1 + x + x^2\}$ and, hence, $f_2 h_2 = 1 + x + x^2$.

iii) The Hensel lift of f_2 is $f = 1$. Indeed, $f_2 h_2 = 1 + x + x^2$, $h_2 = 1 + x + x^2$, and the Hensel lift of h_2 is $h = 1 + x + x^2$.

iv) $(\mathcal{C}_0)_X = \{\mathbf{0}\}$, and then $b = x^3 - 1$.

v) $fh + 2f = 3 + x + x^2$, and $\theta^{-1}(3 + x + x^2) = (1, 0, 1)\mathcal{G}_Y$. Then, we have $\mathbf{v} = (1, 0, 1)$.

vi) $\theta((1, 0, 1)\mathcal{G}_X) = \theta(1, 0, 1) = \ell$. Therefore, $\ell = 1 + x^2 \pmod{x^3 - 1} = 1 + x^2$.

Therefore, considering $b = x^3 - 1 = 0$ in $\mathbb{Z}_2[x]/\langle x^3 - 1\rangle$, we have that

$$\mathcal{C} = \langle (1 + x^2 \mid 3 + x + x^2)\rangle,$$

where the generator polynomials are in standard form. $\triangle$

Example 7.15. By using the MAGMA package for $\mathbb{Z}_2\mathbb{Z}_4$-additive codes [29], we construct the code $\mathcal{C}$ defined in Example 7.3 and used in Examples 7.10 and 7.14. We show that it is cyclic, it can be generated as a cyclic code from the polynomials corresponding to the rows of the generator matrix, or from the generator polynomials in standard form.

```
> Z2 := Integers(2);
> Z4 := Integers(4);
> PR2<x> := PolynomialRing(Z2);
> PR4<y> := PolynomialRing(Z4);
> alpha :=  3;
> beta := 3;
> R2 := RSpace(Integers(2), alpha);
> R4 := RSpace(Integers(4), beta);
> R2R4 := CartesianProduct(R2, R4);
>
> G := Matrix(Z4, [[2,0,2, 2,0,0],
                    [0,2,2, 2,2,0],
                    [0,0,0, 1,1,1]]);
> C := Z2Z4AdditiveCode(G, alpha);
> IsCyclic(C);
true
>
> G2 := [<PR2!Eltseq(v[1]), PR4!Eltseq(v[2])> :
                  v in FromZ4toZ2Z4(G, alpha)];
> G2; // generator polynomials from rows of generator matrix
[
    <x^2 + 1, 2>,
    <x^2 + x, 2*y + 2>,
    <0, y^2 + y + 1>
]
> C2 := Z2Z4CyclicCode(alpha, beta, G2);
> C2 eq C;
true
>
> polyG := GeneratorPolynomials(C);
> b := polyG[1]; l := polyG[2]; f := polyG[3]; h := polyG[4];
> G3 := [ < b, PR4!0 > , < l, f*h + 2*f> ];
> G3; // generator polynomials in standard form
[
    <x^3 + 1, 0>,
    <x^2 + 1, y^2 + y + 3>
]
```

```
> C3 := Z2Z4CyclicCode(alpha, beta, G3);
> C3 eq C;
true
```
△

Note that the generator polynomials given in Example 7.10 are not in standard form. From now on, unless it is stated otherwise, the generator polynomials of a $\mathbb{Z}_2\mathbb{Z}_4$-additive cyclic code are considered to be in standard form.

Proposition 7.16. *Let* $\mathcal{C} = \langle (b \mid 0), (\ell \mid fh + 2f) \rangle$ *be a* $\mathbb{Z}_2\mathbb{Z}_4$-*additive cyclic code. Then,*

   i) $\mathcal{C}_X$ *is a binary cyclic code generated by* $\gcd(b, \ell)$.

   ii) $\mathcal{C}_Y$ *is a quaternary cyclic code generated by* $fh + 2f$.

Proof. The generator polynomial of $\mathcal{C}_Y$ is clear. However, in the case of $\mathcal{C}_X$, it is generated by b and ℓ. Since a binary cyclic code is generated by the monic polynomial of minimum degree in the code, we can easily check that $\mathcal{C}_X$ is generated by $\gcd(b, \ell)$.   □

Note that if $\mathcal{C}$ is a $\mathbb{Z}_2\mathbb{Z}_4$-additive cyclic code, then $\mathcal{C}_X$ and $\mathcal{C}_Y$ are binary and quaternary cyclic codes, respectively. The following example illustrates that the converse is not true in general. Nevertheless, Theorem 7.18 shows that if a $\mathbb{Z}_2\mathbb{Z}_4$-additive cyclic code is separable, then the converse is satisfied.

Example 7.17. Let $\mathcal{C}$ be a $\mathbb{Z}_2\mathbb{Z}_4$-additive code generated by

$$\begin{pmatrix} 1 & 0 & 1 & 0 & 0 \\ 0 & 1 & 0 & 1 & 0 \\ 0 & 0 & 0 & 0 & 1 \end{pmatrix}. \tag{7.2}$$

Clearly, $\mathcal{C}_X$ and $\mathcal{C}_Y$ are cyclic codes. However, note that $(0, 0 \mid 0, 0, 1)^{(1)} = (0, 0 \mid 1, 0, 0)$ does not belong to $\mathcal{C}$ and hence $\mathcal{C}$ is not cyclic.   △

Theorem 7.18. *A separable* $\mathbb{Z}_2\mathbb{Z}_4$-*additive code* $\mathcal{C}$ *is cyclic if and only if* $\mathcal{C}_X$ *is a binary cyclic code and* $\mathcal{C}_Y$ *is a quaternary cyclic code. Moreover,*

$$\mathcal{C} = \langle (b \mid 0), (0 \mid fh + 2f) \rangle.$$

Proof. Let $\mathcal{C}$ be a separable $\mathbb{Z}_2\mathbb{Z}_4$-additive cyclic code. Let $u \in \mathcal{C}_X$ and $u' \in \mathcal{C}_Y$. Then, $(u \mid \mathbf{0}) \in \mathcal{C}$ and $(u \mid \mathbf{0})^{(1)} = (u^{(1)} \mid \mathbf{0}) \in \mathcal{C}$ which gives that $u^{(1)} \in \mathcal{C}_X$. Therefore, $\mathcal{C}_X$ is cyclic. Similarly, $(\mathbf{0} \mid u') \in \mathcal{C}$ and $(\mathbf{0} \mid u')^{(1)} = (\mathbf{0} \mid (u')^{(1)}) \in \mathcal{C}$ which gives that $(u')^{(1)} \in \mathcal{C}_Y$. Therefore, $\mathcal{C}_Y$ is cyclic.

If both $\mathcal{C}_X$ and $\mathcal{C}_Y$ are cyclic, then $(u \mid u')^{(1)} = (u^{(1)} \mid (u')^{(1)}) \in \mathcal{C}_X \times \mathcal{C}_Y = \mathcal{C}$ and so $\mathcal{C}$ is cyclic. The polynomial representation follows immediately from the standard form of the generator polynomials given in (7.1) since the code is separable; that is, $\mathcal{C} = \mathcal{C}_X \times \mathcal{C}_Y$.   □

Corollary 7.19. *The $\mathbb{Z}_2\mathbb{Z}_4$-additive cyclic code $\mathcal{C} = \langle (b \mid 0), (\ell \mid fh + 2f) \rangle$ is separable if and only if $\ell = 0$.*

Example 7.20. By using the MAGMA package for $\mathbb{Z}_2\mathbb{Z}_4$-additive codes [29], we construct a separable $\mathbb{Z}_2\mathbb{Z}_4$-additive cyclic code. We check that $\mathcal{C}_X$ and $\mathcal{C}_Y$ are cyclic codes, and that $\ell = 0$.

```
> Z2 := Integers(2);
> Z4 := Integers(4);
> PR2<x> := PolynomialRing(Z2);
> PR4<y> := PolynomialRing(Z4);
> alpha := 3;
> beta := 5;
> G := Matrix(Z4,[ [2,0,2,0,0,0,0,0],
                    [0,2,2,0,0,0,0,0],
                    [0,0,0,1,0,0,0,3],
                    [0,0,0,0,1,0,0,3],
                    [0,0,0,0,0,1,0,3],
                    [0,0,0,0,0,0,1,3] ]);
> C := Z2Z4AdditiveCode(G, alpha);
> IsSeparable(C) and IsCyclic(C);
true
>
> CX := LinearBinaryCode(C);
> CY := LinearQuaternaryCode(C);
> IsCyclic(CX) and IsCyclic(CY);
true
> polyG := GeneratorPolynomials(C);
> b := polyG[1]; l := polyG[2]; f := polyG[3]; h := polyG[4];
> IsZero(l);
true                                                                    △
```

Recall that, for a $\mathbb{Z}_2\mathbb{Z}_4$-additive code $\mathcal{C}$, $\mathcal{C}_b$ is the subcode of $\mathcal{C}$ which contains all codewords of order at most two.

Lemma 7.21 ([39]). *Let $\mathcal{C} = \langle (b \mid 0), (\ell \mid fh + 2f) \rangle$ be a $\mathbb{Z}_2\mathbb{Z}_4$-additive cyclic code with $fgh = x^\beta - 1$. Then,*

$$\mathcal{C}_b = \langle (b \mid 0), (\ell\tilde{g} \mid 2fg), (0 \mid 2fh) \rangle,$$

where the generator polynomials are not necessarily in standard form.

Proof. Since $\mathcal{C} = \langle (b \mid 0), (\ell \mid fh + 2f) \rangle$, then $(b \mid 0)$, $g \star (\ell \mid fh + 2f) = (\ell\tilde{g} \mid 2fg)$ and $2 \star (\ell \mid fh + 2f) = (0 \mid 2fh)$ are codewords of order two and therefore are elements in $\mathcal{C}_b$. Then, $\langle (b \mid 0), (\ell\tilde{g} \mid 2fg), (0 \mid 2fh) \rangle \subseteq \mathcal{C}_b$.

Let $(u \mid v) \in \mathcal{C}_b \subseteq \mathcal{C}$. We have that $(u \mid v) = p \star (b \mid 0) + q \star (\ell \mid fh + 2f)$, where $q = 2q_1 + q_2$ and all the coefficients of q_2 are zeros or ones. Then,

$q \star (\ell \mid fh+2f) = (2q_1+q_2) \star (\ell \mid fh+2f) = (0 \mid 2q_1fh) + (\tilde{q}_2\ell \mid q_2fh+q_22f)$. Since $(u \mid v)$ is of order *two*, $q_2fh = 0$ and hence $q_2 = q_3g$. Therefore, $(\tilde{q}_2\ell \mid q_2fh + q_22f) = (\tilde{q}_3\tilde{g}\ell \mid q_3g2f)$ and $(u \mid v) = p \star (b \mid 0) + q_1 \star (0 \mid 2fh) + q_3 \star (\ell\tilde{g} \mid 2fg)$. Thus, $\mathcal{C}_b \subseteq \langle(b \mid 0),(\ell\tilde{g} \mid 2f),(0 \mid 2fh)\rangle$, and we get the equality. $\qquad\square$

Corollary 7.22. *Let* $\mathcal{C} = \langle(b \mid 0),(\ell \mid fh + 2f)\rangle$ *be a* $\mathbb{Z}_2\mathbb{Z}_4$-*additive cyclic code with* $fgh = x^\beta - 1$. *Then*, $(\mathcal{C}_b)_X = \langle \gcd(b,\ell\tilde{g})\rangle$.

Lemma 7.23. *Let* $fgh = x^\beta - 1$ *in* $\mathbb{Z}_4[x]$ *and* β *odd, where* f, g *and* h *are pairwise coprime polynomials. Then, there exist polynomials* λ *and* μ *in* $\mathbb{Z}_4[x]$ *such that*

$$\lambda h + \mu g = 1. \tag{7.3}$$

Proof. By Lemma 7.6 and, since β is odd, the uniqueness of the factorization of $x^\beta - 1$ in $\mathbb{Z}_2[x]$, we have that $\tilde{h}$ and $\tilde{g}$ are coprime in $\mathbb{Z}_2[x]$. Therefore, by Bezout's identity, there exist $\lambda_2, \mu_2 \in \mathbb{Z}_2[x]$, such that

$$\lambda_2\tilde{h} + \mu_2\tilde{g} = 1.$$

Now, applying the Hensel lift to the previous equation, we obtain that $\lambda h + \mu g = 1$, where λ and μ are the Hensel lift of λ_2 and μ_2, respectively. $\qquad\square$

Proposition 7.24 ([31]). *Let* $\mathcal{C} = \langle(b \mid 0),(\ell \mid fh+2f)\rangle$ *be a* $\mathbb{Z}_2\mathbb{Z}_4$-*additive cyclic code with* $fgh = x^\beta - 1$. *Then*,

$$\mathcal{C}_b = \langle(b \mid 0),(\tilde{\mu}\ell\tilde{g} \mid 2f)\rangle,$$

where μ *is as in* (7.3).

Proof. By Lemma 7.21, $\mathcal{C}_b = \langle(b \mid 0),(\ell\tilde{g} \mid 2fg),(0 \mid 2fh)\rangle$ with $\gcd(h,g) = 1$. Note that $(b \mid 0) \in \mathcal{C}$ and $g \star (\ell \mid fh+2f) = (\ell\tilde{g} \mid 2fg) \in \mathcal{C}$. Since $(0 \mid 2fh) \in \mathcal{C}$, we obtain $\mu \star (\ell\tilde{g} \mid 2fg) + \lambda \star (0 \mid 2fh) = (\tilde{\mu}\ell\tilde{g} \mid 2f) \in \mathcal{C}$ for λ, μ as in (7.3). Hence, $(\tilde{\mu}\ell\tilde{g} \mid 2f) \in \mathcal{C}_b$.

Let $\mathcal{C}'_b = \langle(b \mid 0),(\tilde{\mu}\ell\tilde{g} \mid 2f)\rangle$. Since $(b \mid 0),(\tilde{\mu}\ell\tilde{g} \mid 2f) \in \mathcal{C}_b$, we have that $\mathcal{C}'_b \subseteq \mathcal{C}_b$. We shall prove that $\mathcal{C}_b \subseteq \mathcal{C}'_b$. It is evident that $(b \mid 0) \in \mathcal{C}'_b$. Since $\lambda h + \mu g = 1$, we have that $\tilde{\lambda}\tilde{h}\ell\tilde{g} + \tilde{\mu}\tilde{g}\ell\tilde{g} = \ell\tilde{g}$. By Theorem 7.11, we know that b divides $\tilde{h}\ell\tilde{g}$. So $\tilde{\mu}\tilde{g}\ell\tilde{g} = \ell\tilde{g} \pmod{b}$. Therefore, $g \star (\tilde{\mu}\ell\tilde{g} \mid 2f) = (\tilde{g}\tilde{\mu}\ell\tilde{g} \mid 2fg) = (\ell\tilde{g} + pb \mid 2fg) = (\ell\tilde{g} \mid 2fg) + (pb \mid 0) \in \mathcal{C}'_b$ for some $p \in \mathbb{Z}_2[x]$. Hence, $(\ell\tilde{g} \mid 2fg) \in \mathcal{C}'_b$. Finally, applying Corollary 7.12, $h \star (\tilde{\mu}\ell\tilde{g} \mid 2f) = (\tilde{h}\tilde{\mu}\ell\tilde{g} \mid 2fh) = (p'b \mid 2fh) \in \mathcal{C}'_b$. Thus, $(0 \mid 2fh) \in \mathcal{C}'_b$, and hence $\mathcal{C}_b \subseteq \mathcal{C}'_b$. $\qquad\square$

Given a code $\mathcal{C}$, we say that a set of codewords S is a *spanning set* of $\mathcal{C}$ if any codeword in $\mathcal{C}$ can be written as a linear combination of codewords in S. We say that a spannning set of $\mathcal{C}$ is minimal if there is not any proper subset of S that is also a spanning set of $\mathcal{C}$.

The following results show the close relation of the type of a $\mathbb{Z}_2\mathbb{Z}_4$-additive cyclic code and the degrees of the generator polynomials of the code. First, the next theorem gives a minimal spanning set in terms of the generator polynomials.

Theorem 7.25 ([1]). *Let $\mathcal{C} = \langle (b \mid 0), (\ell \mid fh + 2f) \rangle$ be a $\mathbb{Z}_2\mathbb{Z}_4$-additive cyclic code of type $(\alpha, \beta; \gamma, \delta; \kappa)$, with $fgh = x^\beta - 1$. Let*

$$S_1 = \bigcup_{i=0}^{\alpha - \deg(b) - 1} \{x^i \star (b \mid 0)\},$$

$$S_2 = \bigcup_{i=0}^{\deg(g) - 1} \{x^i \star (\ell \mid fh + 2f)\}, \text{ and}$$

$$S_3 = \bigcup_{i=0}^{\deg(h) - 1} \{x^i \star (\ell\tilde{g} \mid 2fg)\}.$$

Then, $S_1 \cup S_2 \cup S_3$ forms a minimal spanning set for $\mathcal{C}$ as a $\mathbb{Z}_4[x]$-submodule. Moreover, $\mathcal{C}$ has $2^{\alpha - \deg(b)} 4^{\deg(g)} 2^{\deg(h)}$ codewords.

Note that S_2 generates all order four codewords, and the subcode of codewords of order two, $\mathcal{C}_b$, is generated by $\{S_1, 2S_2, S_3\}$. Hence, in the following theorem, by using these spanning sets, we can obtain the type $(\alpha, \beta; \gamma, \delta; \kappa)$ of the code.

Theorem 7.26 ([39]). *Let $\mathcal{C} = \langle (b \mid 0), (\ell \mid fh + 2f) \rangle$ be a $\mathbb{Z}_2\mathbb{Z}_4$-additive cyclic code of type $(\alpha, \beta; \gamma, \delta; \kappa)$, with $fgh = x^\beta - 1$. Then,*

$$\gamma = \alpha - \deg(b) + \deg(h),$$
$$\delta = \deg(g),$$
$$\kappa = \alpha - \deg(\gcd(\ell\tilde{g}, b)).$$

Proof. The parameters γ and δ are known from Theorem 7.25 and the parameter κ is the dimension of $(\mathcal{C}_b)_X$. By Lemma 7.21, $(\mathcal{C}_b)_X$ is generated by the polynomials b and $\ell\tilde{g}$. Since the ring is a polynomial ring and thus a principal ideal ring, it is generated by the greatest common divisor of the two polynomials. Then, $\kappa = \alpha - \deg(\gcd(\ell\tilde{g}, b))$. $\square$

Corollary 7.27 ([39]). *Let* $\mathcal{C} = \langle (b \mid 0), (\ell \mid fh + 2f) \rangle$ *be a* $\mathbb{Z}_2\mathbb{Z}_4$-*additive cyclic code of type* $(\alpha, \beta; \gamma, \delta; \kappa)$, *with* $fgh = x^\beta - 1$. *Then,*

$$|\mathcal{C}| = 2^{\alpha - \deg(b)} 4^{\deg(g)} 2^{\deg(h)}.$$

Proposition 7.28 ([39]). *Let* $\mathcal{C} = \langle (b \mid 0), (\ell \mid fh + 2f) \rangle$ *be a* $\mathbb{Z}_2\mathbb{Z}_4$-*additive cyclic code of type* $(\alpha, \beta; \gamma, \delta = \delta_1 + \delta_2; \kappa = \kappa_1 + \kappa_2)$, *with* $fgh = x^\beta - 1$ *and* κ_1, δ_2 *defined in (2.3). Then,*

$$\kappa_1 = \alpha - \deg(b), \quad \kappa_2 = \deg(b) - \deg(\gcd(b, \ell\tilde{g})),$$

$$\delta_1 = \deg(\gcd(b, \ell\tilde{g})) - \deg(\gcd(b, \ell)) \text{ and } \delta_2 = \deg(g) - \delta_1.$$

Proof. The result follows from Proposition 3.17 and considering that the generator polynomials of $\mathcal{C}_X$ and $(\mathcal{C}_b)_X$ are $\gcd(b, \ell)$ and $\gcd(b, \ell\tilde{g})$ by Proposition 7.16 and Corollary 7.22, respectively. $\square$

Example 7.29. Let $\mathcal{C} = \langle (1 + x^2 \mid 3 + x + x^2) \rangle$ be the $\mathbb{Z}_2\mathbb{Z}_4$-additive code of type $(3, 3; 2, 1; 2)$ given in Example 7.14. From the generator matrix given in Example 7.3, it is easy to check that $\kappa_1 = 0, \kappa_2 = 2, \delta_1 = 0$ and $\delta_2 = 1$. We have seen that $b = 1 + x^3$, $f = 1$, $h = 1 + x + x^2$, $\ell = 1 + x^2$. Then, $g = \frac{x^3 - 1}{1 + x + x^2} = 3 + x$, $\gcd(b, \ell) = 1 + x$, and $\gcd(b, \ell\tilde{g}) = 1 + x$. By applying Proposition 7.28, $\kappa_1 = 3 - 3 = 0$, $\kappa_2 = 3 - 1 = 2$, $\delta_1 = 1 - 1 = 0$ and $\delta_2 = 1 - 0 = 1$. $\triangle$

7.2 Duality of $\mathbb{Z}_2\mathbb{Z}_4$-Additive Cyclic Codes

In this section, we consider the dual $\mathcal{C}^\perp$ of a $\mathbb{Z}_2\mathbb{Z}_4$-additive cyclic code $\mathcal{C}$. We see that it is also a $\mathbb{Z}_2\mathbb{Z}_4$-additive cyclic code, and we determine the generator polynomials of $\mathcal{C}^\perp$ in terms of the generator polynomials of $\mathcal{C}$.

The *reciprocal polynomial* of a polynomial p is $x^{\deg(p)}p(x^{-1})$ and is denoted by p^*. As in the theory of cyclic codes over $\mathbb{Z}_2$ and $\mathbb{Z}_4$ [112], [126], reciprocal polynomials have an important role on duality.

Proposition 7.30 ([1]). *Let* $\mathcal{C}$ *be a* $\mathbb{Z}_2\mathbb{Z}_4$-*additive cyclic code. Then,* $\mathcal{C}^\perp$ *is also a* $\mathbb{Z}_2\mathbb{Z}_4$-*additive cyclic code.*

Proof. We have that $\mathcal{C}^\perp$ is a $\mathbb{Z}_2\mathbb{Z}_4$-additive code. Let $\mathbf{u} = (u \mid u') \in \mathcal{C}^\perp$. For all $\mathbf{v} = (v \mid v') \in \mathcal{C}$, $i \geq 0$, $\mathbf{v}^{(-i)} \in \mathcal{C}$, and we have $\mathbf{u}^{(i)} \cdot \mathbf{v} = \mathbf{u} \cdot \mathbf{v}^{(-i)} = \mathbf{0}$. Therefore, $\mathbf{u}^{(i)} \in \mathcal{C}^\perp$, and then $\mathcal{C}^\perp$ is cyclic. $\square$

If $\mathcal{C} = \langle (b \mid 0), (\ell \mid fh + 2f) \rangle$, then we denote

$$\mathcal{C}^\perp = \langle (\bar{b} \mid 0), (\bar{\ell} \mid \bar{f}\bar{h} + 2\bar{f}) \rangle,$$

where the generator polynomials are in standard form.

The following proposition determines the degrees of the generator polynomials of $\mathcal{C}^\perp$ in terms of the degrees of the generator polynomials of $\mathcal{C}$. This result is helpful to determine the generator polynomials of the dual code.

Proposition 7.31 ([39]). *Let* $\mathcal{C} = \langle (b \mid 0), (\ell \mid fh + 2f) \rangle$ *be a* $\mathbb{Z}_2\mathbb{Z}_4$-*additive cyclic code of type* $(\alpha, \beta; \gamma, \delta; \kappa)$, *with* $fgh = x^\beta - 1$, *and dual code* $\mathcal{C}^\perp = \langle (\bar{b} \mid 0), (\bar{\ell} \mid \bar{f}\bar{g}\bar{h} + 2\bar{f}) \rangle$, *with* $\bar{f}\bar{g}\bar{h} = x^\beta - 1$. *Then,*

$$\deg(\bar{b}) = \alpha - \deg(\gcd(b, \ell)),$$
$$\deg(\bar{f}) = \deg(g) + \deg(\gcd(b, \ell)) - \deg(\gcd(b, \ell\tilde{g})),$$
$$\deg(\bar{h}) = \deg(h) - \deg(b) - \deg(\gcd(b, \ell)) + 2\deg(\gcd(b, \ell\tilde{g})),$$
$$\deg(\bar{g}) = \deg(f) + \deg(b) - \deg(\gcd(b, \ell\tilde{g})).$$

Proof. Let $\mathcal{C}^\perp$ be a code of type $(\alpha, \beta; \bar{\gamma}, \delta; \bar{\kappa})$. It is easy to prove that $(\mathcal{C}_X)^\perp$ is a binary cyclic code generated by $\bar{b}$, so $|(\mathcal{C}_X)^\perp| = 2^{\alpha - \deg(\bar{b})}$. Moreover, by Proposition 3.17, $|(\mathcal{C}_X)^\perp| = 2^{\alpha - \kappa - \delta_1}$ and, by Proposition 7.28, we obtain that $\deg(\bar{b}) = \alpha - \deg(\gcd(b, \ell))$. Finally, from Theorem 3.16, we have that

$$\bar{\gamma} = \alpha + \gamma - 2\kappa,$$
$$\delta = \beta - \gamma - \delta + \kappa,$$
$$\bar{\kappa} = \alpha - \kappa,$$

and applying Theorem 7.26 to the parameters of $\mathcal{C}$ and $\mathcal{C}^\perp$, we obtain the result. $\square$

By Corollary 3.18, we know that a $\mathbb{Z}_2\mathbb{Z}_4$-additive code $\mathcal{C}$ is separable if and only if $\mathcal{C}^\perp$ is separable. If a $\mathbb{Z}_2\mathbb{Z}_4$-additive cyclic code is separable, then it is easy to find the generator polynomials of its dual, which are given in the following proposition.

Proposition 7.32 ([39]). *Let* $\mathcal{C} = \langle (b \mid 0), (0 \mid fh + 2f) \rangle$ *be a separable* $\mathbb{Z}_2\mathbb{Z}_4$-*additive cyclic code of type* $(\alpha, \beta; \gamma, \delta; \kappa)$, *with* $fgh = x^\beta - 1$. *Then,*

$$\mathcal{C}^\perp = \langle (\frac{x^\alpha - 1}{b^*} \mid 0), (0 \mid g^*h^* + 2g^*) \rangle.$$

Proof. If $\mathcal{C}$ is separable, by Corollary 3.18, $\mathcal{C}^\perp = (\mathcal{C}_X)^\perp \times (\mathcal{C}_Y)^\perp$. Since $\mathcal{C}_X = \langle b \rangle$, $(\mathcal{C}_X)^\perp = \langle \frac{x^\alpha - 1}{b^*} \rangle$ [112]. Moreover, $\mathcal{C}_Y = \langle fh + 2f \rangle$ and $(\mathcal{C}_Y)^\perp = \langle g^*h^* + 2g^* \rangle$ [126]. Therefore, the statement follows. $\square$

Example 7.33. We continue with the separable $\mathbb{Z}_2\mathbb{Z}_4$-additive code $\mathcal{C}$ generated by using MAGMA in Example 7.20. Now, we compute the additive dual code $\mathcal{C}^\perp$ and check that it can be generated by using the generator polinomials of $\mathcal{C}$, as shown in Proposition 7.32.

```
> C;
(8, 1024) Z2Z4-additive code of type (3, 5; 2, 4; 2)
Generator matrix:
[2 0 2 0 0 0 0 0]
[0 2 2 0 0 0 0 0]
[0 0 0 1 0 0 0 3]
[0 0 0 0 1 0 0 3]
[0 0 0 0 0 1 0 3]
[0 0 0 0 0 0 1 3]
> polyG := GeneratorPolynomials(C);
> polyG;
<x + 1, 0, y + 3, 1>
> b := polyG[1]; l := polyG[2]; f := polyG[3]; h := polyG[4];
>
> g := (y^beta -1) div (f*h);
>
> bd := (x^alpha-1) div PR2!Reverse(b);
> fd := PR4!Reverse(Eltseq(g));
> hd := PR4!Reverse(Eltseq(h));
> Cd := Z2Z4CyclicCode(alpha, beta, bd, PR2!0, fd, hd);
> Cd eq Dual(C);
true                                                        △
```

Now, we define the generator polynomials of $\mathcal{C}^\perp$ in terms of the generator polynomials of $\mathcal{C}$, when $\mathcal{C}$ is a non-separable $\mathbb{Z}_2\mathbb{Z}_4$-additive cyclic code. First, we denote the polynomial $\sum_{i=0}^{m-1} x^i$ by $\theta_m(x)$. Using this notation, we have the following proposition. The proof follows from the fact that $y^m - 1 = (y - 1)\theta_m(y)$, replacing y by x^n.

Proposition 7.34 ([39]). *Let* $n, m \in \mathbb{N}$. *Then,*

$$x^{nm} - 1 = (x^n - 1)\theta_m(x^n).$$

From now on, $\mathfrak{m}$ denotes the least common multiple of α and β. The following map $\circ$ is related to the orthogonality of codewords.

Definition 7.35. *Let* $\mathbf{u}(x) = (u(x) \mid u'(x))$ *and* $\mathbf{v}(x) = (v(x) \mid v'(x))$ *be elements in* $R_{\alpha,\beta}$. *We define the map*

$$\circ : R_{\alpha,\beta} \times R_{\alpha,\beta} \longrightarrow \mathbb{Z}_4[x]/\langle x^{\mathfrak{m}} - 1\rangle,$$

by

$$\mathbf{u}(x) \circ \mathbf{v}(x) = 2u(x)\theta_{\frac{\mathfrak{m}}{\alpha}}(x^\alpha)x^{\mathfrak{m}-1-\deg(v(x))}v^*(x) +$$
$$+ u'(x)\theta_{\frac{\mathfrak{m}}{\beta}}(x^\beta)x^{\mathfrak{m}-1-\deg(v'(x))}v'^*(x) \mod (x^{\mathfrak{m}} - 1),$$

where the computations are made taking the binary zeros and ones in $u(x)$ and $v(x)$ as quaternary zeros and ones, respectively.

Note that the map $\circ$ is a bilinear map between $\mathbb{Z}_4[x]$-modules and $\mathbf{u}(x) \circ \mathbf{v}(x)$ belongs to $\mathbb{Z}_4[x]/\langle x^{\mathfrak{m}} - 1 \rangle$.

Proposition 7.36 ([39]). *Let $\mathbf{u}$ and $\mathbf{v}$ be vectors in $\mathbb{Z}_2^\alpha \times \mathbb{Z}_4^\beta$ with associated polynomials $\mathbf{u}(x) = (u(x) \mid u'(x))$ and $\mathbf{v}(x) = (v(x) \mid v'(x))$. Then, $\mathbf{u}$ is orthogonal to $\mathbf{v}$ and all its shifts if and only if*

$$\mathbf{u}(x) \circ \mathbf{v}(x) = 0.$$

Proof. Let $\mathbf{u}(x) = (u(x) \mid u'(x)), \mathbf{v}(x) = (v(x) \mid v'(x)) \in R_{\alpha,\beta}$. By definition, $\mathbf{u} \cdot \mathbf{v}^{(i)} = 0$ if and only if $2\sum_{j=0}^{\alpha-1} u_j v_{j+i} + \sum_{k=0}^{\beta-1} u'_k v'_{k+i} = 0$. Arranging the terms in $\mathbf{u}(x) \circ \mathbf{v}(x)$, we have that $\mathbf{u}(x) \circ \mathbf{v}(x) = 0$ if and only if $2\sum_{j=0}^{\alpha-1} u_j v_{j+i} + \sum_{k=0}^{\beta-1} u'_k v'_{k+i} = 0$ for $0 \le i \le \mathfrak{m} - 1$; that is, $\mathbf{u}$ is orthogonal to $\mathbf{v}$ and all its shifts. $\qquad\qquad\square$

In order to describe the generator polynomials of the dual of a non-separable $\mathbb{Z}_2\mathbb{Z}_4$-additive cyclic code, we use the following technical lemma proved in [39].

Lemma 7.37 ([39]). *Let $\mathbf{u} = (u(x) \mid u'(x))$ and $\mathbf{v}(x) = (v(x) \mid v'(x))$ be elements in $R_{\alpha,\beta}$ such that $\mathbf{u}(x) \circ \mathbf{v}(x) = 0$. If $u'(x)$ or $v'(x)$ equals 0, then $u(x)v^*(x) \equiv 0 \pmod{(x^\alpha - 1)}$ over $\mathbb{Z}_2$. If $u(x)$ or $v(x)$ equals 0, then $u'(x)v'^*(x) \equiv 0 \pmod{(x^\beta - 1)}$ over $\mathbb{Z}_4$.*

Proposition 7.38 ([39]). *Let $\mathcal{C} = \langle (b \mid 0), (\ell \mid fh + 2f) \rangle$ be a $\mathbb{Z}_2\mathbb{Z}_4$-additive cyclic code of type $(\alpha, \beta; \gamma, \delta; \kappa)$ with dual code $\mathcal{C}^\perp = \langle (\bar{b} \mid 0), (\bar{\ell} \mid \bar{f}\bar{h} + 2\bar{f}) \rangle$. Then,*

$$\bar{b} = \frac{x^\alpha - 1}{(\gcd(b, \ell))^*} \in \mathbb{Z}_2[x].$$

Proof. Since $(\bar{b} \mid 0) \in \mathcal{C}^\perp$, $(b \mid 0) \circ (\bar{b} \mid 0) = 0$ and $(\ell \mid fh + 2f) \circ (\bar{b} \mid 0) = 0$. By Lemma 7.37, we have $b\bar{b}^* \equiv 0 \pmod{(x^\alpha - 1)}$ and $\ell\bar{b}^* \equiv 0 \pmod{(x^\alpha - 1)}$ over $\mathbb{Z}_2$. Therefore, $\gcd(b, \ell)\bar{b}^* \equiv 0 \pmod{(x^\alpha - 1)}$, and there exists $\mu \in \mathbb{Z}_2[x]$ such that $\gcd(b, \ell)\bar{b}^* = \mu(x^\alpha - 1)$. Since $\gcd(b, \ell)$ and $\bar{b}^*$ divide $(x^\alpha - 1)$ and, by Proposition 7.31, $\deg(\bar{b}) = \alpha - \deg(\gcd(b, \ell))$, we obtain $\gcd(b, \ell)\bar{b}^* = (x^\alpha - 1)$ and hence $\bar{b} = \frac{x^\alpha - 1}{(\gcd(b, \ell))^*} \in \mathbb{Z}_2[x]$. $\qquad\square$

Proposition 7.39 ([39]). *Let $\mathcal{C} = \langle (b \mid 0), (\ell \mid fh + 2f) \rangle$ be a $\mathbb{Z}_2\mathbb{Z}_4$-additive cyclic code of type $(\alpha, \beta; \gamma, \delta; \kappa)$, with $fgh = x^\beta - 1$, and with dual code $\mathcal{C}^\perp = \langle (\bar{b} \mid 0), (\bar{\ell} \mid \bar{f}\bar{h} + 2\bar{f}) \rangle$, with $\bar{f}\bar{g}\bar{h} = x^\beta - 1$. Then, $\bar{f}\bar{h}$ is the Hensel lift of*

$$\frac{(x^\beta - 1)\gcd(b, \ell\tilde{g})^*}{f^* b^*} \in \mathbb{Z}_2[x].$$

Proof. The polynomials h and g are coprime, $fgh = x^\beta - 1$. By Lemma 7.23, we have that $p_1 fh + p_2 fg = f$ for some $p_1, p_2 \in \mathbb{Z}_4[x]$. Since $(b \mid 0)$, $(0 \mid 2fh)$ and $(\ell\tilde{g} \mid 2fg)$ belong to $\mathcal{C}$,

$$(0 \mid \frac{b}{\gcd(b, \ell\tilde{g})}(2p_1 fh + 2p_2 fg)) = (0 \mid \frac{b}{\gcd(b, \ell\tilde{g})}2f) \in \mathcal{C}.$$

Therefore,

$$(\bar{\ell} \mid \bar{f}h + 2\bar{f}) \circ (0 \mid \frac{b}{\gcd(b, \ell\tilde{g})}2f) = 0.$$

Thus, by Lemma 7.37,

$$(\overline{fh} + 2\bar{f})\left(\frac{b^*2f^*}{\gcd(b, \ell\tilde{g})^*}\right) \equiv 0 \pmod{(x^\beta - 1)},$$

and

$$(2\overline{fh})\left(\frac{b^* f^*}{\gcd(b, \ell\tilde{g})^*}\right) = 2\mu(x^\beta - 1) \tag{7.4}$$

for some $\mu \in \mathbb{Z}_4[x]$. If (7.4) holds over $\mathbb{Z}_4$, then it is equivalent to

$$(\overline{fh})\left(\frac{b^* f^*}{\gcd(b, \ell\tilde{g})^*}\right) = \mu(x^\beta - 1) \in \mathbb{Z}_2[x].$$

It is known that $\overline{fh}$ is a divisor of $x^\beta - 1$ and, by Corollary 7.12, we have that $\left(\frac{b^* f^*}{\gcd(b, \ell\tilde{g})^*}\right)$ divides $(x^\beta - 1)$ over $\mathbb{Z}_2$. By Proposition 7.31, $\deg(\overline{fh}) = \beta - \deg(f) - \deg(b) + \deg(\gcd(b, \ell\tilde{g}))$, so

$$\beta = \deg\left(\overline{fh}\frac{b^* f^*}{\gcd(b, \ell\tilde{g})^*}\right) = \deg(x^\beta - 1).$$

Hence, we obtain that $\mu = 1 \in \mathbb{Z}_2$ and

$$\overline{fh} = \frac{(x^\beta - 1)\gcd(b, \ell\tilde{g})^*}{f^* b^*} \in \mathbb{Z}_2[x]. \tag{7.5}$$

Since β is odd, by the uniqueness of the Hensel lift given by Lemma 7.6, $\overline{fh}$ is the unique monic polynomial in $\mathbb{Z}_4[x]$ dividing $(x^\beta - 1)$ and satisfying (7.5). $\square$

Proposition 7.40 ([39]). *Let $\mathcal{C} = \langle(b \mid 0), (\ell \mid fh + 2f)\rangle$ be a $\mathbb{Z}_2\mathbb{Z}_4$-additive cyclic code of type $(\alpha, \beta; \gamma, \delta; \kappa)$, with $fgh = x^\beta - 1$, and with dual code $\mathcal{C}^\perp = \langle(\bar{b} \mid 0), (\bar{\ell} \mid \bar{f}h + 2\bar{f})\rangle$, with $\bar{f}\bar{g}h = x^\beta - 1$. Then, $\bar{f}$ is the Hensel lift of*

$$\frac{(x^\beta - 1)\gcd(b, \ell)^*}{f^* h^* \gcd(b, \ell\tilde{g})^*} \in \mathbb{Z}_2[x].$$

Proof. In $\mathbb{Z}_2[x]$ the polynomials $b, \ell, \ell\tilde{g}$ can be written as

$$\ell = \gcd(b, \ell)\rho,$$
$$\ell\tilde{g} = \gcd(b, \ell\tilde{g})\rho\tau_1,$$
$$b = \gcd(b, \ell\tilde{g})\tau_2,$$

where τ_1 and τ_2 are coprime polynomials. Hence, there exist $t_1, t_2 \in \mathbb{Z}_2[x]$ such that $t_1\tau_1 + t_2\tau_2 = 1$. Then, $t_1\ell\tilde{g} + \rho t_2 b = \gcd(b, \ell\tilde{g})\rho(t_1\tau_1 + t_2\tau_2) = \gcd(b, \ell\tilde{g})\rho = \frac{\gcd(b,\ell\tilde{g})}{\gcd(b,\ell)}\ell$. Therefore,

$$\frac{\gcd(b, \ell\tilde{g})}{\gcd(b, \ell)} \star (\ell \mid fh + 2f) + t_1 \star (\ell\tilde{g} \mid 2fg) + \rho t_2 \star (b \mid 0) =$$
$$\left(0 \mid \frac{\gcd(b, \ell\tilde{g})}{\gcd(b, \ell)}(fh + 2f) + t_1 2fg \right) \in \mathcal{C}.$$

Since $\bar{h}$ and $\bar{g}$ are coprime, there exist $\bar{p}_1, \bar{p}_2 \in \mathbb{Z}_4[x]$ such that $2\bar{p}_1\bar{f}\bar{h} + 2\bar{p}_2\bar{f}\bar{g} = 2\bar{f}$ by Lemma 7.23. Thus, $(2\bar{p}_1 + \bar{p}_2\bar{g}) \star (\bar{\ell} \mid \bar{f}\bar{h} + 2\bar{f}) = (\bar{p}_2\bar{\ell}\bar{g} \mid 2\bar{f}) \in \mathcal{C}^\perp$. Therefore, $(\bar{p}_2\bar{\ell}\bar{g} \mid 2\bar{f}) \circ \left(0 \mid \frac{\gcd(b,\ell\tilde{g})}{\gcd(b,\ell)}(fh + 2f) + t_1 2fg \right) = 0$. By Lemma 7.37, arranging properly, we obtain that

$$2\bar{f} \left(\frac{\gcd(b, \ell\tilde{g})^*}{\gcd(b, \ell)^*} \right) f^* h^* \equiv 0 \pmod{(x^\beta - 1)}$$

and hence

$$2\bar{f} \left(\frac{\gcd(b, \ell\tilde{g})^*}{\gcd(b, \ell)^*} \right) f^* h^* = 2\mu(x^\beta - 1) \qquad (7.6)$$

for some $\mu \in \mathbb{Z}_4[x]$. If (7.6) holds over $\mathbb{Z}_4$, then it is equivalent to

$$\bar{f} \left(\frac{\gcd(b, \ell\tilde{g})^*}{\gcd(b, \ell)^*} \right) f^* h^* = \mu(x^\beta - 1) \in \mathbb{Z}_2[x].$$

It is easy to prove that $\left(\frac{\gcd(b,\ell\tilde{g})^*}{\gcd(b,\ell)^*} \right) f^* h^*$ divides $(x^\beta - 1)$ in $\mathbb{Z}_2[x]$. By Proposition 7.31, $\deg(\bar{f}) = \beta - \deg(f) - \deg(h) + \deg(\gcd(b, \ell)) - \deg(\gcd(b, \ell\tilde{g}))$, so

$$\beta = \deg \left(\bar{f} \left(\frac{\gcd(b, \ell\tilde{g})^*}{\gcd(b, \ell)^*} \right) f^* h^* \right) = \deg(x^\beta - 1).$$

Hence, we obtain that $\mu = 1$ and

$$\bar{f} = \frac{(x^\beta - 1)\gcd(b, \ell)^*}{\gcd(b, \ell\tilde{g})^* f^* h^*} \in \mathbb{Z}_2[x]. \qquad (7.7)$$

Since β is odd, by the uniqueness of the Hensel lift given by Lemma 7.6, $\bar{f}$ is the unique monic polynomial in $\mathbb{Z}_4[x]$ dividing $(x^\beta - 1)$ and holding (7.7). $\qquad \square$

Lemma 7.41 ([39]). *Let $\mathcal{C} = \langle (b \mid 0), (\ell \mid fh + 2f) \rangle$ be a $\mathbb{Z}_2\mathbb{Z}_4$-additive cyclic code of type $(\alpha, \beta; \gamma, \delta; \kappa)$, where $fgh = x^\beta - 1$. Then, the Hensel lift of $\frac{b}{\gcd(b, \ell \tilde{g})}$ divides h.*

Proof. In general, if $a \mid b \mid x^\beta - 1$ over $\mathbb{Z}_2[x]$ with odd β, then the Hensel lift of a divides the Hensel lift of b that divides $x^\beta - 1$ over $\mathbb{Z}_4[x]$. Then, by Corollary 7.12, the result follows. □

Proposition 7.42 ([39]). *Let $\mathcal{C} = \langle (b \mid 0), (\ell \mid fh + 2f) \rangle$ be a non-separable $\mathbb{Z}_2\mathbb{Z}_4$-additive cyclic code of type $(\alpha, \beta; \gamma, \delta; \kappa)$, with $fgh = x^\beta - 1$, and with dual code $\mathcal{C}^\perp = \langle (\bar{b} \mid 0), (\bar{\ell} \mid \bar{f}\bar{h} + 2\bar{f}) \rangle$, with $\bar{f}\bar{g}\bar{h} = x^\beta - 1$. Let $\rho = \frac{\ell}{\gcd(b, \ell)}$. Then,*

$$\bar{\ell} = \frac{x^\alpha - 1}{b^*} \left(\frac{\gcd(b, \ell\tilde{g})^*}{\gcd(b, \ell)^*} x^{\mathrm{m} - \deg(f)} \mu_1 \right.$$

$$\left. + \frac{b^*}{\gcd(b, \ell\tilde{g})^*} x^{\mathrm{m} - \deg(fh)} \mu_2 \right),$$

where

$$\begin{cases} \mu_1 = x^{\deg(\ell)}(\rho^*)^{-1} \mod \left(\frac{b^*}{\gcd(b, \ell\tilde{g})^*} \right), \\ \mu_2 = x^{\deg(\ell)}(\rho^*)^{-1} \mod \left(\frac{b^*}{\gcd(b, \ell)^*} \right). \end{cases}$$

Proof. In order to calculate $\bar{\ell}$, by using $\circ$, we are going to operate $(\bar{\ell} \mid \bar{f}\bar{h} + 2\bar{f})$ by three different codewords of $\mathcal{C}$. The result of these operations is 0 modulo $x^{\mathrm{m}} - 1$.

First, consider $(\bar{\ell} \mid \bar{f}\bar{h} + 2\bar{f}) \circ (b \mid 0) = 0$. By Lemma 7.37, $\bar{\ell}b^* \equiv 0$ (mod $(x^\alpha - 1)$) and, for some $\lambda \in \mathbb{Z}_2[x]$, we have that $\bar{\ell} = \frac{x^\alpha - 1}{b^*}\lambda$.

Second, consider $\tau = \frac{\gcd(b, \ell\tilde{g})}{\gcd(b, \ell)}$ and compute $(\bar{\ell} \mid \bar{f}\bar{h} + 2\bar{f}) \circ (\tau\ell \mid \tau fh + 2\tau f)$. Let $t = \deg(\tau)$ and note that $(fh + 2f)^* = f^*h^* + 2x^{\deg(h)}f^*$. We obtain that

$$0 = (\bar{\ell} \mid \bar{f}\bar{h} + 2\bar{f}) \circ (\tau\ell \mid \tau fh + 2\tau f) =$$
$$2\bar{\ell}\theta_{\frac{\mathrm{m}}{\alpha}}(x^\alpha)x^{\mathrm{m} - \deg(\ell) - 1 - t}\tau^*\ell^*$$
$$+ \bar{f}\bar{h}\theta_{\frac{\mathrm{m}}{\beta}}(x^\beta)x^{\mathrm{m} - \deg(fh) - 1 - t}\tau^*f^*h^* \qquad (7.8)$$
$$+ 2\bar{f}\bar{h}\theta_{\frac{\mathrm{m}}{\beta}}(x^\beta)x^{\mathrm{m} - \deg(f) - 1 - t}\tau^*f^*$$
$$+ 2\bar{f}\theta_{\frac{\mathrm{m}}{\beta}}(x^\beta)x^{\mathrm{m} - \deg(fh) - 1 - t}\tau^*f^*h^* \quad \mod (x^{\mathrm{m}} - 1).$$

Apply Proposition 7.34 to each addend and $\bar{\ell} = \frac{x^\alpha - 1}{b^*}\lambda$. In addend (7.8), by Proposition 7.39, we may replace $\bar{f}\bar{h}$ by the Hensel lift of $\frac{(x^\beta - 1)\gcd(b, \ell\tilde{g})^*}{f^*b^*}$. The Hensel lift of $(x^\beta - 1)$ and f^* (mod 2) are the same polynomials $(x^\beta - 1)$

and f^*. Moreover, by Lemma 7.41, the addend (7.8) is 0 modulo $(x^m - 1)$. Therefore, by Propositions 7.39 and 7.40, we get that

$$0 = (\bar{\ell} \mid \bar{f}h + 2\bar{f}) \circ (\tau\ell \mid \tau fh + 2\tau f) =$$
$$2\frac{(x^m - 1)}{b^*}\lambda x^{m-\deg(\ell)-1-t}\tau^*\ell^*$$
$$+ 2\frac{(x^m - 1)\gcd(b, \ell)^*}{f^*h^*\gcd(b, \ell\tilde{g})^*}x^{m-\deg(fh)-1-t}\tau^*f^*h^* \tag{7.9}$$
$$+ 2\frac{(x^m - 1)\gcd(b, \ell\tilde{g})^*}{f^*b^*}x^{m-\deg(f)-1-t}\tau^*f^* \quad \bmod (x^m - 1).$$

Clearly, the addend (7.9) is 0 modulo $(x^m - 1)$. Since $\tau = \frac{\gcd(b,\ell\tilde{g})}{\gcd(b,\ell)}$, we have that $(\bar{\ell} \mid \bar{f}h + 2\bar{f}) \circ (\tau\ell \mid \tau fh + 2\tau f)$ is equal to

$$2\frac{(x^m - 1)\gcd(b, \ell\tilde{g})^*}{b^*}\left(\lambda x^{m-\deg(\ell)-1-t}\rho^*\right.$$
$$\left. + x^{m-\deg(f)-1-t}\tau^*\right) \equiv 0 \pmod{(x^m - 1)}. \tag{7.10}$$

This is equivalent, over $\mathbb{Z}_2$, to

$$\frac{(x^m - 1)\gcd(b, \ell\tilde{g})^*}{b^*}\left(\lambda x^{m-\deg(\ell)-1-t}\rho^*\right.$$
$$\left. + x^{m-\deg(f)-1-t}\tau^*\right) \equiv 0 \pmod{(x^m - 1)}.$$

Then,

$$\left(\lambda x^{m-\deg(\ell)-1-t}\rho^*\right.$$
$$\left. + x^{m-\deg(f)-1-t}\tau^*\right) \equiv 0 \pmod{(x^m - 1)}, \tag{7.11}$$

or

$$\left(\lambda x^{m-\deg(\ell)-1-t}\rho^*\right.$$
$$\left. + x^{m-\deg(f)-1-t}\tau^*\right) \equiv 0 \pmod{\left(\frac{b^*}{\gcd(b, \ell\tilde{g})^*}\right)}. \tag{7.12}$$

Since $\left(\frac{b^*}{\gcd(b,\ell\tilde{g})^*}\right)$ divides $(x^m - 1)$, then (7.11) implies (7.12).

The greatest common divisor between ρ and $\left(\frac{b}{\gcd(b,\ell\tilde{g})}\right)$ is 1, then ρ^* is invertible modulo $\left(\frac{b^*}{\gcd(b,\ell\tilde{g})^*}\right)$. Thus,

$$\lambda = \tau^* x^{m-\deg(f)+\deg(\ell)}(\rho^*)^{-1} \quad \bmod \left(\frac{b^*}{\gcd(b, \ell\tilde{g})^*}\right).$$

Let $\lambda_1 = \tau^* x^{m-\deg(f)+\deg(\ell)}(\rho^*)^{-1} \mod \left(\frac{b^*}{\gcd(b,\ell\tilde{g})^*}\right)$. Then, $\lambda = \lambda_1 + \lambda_2$ with $\lambda_2 \equiv 0 \pmod{\left(\frac{b^*}{\gcd(b,\ell\tilde{g})^*}\right)}$.

Finally, we compute $(\bar{\ell} \mid \bar{f}\bar{h} + 2\bar{f}) \circ (\ell \mid fh + 2f)$.

$$
\begin{aligned}
0 = (\bar{\ell} \mid \bar{f}\bar{h} + 2\bar{f}) \circ (\ell \mid fh + 2f) = & \\
2\bar{\ell}\ell\theta_{\frac{m}{\alpha}}(x^\alpha)x^{m-\deg(\ell)-1}\ell^* & \\
+ \bar{f}\bar{h}\theta_{\frac{m}{\beta}}(x^\beta)x^{m-\deg(fh)-1}f^*h^* & \qquad (7.13)\\
+ 2\bar{f}\bar{h}\theta_{\frac{m}{\beta}}(x^\beta)x^{m-\deg(f)-1}f^* & \\
+ 2\bar{f}\theta_{\frac{m}{\beta}}(x^\beta)x^{m-\deg(fh)-1}f^*h^* & \quad \mod (x^m - 1).
\end{aligned}
$$

Apply Proposition 7.34 to each addend. By Lemma 7.41 and replacing $\bar{f}\bar{h}$ by the Hensel lift of $\frac{(x^\beta-1)\gcd(b,\ell\tilde{g})^*}{f^*b^*}$, then the addend (7.13) is $0 \mod (x^m-1)$ and, by Propositions 7.39 and 7.40, $(\bar{\ell} \mid \bar{f}\bar{h} + 2\bar{f}) \circ (\ell \mid fh + 2f)$ is equal to

$$
\begin{aligned}
& 2\frac{(x^m - 1)}{b^*}(\lambda_1 + \lambda_2)x^{m-\deg(\ell)-1}\ell^* \\
& + 2\frac{(x^m - 1)\gcd(b, \ell\tilde{g})^*}{b^*}x^{m-\deg(f)-1} \\
& + 2\frac{(x^m - 1)\gcd(b, \ell\tilde{g})^*}{\gcd(b, \ell\tilde{g})^*}x^{m-\deg(fh)-1} \equiv 0 \pmod{(x^m - 1)}.
\end{aligned}
$$

Since $\lambda_1 = \tau^* x^{m-\deg(f)+\deg(\ell)}(\rho^*)^{-1} \mod \left(\frac{b^*}{\gcd(b,\ell\tilde{g})^*}\right)$, we have that

$$
\begin{aligned}
& 2\frac{(x^m - 1)}{b^*}\lambda_1 x^{m-\deg(\ell)-1}\ell^* \\
& + 2\frac{(x^m - 1)\gcd(b, \ell\tilde{g})^*}{b^*}x^{m-\deg(f)-1} \equiv 0 \pmod{(x^m - 1)}.
\end{aligned}
$$

Therefore, we obtain that

$$
\begin{aligned}
& 2\frac{(x^m - 1)}{b^*}\lambda_2 x^{m-\deg(\ell)-1}\ell^* \\
& + 2\frac{(x^m - 1)\gcd(b, \ell)^*}{\gcd(b, \ell\tilde{g})^*}x^{m-\deg(fh)-1} \equiv 0 \pmod{(x^m - 1)},
\end{aligned}
$$

and then

$$
\begin{aligned}
& 2\frac{(x^m - 1)\gcd(b, \ell)^*}{b^*}\Big(\lambda_2 x^{m-\deg(\ell)-1}\rho^* \\
& + \frac{b^*}{\gcd(b, \ell\tilde{g})^*}x^{m-\deg(fh)-1}\Big) \equiv 0 \pmod{(x^m - 1)}.
\end{aligned}
$$

Arguing similarly to the calculation of λ in (7.10), we obtain that

$$\lambda_2 = \frac{b^*}{\gcd(b, \ell\tilde{g})^*} x^{m-\deg(fh)+\deg(\ell)}(\rho^*)^{-1} \quad \mod \left(\frac{b^*}{\gcd(b, \ell)^*}\right).$$

Now, considering the values of λ_1 and λ_2 and defining properly μ_1 and μ_2, we obtain the expected result.   □

We summarize the previous results in the next theorem.

Theorem 7.43 ([39]). *Let* $\mathcal{C} = \langle (b \mid 0), (\ell \mid fh + 2f) \rangle$ *be a* $\mathbb{Z}_2\mathbb{Z}_4$-*additive cyclic code of type* $(\alpha, \beta; \gamma, \delta; \kappa)$, *with* $fgh = x^\beta - 1$, *and with dual code* $\mathcal{C}^\perp = \langle (\bar{b} \mid 0), (\bar{\ell} \mid \bar{f}\bar{h} + 2\bar{f}) \rangle$, *with* $\bar{f}\bar{g}\bar{h} = x^\beta - 1$. *Let* $\rho = \frac{\ell}{\gcd(b,\ell)}$. *Then,*

i) $\bar{b} = \frac{x^\alpha - 1}{(\gcd(b,\ell))^*} \in \mathbb{Z}_2[x]$,

ii) $\bar{f}\bar{h}$ *is the Hensel lift of the polynomial* $\frac{(x^\beta-1)\gcd(b,\ell\tilde{g})^*}{f^* b^*} \in \mathbb{Z}_2[x]$,

iii) $\bar{f}$ *is the Hensel lift of the polynomial* $\frac{(x^\beta-1)\gcd(b,\ell)^*}{f^* h^* \gcd(b,\ell\tilde{g})^*} \in \mathbb{Z}_2[x]$,

iv)

$$\bar{\ell} = \frac{x^\alpha - 1}{b^*} \left(\frac{\gcd(b, \ell\tilde{g})^*}{\gcd(b, \ell)^*} x^{m-\deg(f)}\mu_1 \right.$$
$$\left. + \frac{b^*}{\gcd(b, \ell\tilde{g})^*} x^{m-\deg(fh)}\mu_2 \right) \in \mathbb{Z}_2[x],$$

where

$$\begin{cases} \mu_1 = x^{\deg(\ell)}(\rho^*)^{-1} \quad \mod \left(\frac{b^*}{\gcd(b,\ell\tilde{g})^*}\right), \\ \mu_2 = x^{\deg(\ell)}(\rho^*)^{-1} \quad \mod \left(\frac{b^*}{\gcd(b,\ell)^*}\right). \end{cases}$$

Note that from Theorems 7.25 and 7.43 one can easily compute the minimal spanning set of the dual code $\mathcal{C}^\perp$ as a $\mathbb{Z}_4$-module, and use the encoding method for $\mathbb{Z}_2\mathbb{Z}_4$-additive cyclic codes described in [1].

Example 7.44. Let $\mathcal{C} = \langle (1+x^2 \mid (1+x+x^2)+2) \rangle$ be the $\mathbb{Z}_2\mathbb{Z}_4$-additive cyclic code of type $(3, 3; 2, 1; 2)$ given in Example 7.14. We have that $b = x^3 - 1, \ell = 1+x^2, f = 1$ and $h = 1+x+x^2$. Then, applying the formulas of Theorem 7.43, we have that $\bar{b} = 1 + x + x^2, \bar{\ell} = 1, \bar{f}\bar{h} = x - 1$, and $\bar{f} = x - 1$. Therefore, $\bar{f}\bar{h}+2\bar{f} = (x-1)+2(x-1) = 1+3x$, and $\mathcal{C}^\perp = \langle (1+x+x^2 \mid 0), (1 \mid 1+3x) \rangle$ is of type $(3, 3; 1, 2; 1)$ and has generator matrix

$$\mathcal{H} = \begin{pmatrix} 1 & 1 & 1 & 0 & 0 & 0 \\ 1 & 0 & 0 & 1 & 3 & 0 \\ 0 & 0 & 1 & 3 & 0 & 1 \end{pmatrix}.$$

△

Example 7.45. We continue with the $\mathbb{Z}_2\mathbb{Z}_4$-additive cyclic code $\mathcal{C}$, considered in Examples 7.3, 7.10 and 7.14, which has been generated by using MAGMA in Example 7.15. Now, we compute the generator polynomials of the additive dual code $\mathcal{C}^\perp$ and check some statements given in Theorem 7.43.

```
> C;
(6, 16) Z2Z4-additive code of type (3, 3; 2, 1; 2)
Generator matrix:
[2 0 2 0 2 2]
[0 2 2 0 0 2]
[0 0 0 1 1 1]
> polyG := GeneratorPolynomials(C);
> polyG; // generator polynomials of the code
<x^3 + 1, x^2 + 1, 1, y^2 + y + 1>
> b := polyG[1]; l := polyG[2]; f := polyG[3]; h := polyG[4];
> g := (y^beta - 1) div (f*h);
>
> Cd := Dual(C);
> polyGd := GeneratorPolynomials(Cd);
> polyGd; // generator polynomials of the additive dual code
<x^2 + x + 1, 1, y + 3, 1>
> bd := polyGd[1]; ld := polyGd[2]; fd := polyGd[3]; hd := polyGd[4];
> gd := (y^beta - 1) div (fd*hd);
>
> bd eq (x^alpha - 1) div PR2!Reverse(Eltseq(Gcd(b, l)));
true
> g2 := PR2!g;
> PR2!fd eq ((x^beta-1) * PR2!Reverse(Eltseq(Gcd(b, l)))) div
             (PR2!Reverse(Eltseq(Gcd(b, l*g2)))) *
             PR2!Reverse(Eltseq(f)) *
             PR2!Reverse(Eltseq(h))));
true
> PR2!(fd*hd) eq ((x^beta-1) * PR2!Reverse(Eltseq(Gcd(b, l)))) div
             (PR2!Reverse(Eltseq(f)) * PR2!Reverse(Eltseq(b))));
true                                                              △
```

7.3 Remarkable $\mathbb{Z}_2\mathbb{Z}_4$-Additive Cyclic Codes

In this section, we show that some remarkable $\mathbb{Z}_2\mathbb{Z}_4$-additive codes are cyclic. First, we determine the existence of perfect and extended perfect $\mathbb{Z}_2\mathbb{Z}_4$-additive cyclic codes. Next, we consider optimal codes with respect to the minimum distance, called MDSS codes, and we show that all of them are cyclic. Finally, we give some examples of self-dual $\mathbb{Z}_2\mathbb{Z}_4$-additive codes, including an infinite family, which are cyclic.

The first family of codes we are considering are the $\mathbb{Z}_2\mathbb{Z}_4$-additive (extended) perfect codes described in Section 6.1.2. Recall, from Theorem 6.11, that for any integer $t \geq 4$ and each $\delta \in \{0, \ldots, \lfloor t/2 \rfloor\}$, there exists a unique (up to equivalence) perfect $\mathbb{Z}_2\mathbb{Z}_4$-linear code C of length $n = 2^t - 1 \geq 15$ such that its $\mathbb{Z}_2\mathbb{Z}_4$-dual is of type $(\alpha, \beta; \gamma, \delta; \kappa)$. By Theorem 3.16, note that C is of type $(\alpha, \beta; \bar{\gamma}, \bar{\delta}; \bar{\kappa})$, where

$$\alpha = 2^{t-\delta} - 1$$
$$\beta = 2^{t-1} - 2^{t-\delta-1}$$
$$\bar{\gamma} = 2^{t-\delta} - t + 2\delta - 1$$
$$\bar{\delta} = 2^{t-1} - 2^{t-\delta-1} - \delta$$
$$\bar{\kappa} = 2^{t-\delta} - t + 2\delta - 1.$$

In the case of $\mathbb{Z}_2\mathbb{Z}_4$-additives extended perfect codes, we have from Theorem 6.10 that, for any integer $t \geq 4$ and each $\delta \in \{0, \ldots, \lfloor t/2 \rfloor\}$, there exists a unique (up to equivalence) extended perfect $\mathbb{Z}_2\mathbb{Z}_4$-linear code C^* of length $2^t \geq 16$ such that its $\mathbb{Z}_2\mathbb{Z}_4$-dual is of type $(\alpha, \beta; \gamma, \delta; \kappa)$. Again, by Theorem 3.16, C^* is of type $(\alpha, \beta; \bar{\gamma}, \bar{\delta}; \bar{\kappa})$, where

$$\alpha = 2^{t-\delta}$$
$$\beta = 2^{t-1} - 2^{t-\delta-1}$$
$$\bar{\gamma} = 2^{t-\delta} - t + 2\delta - 1$$
$$\bar{\delta} = 2^{t-1} - 2^{t-\delta-1} - \delta$$
$$\bar{\kappa} = \bar{\gamma}.$$

In the next theorem, the possible parameters t and δ for a $\mathbb{Z}_2\mathbb{Z}_4$-additive perfect code to be cyclic are given.

Proposition 7.46 ([30]). *Let $t \geq 1$, and let C be a $\mathbb{Z}_2\mathbb{Z}_4$-additive perfect code such that its dual is of type $(\alpha, \beta; \gamma, \delta; \kappa)$. If C is cyclic, then $\delta = 0$ or $t = 2\delta$.*

Proof: Since C is perfect, let $\mathbf{u} = (u \mid u')$ be the codeword in C at distance 1 from the vector $(0, \ldots, 0 \mid 2, 0, \ldots, 0)$. The number of order four coordinates in u' cannot be 1; otherwise, $\mathrm{wt}(2\mathbf{u}) = 2$, which is not possible. Then, $\mathbf{u} = (u \mid 2, 0, \ldots, 0)$ with $\mathrm{wt}_H(u) = 1$. Now, consider the codeword $\mathbf{v} = \mathbf{u}^{(\beta)}$. If $\mathbf{v} \neq \mathbf{u}$, then $\mathbf{v} + \mathbf{u}$ would have weight 2, that is not possible. Consequently, $\mathbf{v}$ must be equal to $\mathbf{u}$ implying that β is a multiple of α, that is, $2^{t-1} - 2^{t-\delta-1}$ is a multiple of $2^{t-\delta} - 1$. Thus,

$$\frac{2^{t-\delta-1}(2^\delta - 1)}{2^{t-\delta} - 1} \in \mathbb{N} \implies \frac{2^\delta - 1}{2^{t-\delta} - 1} \in \mathbb{N}.$$

Therefore, $t - \delta$ divides δ. Since $2\delta \leq t$, $t - \delta \geq \delta$ and the only possibilities are $\delta = 0$ or $t = 2\delta$. □

Let $\mathcal{C}$ be a $\mathbb{Z}_2\mathbb{Z}_4$-additive perfect cyclic code such that its dual is of type $(\alpha, \beta; \gamma, \delta; \kappa)$. On the one hand, if $\delta = 0$, then $\bar{\delta} = 0$ by Theorem 3.16. Hence $\Phi(\mathcal{C})$ is linear by Proposition 5.8. Therefore, $\Phi(\mathcal{C})$ is a Hamming code, and it is well known that the coordinates in $\mathcal{C}$ can be arranged so that it is cyclic. On the other hand, if $t = 2\delta$, then $\mathcal{C}^\perp$ is of type $(2^\delta - 1, 2^{\delta-1}(2^\delta - 1); 0, \delta; 0)$ and hence, by Theorem 3.16, $\mathcal{C}$ is of type $(2^\delta - 1, 2^{\delta-1}(2^\delta - 1); 2^\delta - 1, 2^{\delta-1}(2^\delta - 1) - \delta; 2^\delta - 1)$.

Example 7.47. Let $\mathcal{D}$ be the $\mathbb{Z}_2\mathbb{Z}_4$-additive code of type $(3, 6; 3, 4; 3)$ with parity check matrix

$$\mathcal{H} = \begin{pmatrix} 1 & 1 & 0 & | & 1 & 1 & 2 & 3 & 1 & 0 \\ 0 & 1 & 1 & | & 0 & 1 & 1 & 2 & 3 & 1 \end{pmatrix}.$$

First note that $\mathcal{H}$ generates a $\mathbb{Z}_2\mathbb{Z}_4$-additive code of type $(3, 6; 0, 2; 0)$. Since in $\mathcal{H}$ there is not the zero column and no column is multiple of another one, $\mathcal{D}$ has minimum distance at least 3. Since its type is $(3, 6; 3, 4; 3)$, it has size 2^{11}. Therefore, $\mathcal{D}$ is a $\mathbb{Z}_2\mathbb{Z}_4$-additive perfect code. Note that the second row of $\mathcal{H}$ is the shift of the first one. Also, the first row minus the second one gives the shift of the second row. By Proposition 7.4, $\mathcal{D}^\perp$ is cyclic and so is $\mathcal{D}$. Hence, $\mathcal{D}$ is a $\mathbb{Z}_2\mathbb{Z}_4$-additive perfect cyclic code. △

In [30], it is proved that there are no other $\mathbb{Z}_2\mathbb{Z}_4$-additive perfect cyclic codes with $t = 2\delta$ that the code $\mathcal{D}$ given in Example 7.47. As a result, we have the following statement.

Theorem 7.48 ([30]). *Let $t \geq 1$, and let $\mathcal{C}$ be a $\mathbb{Z}_2\mathbb{Z}_4$-additive perfect cyclic code such that its dual is of type $(\alpha, \beta; \gamma, \delta; \kappa)$. Then,*

i) $\delta = 0$ and $\Phi(\mathcal{C})$ is a Hamming code, or

ii) $t = 2\delta$ and $\mathcal{C}$ is the code given in Example 7.47.

Let $\mathcal{C}^*$ be a $\mathbb{Z}_2\mathbb{Z}_4$-additive extended perfect code such that its dual is of type $(\alpha, \beta; \gamma, \delta; \kappa)$. If $t = 2$ and $\delta = 0$, then $\mathcal{C}^*$ is of type $(4, 0; 1, 0; 1)$, and it is, in fact, the binary repetition code of length 4, that is certainly cyclic. In Example 7.49, we see that this code is the only $\mathbb{Z}_2\mathbb{Z}_4$-additive extended perfect cyclic code with $\alpha = 4$. Moreover, Theorem 7.50 shows that this is the only one for $t \geq 2$, that is, there is not any $\mathbb{Z}_2\mathbb{Z}_4$-additive extended perfect cyclic code for $t \geq 3$.

Example 7.49. In this example, by using MAGMA, it is checked whose $\mathbb{Z}_2\mathbb{Z}_4$-additive extended perfect codes with $\alpha = 4$ are cyclic. If $\alpha = 2^{t-\delta} = 4$, then we have three options: $t = 2, \delta = 0$; $t = 3, \delta = 1$; and $t = 4, \delta = 2$.

Note that the code with $t = 2$ given by MAGMA is cyclic. However, the codes for $t = 3$ and $t = 4$ are not cyclic. Moreover, these two codes are not permutation equivalent to any cyclic code. Therefore, there does not exist any $\mathbb{Z}_2\mathbb{Z}_4$-additive extended perfect codes with $\alpha = 4$ and $t > 2$.

```
> C0 := Z2Z4ExtendedPerfectCode(0, 2);
> IsCyclic(C0);
true
>
> Z4 := Integers(4);
> C1, G1 := Z2Z4ExtendedPerfectCode(1, 3);
> typeC1 := Z2Z4Type(C1);
> alpha := typeC1[1]; beta := typeC1[2];
> G1alpha := ColumnSubmatrix(G1, alpha);
> G1beta := ColumnSubmatrix(G1, alpha + 1, beta);
> isCyclic := false;
> for permAlpha in Sym(alpha) do
for> for permBeta in Sym(beta) do
for|for> newG1 := HorizontalJoin(G1alpha*PermutationMatrix(Z4, permAlpha),
for|for>                          G1beta*PermutationMatrix(Z4, permBeta));
for|for> newC1 := Z2Z4AdditiveCode(newG1, alpha );
for|for> if IsCyclic(newC1) then break; end if;
for|for> end for;
for> end for;
> isCyclic;
false
>
> C2, G2 := Z2Z4ExtendedPerfectCode(2, 4);
> typeC2 := Z2Z4Type(C2);
> alpha := typeC2[1]; beta := typeC2[2];
> G2alpha := ColumnSubmatrix(G2, alpha);
> G2beta := ColumnSubmatrix(G2, alpha + 1, beta);
> isCyclic := false;
> for permAlpha in Sym(alpha) do
for> for permBeta in Sym(beta) do
for|for> newG2 := HorizontalJoin(G2alpha*PermutationMatrix(Z4, permAlpha),
for|for>                          G2beta*PermutationMatrix(Z4, permBeta));
for|for> newC2 := Z2Z4AdditiveCode(newG2, alpha );
for|for> if IsCyclic(newC2) then break; end if;
for|for> end for;
for> end for;
> isCyclic;
false                                                                    △
```

Theorem 7.50 ([30]). *Let $t \geq 3$, and let $\mathcal{C}^*$ be an $\mathbb{Z}_2\mathbb{Z}_4$-additive extended perfect code such that its dual is of type $(\alpha, \beta; \gamma, \delta; \kappa)$. Then, $\mathcal{C}^*$ is not cyclic.*

Now, we consider MDSS $\mathbb{Z}_2\mathbb{Z}_4$-additive cyclic codes. Applying the classical Singleton bound [149] to a $\mathbb{Z}_2\mathbb{Z}_4$-additive code $\mathcal{C}$ of type $(\alpha, \beta; \gamma, \delta; \kappa)$ and minimum distance $d(\mathcal{C})$, as seen in Section 6.4, the following bound is obtained:

$$\frac{d(\mathcal{C}) - 1}{2} \leq \frac{\alpha}{2} + \beta - \frac{\gamma}{2} - \delta. \tag{7.14}$$

According to [28], a code meeting the bound (7.14) is called maximum distance separable with respect to the Singleton bound, briefly MDSS.

By [1, Theorem 19], it is known that $\mathcal{C} = \langle (b \mid 0), (\ell \mid fh + 2f) \rangle$ with $b = x - 1$, $\ell = 1$ and $f = h = 1$ is an MDSS code of type $(\alpha, \beta; \alpha - 1, \beta; \alpha - 1)$. Applying Theorem 7.43 to compute the dual code of $\mathcal{C}$, one obtain that $\mathcal{C}^\perp = \langle (\bar{b} \mid 0), (\bar{\ell} \mid \bar{f}\bar{h} + 2\bar{f}) \rangle$ with $\bar{b} = x^\alpha - 1$, $\bar{\ell} = \theta_\alpha(x)$, $\bar{f} = \theta_\beta(x)$ and $\bar{h} = x - 1$, which is also an MDSS code. In fact, the binary Gray map image of $\mathcal{C}$ is the set of all even weight vectors and the binary Gray map image of $\mathcal{C}^\perp$ is the repetition code. Moreover, these are the only MDSS $\mathbb{Z}_2\mathbb{Z}_4$-additive codes with more than one codeword and minimum distance $d > 1$, as can be seen by Theorem 6.39. Therefore, we obtain the following theorem.

Theorem 7.51. *If $\mathcal{C}$ is an MDSS $\mathbb{Z}_2\mathbb{Z}_4$-additive code, then $\mathcal{C}$ is cyclic.*

Proof: Let $\mathcal{C}$ be an MDSS $\mathbb{Z}_2\mathbb{Z}_4$-additive code of type $(\alpha, \beta; \gamma, \delta; \kappa)$. If $|\mathcal{C}| = 1$, then $\mathcal{C} = \{0\}$ is the zero code and it is cyclic. If $|\mathcal{C}| = 2^{\alpha+2\beta}$, then $\mathcal{C} = \mathbb{Z}_2^\alpha \times \mathbb{Z}_4^\beta$ is the universe code and it is also cyclic. Finally, if $\mathcal{C}$ is an MDSS $\mathbb{Z}_2\mathbb{Z}_4$-additive code such that $1 < |\mathcal{C}| < 2^{\alpha+2\beta}$, by Theorem 6.39, it is either the even code or the repetition code. As we mentioned before, both families of codes are cyclic. ☐

Now, we give a family of self-dual $\mathbb{Z}_2\mathbb{Z}_4$-additive codes that are cyclic. First, Table 7.1 shows the generator polynomials and type of two examples of self-dual $\mathbb{Z}_2\mathbb{Z}_4$-additive cyclic codes. The second code from this table belongs, in fact, to an infinite family of self-dual $\mathbb{Z}_2\mathbb{Z}_4$-additive cyclic codes, given by the following proposition.

Proposition 7.52 ([32, Theorem 4]). *Let α be even and β odd. Let $\mathcal{C} = \langle (b \mid 0), (\ell \mid fh + 2f) \rangle$ be a $\mathbb{Z}_2\mathbb{Z}_4$-additive cyclic code with $b = x^{\frac{\alpha}{2}} - 1$, $\ell = 0$, $h = x^\beta - 1$ and $f = 1$. Then, $\mathcal{C}$ is a self-dual code of type $(\alpha, \beta; \beta + \frac{\alpha}{2}, 0; \frac{\alpha}{2})$.*

Proof. By Theorem 7.43, one obtains that $\bar{b} = x^{\frac{\alpha}{2}} - 1$, $\bar{\ell} = 0$, $\bar{h} = x^\beta - 1$ and $\bar{f} = 1$. Hence $\mathcal{C}$ is self-dual and, by Theorem 7.26, it is of type $(\alpha, \beta; \beta + \frac{\alpha}{2}, 0; \frac{\alpha}{2})$. ☐

Example 7.53. By using MAGMA, the first $\mathbb{Z}_2\mathbb{Z}_4$-additive cyclic code given in Table 7.1 is constructed, and we check that it is indeed self-dual.

generator polynomials	type
$b = x^{10} + x^8 + x^7 + x^3 + x + 1$,	$(14, 7; 8, 3; 7)$
$\ell = x^6 + x^4 + x + 1$,	
$fh = x^4 + 2x^3 + 3x^2 + x + 1, f = 1$	
$b = x^5 - 1, \ell = 0$,	$(10, 5; 10, 0; 5)$
$fh = x^5 - 1, f = 1$	

Table 7.1: Generator polynomials and type of some self-dual $\mathbb{Z}_2\mathbb{Z}_4$-additive cyclic codes

```
> PR2<x> := PolynomialRing(Integers(2));
> PR4<y> := PolynomialRing(Integers(4));
> alpha := 14;
> beta := 7;
> b := x^10 + x^8 + x^7 + x^3 + x + 1;
> l := x^6 + x^4 + x + 1;
> f := PR4!1;
> h := y^4 + 2*y^3 + 3*y^2 + y + 1;
> C := Z2Z4CyclicCode(alpha, beta, b, l, f, h);
> IsSelfDual(C);
true                                                            △
```

7.4   Binary Images of $\mathbb{Z}_2\mathbb{Z}_4$-Additive Cyclic Codes

As we have seen in Chapter 5, if $\mathcal{C}$ is a $\mathbb{Z}_2\mathbb{Z}_4$-additive code, then the binary code $\Phi(\mathcal{C})$ may not be linear. The aim in this section is to give a classification of all $\mathbb{Z}_2\mathbb{Z}_4$-additive cyclic codes with odd β whose Gray map images are binary linear codes.

Let p be a divisor of $x^n - 1$ in $\mathbb{Z}_2[x]$ with odd n and let ζ be a primitive nth root of unity over $\mathbb{Z}_2$. The polynomial $(p \otimes p)$ is defined as the divisor of $x^n - 1$ in $\mathbb{Z}_2[x]$ whose roots are the products $\zeta^i \zeta^j$ such that ζ^i and ζ^j are roots of p.

7.4.1   Images Under the Gray Map

In [161], the author characterizes all quaternary cyclic codes of odd length whose Gray map images are binary linear codes, as it is shown in the following theorem.

Theorem 7.54 ([161, Theorem 20]). Let $\mathcal{D} = \langle fh + 2f \rangle$ be a $\mathbb{Z}_4$-additive cyclic code of odd length β, where $fgh = x^\beta - 1$. The following properties are equivalent:

i) $\gcd(\tilde{f}, (\tilde{g} \otimes \tilde{g})) = 1$ in $\mathbb{Z}_2[x]$,

ii) $\phi(\mathcal{D})$ is a binary linear code of length 2β.

We show that Theorem 7.54 gives a necessary but not a sufficient condition for the Gray map image $\Phi(\mathcal{C})$ of a $\mathbb{Z}_2\mathbb{Z}_4$-additive cyclic code $\mathcal{C} = \langle (b \mid 0), (\ell \mid fh + 2f) \rangle$ to be linear, in terms of the generator polynomials.

Corollary 7.55. Let $\mathcal{C} = \langle (b \mid 0), (\ell \mid fh + 2f) \rangle$ be a $\mathbb{Z}_2\mathbb{Z}_4$-additive cyclic code where $fgh = x^\beta - 1$ such that $\Phi(\mathcal{C})$ is linear. Then, $\gcd(\tilde{f}, (\tilde{g} \otimes \tilde{g})) = 1$ in $\mathbb{Z}_2[x]$.

Proof. It follows from Theorem 7.54 and the fact that $\phi(\mathcal{C}_Y)$ is linear by Lemma 5.11. □

Corollary 7.56. Let $\mathcal{C} = \langle (b \mid 0), (\ell \mid fh + 2f) \rangle$ be a $\mathbb{Z}_2\mathbb{Z}_4$-additive cyclic code where $fgh = x^\beta - 1$. If b divides $\ell\tilde{g}$, then $\Phi(\mathcal{C})$ is linear if and only if $\phi(\mathcal{C}_Y)$ is linear.

Proof. By Proposition 7.24, $\mathcal{C}_b = \langle (b \mid 0), (\tilde{\mu}\ell\tilde{g} \mid 2f) \rangle$. If b divides $\ell\tilde{g}$, then $\mathcal{C}_b = \langle (b \mid 0), (0 \mid 2f) \rangle$ is separable and, therefore, by Remark 2.5, $\kappa_2 = 0$. Therefore, by Lemma 5.13, $\Phi(\mathcal{C})$ is linear if and only if $\phi(\mathcal{C}_Y)$ is linear and the statement follows by Theorem 7.54. □

Example 7.57. Let $\mathcal{C} = \langle (b \mid 0), (\ell \mid fh + 2f) \rangle$ be a $\mathbb{Z}_2\mathbb{Z}_4$-additive cyclic code with $\alpha = 2, \beta = 7$, and $b = x - 1$. Since $\deg(\ell) < \deg(b)$, we have that either $\ell = 0$, and the code is separable, or $\ell = 1$. Note that $x^7 - 1 = (x-1)pq$ over $\mathbb{Z}_4$, where $p = 3 + x + 2x^2 + x^3$ and $q = 3 + 2x + 3x^2 + x^3$.

We have that $\mathcal{C}_Y = \langle fh + 2f \rangle$ is a quaternary cyclic code of length 7. The values of f, g and h for which $\phi(\mathcal{C}_Y)$ is linear are given in [72]. First, Table 7.2 shows all possible polynomials f, g, h for which $\phi(\mathcal{C}_Y)$ is not linear. For all these cases and $\ell \in \{0, 1\}$, by Lemma 5.11, we have that $\Phi(\mathcal{C})$ is not linear. Second, we consider the values of f, g and h for which $\phi(\mathcal{C}_Y)$ is linear. Clearly, by Corollary 7.19, if $\ell = 0$, then the code is separable and $\Phi(\mathcal{C})$ is also linear. In the case $\ell = 1$, by using MAGMA [46], we obtain that $\Phi(\mathcal{C})$ is linear in all cases except one, as it is shown in Table 7.3, where $*$ indicates that it takes all possible values. △

Lemma 7.58 ([31]). Let $\mathcal{C} = \langle (b \mid 0), (\ell \mid fh + 2f) \rangle$ be a $\mathbb{Z}_2\mathbb{Z}_4$-additive cyclic code where $fhg = x^\beta - 1$. Then, $\mathcal{C}$ can also be generated by $(b \mid 0), (\ell\tilde{g} \mid 2fg), (\ell' \mid fh)$, where $\ell' = \ell - \tilde{\mu}\ell\tilde{g}$, and μ is as in (7.3).

f	g	h
$(x-1)p$	q	1
$(x-1)q$	p	1
$(x-1)$	pq	1
p	q	$(x-1)$
q	p	$(x-1)$
p	$(x-1)q$	1
q	$(x-1)p$	1

Table 7.2: Values of f, g, h for which $\phi(\mathcal{C}_Y)$ and $\Phi(\mathcal{C})$ are not linear, where $\mathcal{C}$ and $\mathcal{C}_Y = \langle fh + 2f \rangle$ are as considered in Example 7.57

Proof. By Lemma 7.23, there exists λ and μ in $\mathbb{Z}_4[x]$ such that $\lambda h + \mu g = 1$. Let $\ell' = \ell - \tilde{\mu}\ell\tilde{g}$ and $\mathcal{D} = \langle (b \mid 0), (\ell\tilde{g} \mid 2fg), (\ell' \mid fh) \rangle$. We shall prove that $\mathcal{D} = \mathcal{C}$. By Proposition 7.24, $(\tilde{\mu}\ell\tilde{g} \mid 2f) \in \mathcal{C}$. Hence, $(\ell \mid fh + 2f) - (\tilde{\mu}\ell\tilde{g} \mid 2f) = (\ell' \mid fh) \in \mathcal{C}$. Therefore, $\mathcal{D} \subseteq \mathcal{C}$. Finally, since $(0 \mid 2fh) \in \mathcal{D}$ and $(\ell\tilde{g} \mid 2fg) \in \mathcal{D}$, we have that $(\tilde{\mu}\ell\tilde{g} \mid 2f) \in \mathcal{D}$. Therefore, $(\ell \mid fh + 2f) = (\ell' \mid fh) + (\tilde{\mu}\ell\tilde{g} \mid 2f) \in \mathcal{D}$, that implies $\mathcal{C} \subseteq \mathcal{D}$. $\qquad\square$

Theorem 7.59 ([31]). *Let $\mathcal{C} = \langle (b \mid 0), (\ell \mid fh + 2f) \rangle$ be a $\mathbb{Z}_2\mathbb{Z}_4$-additive cyclic code where $fgh = x^\beta - 1$. The following properties are equivalent:*

i) $\gcd(\frac{\tilde{f}b}{\gcd(b,\ell\tilde{g})}, (\tilde{g} \otimes \tilde{g})) = 1$ *in* $\mathbb{Z}_2[x]$,

ii) $\Phi(\mathcal{C})$ *is a binary linear code of length* $\alpha + 2\beta$.

Proof. By Lemma 7.58, $\mathcal{C} = \langle (b \mid 0), (\ell\tilde{g} \mid 2fg), (\ell' \mid fh) \rangle$ for $\ell' = \ell - \tilde{\mu}\ell\tilde{g}$ and, by Lemma 7.21, $\mathcal{C}_b = \langle (b \mid 0), (\ell\tilde{g} \mid 2fg), (0 \mid 2fh) \rangle$. Define the set $S = \{(0 \mid p) \in \mathcal{C}\}$, and the subcode $\mathcal{D} = \langle (\ell' \mid fh), S \rangle \subseteq \mathcal{C}$.

We have that $\mathcal{C} = \langle \mathcal{C}_b, \mathcal{D} \rangle$. The subcode $\mathcal{C}_b$ is generated by all the codewords of $\mathcal{C}$ of order at most two and, therefore $\Phi(\mathcal{C}_b)$ is linear. For $\mathbf{u} = (u \mid u') \in \mathcal{C}$ and $\mathbf{v} = (v \mid v') \in \mathcal{C}_b$, $2\mathbf{u} * \mathbf{v} = \mathbf{0} \in \mathcal{C}$. Therefore, by Corollary 5.4, to check whether $\Phi(\mathcal{C})$ is linear, we only have to consider whether $2\mathbf{u} * \mathbf{v} \in \mathcal{C}$ for $\mathbf{u}, \mathbf{v} \in \mathcal{D}$. Moreover, for $\mathbf{u}, \mathbf{v} \in \mathcal{D}$, we have that $2\mathbf{u} * \mathbf{v} = (\mathbf{0} \mid 2u' * v') \in \mathcal{D}$. Then, $\Phi(\mathcal{C})$ is linear if and only if $2\mathbf{u} * \mathbf{v} \in \mathcal{D}$ for $\mathbf{u}, \mathbf{v} \in \mathcal{D}$. That is, $\Phi(\mathcal{C})$ is linear if and only if $\Phi(\mathcal{D})$ is linear.

Let q_g be the Hensel lift of $\frac{g}{\gcd(b, \ell\tilde{g})}$, q_b the Hensel lift of $\frac{b}{\gcd(b, \ell\tilde{g})}$, and q_ℓ the Hensel lift of ℓ. Now, we prove that $\mathcal{D} = \langle (\ell' \mid fh), (0 \mid 2fgq_b) \rangle$. First, since $(b \mid 0)$ and $(\ell\tilde{g} \mid 2fg)$ belong to $\mathcal{C}$, we have that $(0 \mid 2fgq_b)) = q_g q_\ell * (b \mid 0) + q_b * (\ell\tilde{g} \mid 2fg) \in \mathcal{C}$ and, therefore, $(0 \mid 2fgq_b) \in \mathcal{D}$. Consider $(0 \mid p) \in \mathcal{D}$. Since $\mathcal{D} \subseteq \mathcal{C}$, we have that $(0 \mid p) = a_1 * (b \mid 0) + a_2 * (\ell\tilde{g} \mid 2fg) + a_3 * (\ell' \mid fh)$

$\Phi(\mathcal{C})$ linear		
f	g	h
$*$	1	$*$
$*$	$(x-1)$	$*$
$(x-1)$	p	q
$(x-1)$	q	p
1	$(x-1)p$	q
1	$(x-1)q$	p
1	p	$(x-1)q$
1	q	$(x-1)p$

$\Phi(\mathcal{C})$ not linear		
f	g	h
1	pq	$(x-1)$

Table 7.3: Values of f, g, h and linearity of $\Phi(\mathcal{C})$ for which $\phi(\mathcal{C}_Y)$ is linear, where $\mathcal{C}$ and $\mathcal{C}_Y = \langle fh + 2f \rangle$ are as considered in Example 7.57

for some $a_1, a_2, a_3 \in \mathbb{Z}_4[x]$. Since $\tilde{a}_1 b + \tilde{a}_2 \ell \tilde{g} = 0$, we have that $\tilde{a}_1 \frac{b}{\gcd(b, \ell\tilde{g})} = \tilde{a}_2 \frac{\ell\tilde{g}}{\gcd(b, \ell\tilde{g})}$, where $\frac{\ell\tilde{g}}{\gcd(b, \ell\tilde{g})}$ does not divide $\frac{b}{\gcd(b, \ell\tilde{g})}$. Therefore, $\tilde{a}_2 | \frac{b}{\gcd(b, \ell\tilde{g})}$; that is, $\tilde{a}_2 = \tilde{a} \frac{b}{\gcd(b, \ell\tilde{g})}$ for $\tilde{a} \in \mathbb{Z}_2[x]$. Hence, $a_1 \star (b \mid 0) + a_2 \star (\ell\tilde{g} \mid 2fg) = (0 \mid a_2 2fg) = a \star (0 \mid 2fgq)$, where a is the Hensel lift of $\tilde{a}$.

We have that $\Phi(\mathcal{C})$ is linear if and only if $\Phi(\mathcal{D})$ is linear, where $\mathcal{D} = \langle (\ell' \mid fh), (0 \mid 2fgq_b) \rangle$. We consider the polynomials $f' = fq_b$, $h' = \frac{h}{q_b}$ and $g' = g$, that satisfy $f'g'h' = x^\beta - 1$, with f', g' and h' coprime factors. Then, $\mathcal{D} = \langle (\ell' \mid f'h'), (0 \mid 2f') \rangle = \langle (\ell' \mid f'h' + 2f') \rangle$. By the construction of $\mathcal{D}$, it is easy to check that $\mathcal{D}_b = \langle (0 \mid 2f') \rangle$. Therefore, by Corollary 5.14, $\Phi(\mathcal{D})$ is a binary linear code if and only if $\phi(\mathcal{D}_Y)$ is linear. Finally, by Theorem 7.54, $\phi(\mathcal{D}_Y) = \phi(\langle f'h' + 2f' \rangle)$ is linear if and only if $\gcd(\tilde{f}', (\tilde{g}' \otimes \tilde{g}')) = 1$, this is equivalent to $\gcd(\frac{\tilde{f}b}{\gcd(b, \ell\tilde{g})}, (\tilde{g} \otimes \tilde{g})) = 1$. $\qquad\square$

The following result describes an infinite family of $\mathbb{Z}_2\mathbb{Z}_4$-additive cyclic codes with odd β whose Gray map images are linear. The splitting field of the polynomial $x^\beta - 1 \in \mathbb{Z}_2[x]$ is the extension field K of $\mathbb{Z}_2$ such that $x^\beta - 1$ factorizes in $K[x]$ into linear factors.

Theorem 7.60 ([31]). *Let $\mathcal{C} = \langle (b \mid 0), (\ell \mid fh + 2f) \rangle$ be a $\mathbb{Z}_2\mathbb{Z}_4$-additive cyclic code where $fgh = x^\beta - 1$. If $g = 1$ or $g = x^s - 1$, where s divides β, then $\Phi(\mathcal{C})$ is linear.*

Proof. Clearly, $(\tilde{g} \otimes \tilde{g}) = \tilde{g}$ since the set of roots of $\tilde{g}$ is a multiplicative subgroup of the splitting field of $x^\beta - 1$. Then, $\gcd(\frac{\tilde{f}b}{\gcd(b,\ell\tilde{g})}, (\tilde{g} \otimes \tilde{g})) = 1$. By Theorem 7.59, $\Phi(\mathcal{C})$ is linear. $\qquad\qquad\qquad\qquad\qquad\qquad\square$

Example 7.61. Let $\mathcal{C} = \langle (x-1 \mid 0), (1 \mid fh+2f) \rangle$ be a $\mathbb{Z}_2\mathbb{Z}_4$-additive cyclic code with $\alpha = 2$ and $\beta = 7$. As it is shown in Table 7.3, when $g = 1$ or $g = x - 1$, the code $\Phi(\mathcal{C})$ is linear. $\qquad\qquad\qquad\qquad\qquad\qquad\triangle$

By Theorem 5.28 (or Theorem 5.50), we have that there exists a $\mathbb{Z}_2\mathbb{Z}_4$-additive code for a given type $(\alpha, \beta; \gamma, \delta; \kappa)$ which image is a binary linear code. Considering Theorem 7.59, the next example illustrates that for a given type $(\alpha, \beta; \gamma, \delta; \kappa)$ there does not always exist a $\mathbb{Z}_2\mathbb{Z}_4$-additive cyclic code $\mathcal{C}$ such that $\Phi(\mathcal{C})$ is linear.

Example 7.62. We show that there does not exist a $\mathbb{Z}_2\mathbb{Z}_4$-additive cyclic code $\mathcal{C}$ of type $(2, 7; 2, 3; \kappa)$ such that $\Phi(\mathcal{C})$ is linear, for any possible value of κ. Let $\mathcal{C} = \langle (b \mid 0), (\ell \mid fh + 2f) \rangle$ be a $\mathbb{Z}_2\mathbb{Z}_4$-additive cyclic code of type $(2, 7; 2, 3; \kappa)$, where $fgh = x^7 - 1 = (x-1)(3+x+2x^2+x^3)(3+2x+3x^2+x^3)$ over $\mathbb{Z}_4$.

By Theorem 7.26, $\deg(g) = 3$ and $\deg(b) = \deg(h) \leq 2$. Let $\{p_1, p_2\} = \{3 + x + 2x^2 + x^3, 3 + 2x + 3x^2 + x^3\}$. Without loss of generality, assume that $g = p_1$ and, since $\deg(h) \leq 2$, we have that p_2 divides f. It is easy to see that $\gcd(p_2, (\tilde{p}_1 \otimes \tilde{p}_1)) \neq 1$ and therefore $\gcd(\frac{\tilde{f}b}{\gcd(b,\ell\tilde{g})}, (\tilde{g} \otimes \tilde{g})) \neq 1$. Hence, by Theorem 7.59, there does not exist a $\mathbb{Z}_2\mathbb{Z}_4$-additive cyclic code $\mathcal{C}$ of type $(2, 7; 2, 3; \kappa)$ such that $\Phi(\mathcal{C})$ is linear, for any possible value of κ. From Tables 7.2 and 7.3, we can see that if $\deg(g) = 3$ and $\deg(h) \leq 2$, then the code $\Phi(\mathcal{C})$ is not linear. $\qquad\qquad\qquad\qquad\qquad\qquad\triangle$

7.4.2 Images Under the Neachaev-Gray Map

In this section, we show that the convenient map to obtain a cyclic structure on the binary images from $\mathbb{Z}_2\mathbb{Z}_4$-additive cyclic codes is considering the Nechaev-Gray map instead of the Gray map.

The *Nechaev permutation* is the following permutation:

$$\tau = (2, n+2)(4, n+4) \cdots (2i, n+2i) \cdots (n-1, 2n-1) \in \mathcal{S}_{2n}$$

with odd n. This permutation acts on vectors of $\mathbb{Z}_2^{2n}$ as usual, that is,

$$\tau(v_1, v_2, \ldots, v_{2n}) = (v_{\tau^{-1}(1)}, v_{\tau^{-1}(2)}, \ldots, v_{\tau^{-1}(2n)}),$$

so it defines a map from $\mathbb{Z}_2^{2n}$ into $\mathbb{Z}_2^{2n}$. For this subsection, consider the Gray map ϕ from $\mathbb{Z}_4^n$ into $\mathbb{Z}_2^{2n}$ defined as

$$\phi(w_1, \ldots, w_n) = (z_1, \ldots, z_n, z_1', \ldots, z_n'),$$

where $(z_i, z_i') = \phi(w_i)$ for $i \in \{1, \ldots, n\}$. Let ψ be the map from $\mathbb{Z}_4^n$ into $\mathbb{Z}_2^{2n}$ defined by $\psi = \tau \circ \phi$, with odd n. The map ψ is called the *Nechaev-Gray map* [161]. The *extended Nechaev-Gray map* Ψ is the map from $\mathbb{Z}_2^\alpha \times \mathbb{Z}_4^\beta$ to $\mathbb{Z}_2^{\alpha+2\beta}$ given by

$$\Psi(u \mid u') = (u \mid \psi(u')).$$

Now, we introduce the double cyclic property on binary codes described in [40]. The family of double cyclic codes is a subfamily of generalized quasi-cyclic codes [148]. Let r and s be positive integers. A binary linear code C of length $r + s$ is a $\mathbb{Z}_2$-double cyclic code if the set of coordinates can be partitioned into two subsets, one with the first r coordinates and another one with the last s, such that any cyclic shift of the coordinates of both subsets leaves the code invariant. These codes can be identified as submodules of the $\mathbb{Z}_2[x]$-module $\mathbb{Z}_2[x]/\langle x^r - 1 \rangle \times \mathbb{Z}_2[x]/\langle x^s - 1 \rangle$ [40].

Theorem 7.63 ([40]). *Let C be a $\mathbb{Z}_2$-double cyclic code of length $r+s$. Then, C is generated by*

$$\langle (b \mid 0), (\ell \mid a) \rangle \subseteq \mathbb{Z}_2[x]/\langle x^r - 1 \rangle \times \mathbb{Z}_2[x]/\langle x^s - 1 \rangle,$$

where $a \mid (x^s - 1)$, $b \mid (x^r - 1)$ and we can assume that $\deg(\ell) < \deg(b)$.

From [40], we also have the following useful result.

Theorem 7.64 ([40]). *Let $\mathcal{C}$ be a $\mathbb{Z}_2\mathbb{Z}_4$-additive cyclic code. If $\Phi(\mathcal{C})$ is a binary linear code, then $\Psi(\mathcal{C})$ is a $\mathbb{Z}_2$-double cyclic code.*

If $\mathcal{C}$ is a $\mathbb{Z}_2\mathbb{Z}_4$-additive cyclic code such that $\Phi(\mathcal{C})$ is linear, then the following theorem shows the relation between the generator polynomials of $\mathcal{C}$ and the generators of its $\mathbb{Z}_2$-double cyclic image.

Theorem 7.65 ([31]). *Let $\mathcal{C} = \langle (b \mid 0), (\ell \mid fh + 2f) \rangle$ be a $\mathbb{Z}_2\mathbb{Z}_4$-additive cyclic code, where $fgh = x^\beta - 1$. If $\gcd(\frac{\tilde{f}b}{\gcd(b, \ell\tilde{g})}, (\tilde{g} \otimes \tilde{g})) = 1$ in $\mathbb{Z}_2[x]$, then $\Psi(\mathcal{C})$ is a $\mathbb{Z}_2$-double cyclic code of length $\alpha + 2\beta$ generated by*

$$\Psi(\mathcal{C}) = \langle (b \mid 0), (\ell' \mid \tilde{f}^2\tilde{h}) \rangle,$$

where $\ell' = \tilde{p}\ell \pmod{b} \in \mathbb{Z}_2[x]/\langle x^\alpha - 1 \rangle$ such that $p(fh + 2f) = \psi^{-1}(\tilde{f}^2\tilde{h})$.

Proof. We have that $\gcd(\frac{\tilde{f}b}{\gcd(b,\ell\tilde{g})}, (\tilde{g} \otimes \tilde{g})) = 1$. By Theorems 7.59 and 7.64, $\psi(\mathcal{C}_Y)$ is a $\mathbb{Z}_2$-double cyclic code of length 2β. By a similar argument of [161, Theorem 15], but taking care of the binary part, we obtain that $\Psi(\mathcal{C}) = \langle (b \mid 0), (\ell' \mid \tilde{f}^2\tilde{h}) \rangle$ for some $\ell' \in \mathbb{Z}_2[x]/\langle x^\alpha - 1 \rangle$. Since $\psi^{-1}(\tilde{f}^2\tilde{h}) \in \mathcal{C}_Y = \langle fh + 2f \rangle$, there exists $p \in \mathbb{Z}_4[x]/\langle x^\beta - 1 \rangle$ such that $\psi^{-1}(\tilde{f}^2\tilde{h}) = p(fh+2f)$. Therefore, $(\ell' \mid \psi^{-1}(\tilde{f}^2\tilde{h})) = p' \star (b \mid 0) + p \star (\ell \mid fh + 2f)$ for some p'. Thus, $\ell' = \tilde{p}\ell \pmod{b} \in \mathbb{Z}_2[x]/\langle x^\alpha - 1 \rangle$. $\qquad\square$

$[\alpha, \beta]$	generator polynomials	$(\tilde{g} \otimes \tilde{g})$
[2,3]	$b = x^2 - 1, \ell = x + 1,$ $fh = x^3 - 1, f = 1$	1
[3,3]	$b = x^2 + x + 1, \ell = x,$ $fh = x^2 + x + 1, f = 1$	1
[9,3]	$b = x^8 + x^7 + x^6 + x^5 + x^4 + x^3 + x^2 + x + 1,$ $\ell = x^6 + x^3 + 1,$ $fh = x^3 - 1, f = x - 1$	1
[4,7]	$b = x^3 + x^2 + x + 1, \ell = x^2 + 1,$ $fh = x^4 + 2x^3 + 3x^2 + x + 1, f = 1$	$x^6 + x^5 + x^4 +$ $x^3 + x^2 + x + 1$
[4,7]	$b = x^4 - 1, \ell = x^3 + x^2 + x + 1,$ $fh = x^6 + x^5 + x^4 + x^3 + x^2 + x + 1,$ $f = x^3 + 3x^2 + 2x + 3$	$x + 1$
[4,7]	$b = x^4 - 1, \ell = x^3 + x^2 + x + 1,$ $fh = x^7 - 1, f = x^3 + 3x^2 + 2x + 3$	1
[7,7]	$b = x^7 - 1, \ell = x^6 + x^5 + x^3,$ $fh = x^6 + x^5 + x^4 + x^3 + x^2 + x + 1,$ $f = x^3 + 3x^2 + 2x + 3$	$x + 1$
[7,7]	$b = x^6 + x^5 + x^4 + x^3 + x^2 + x + 1,$ $\ell = x^3 + x + 1,$ $fh = x^7 - 1, f = x - 1$	1
[7,7]	$b = x^3 + x + 1, \ell = x,$ $fh = x^4 + 2x^3 + 3x^2 + x + 1, f = x - 1$	$x^6 + x^5 + x^4 +$ $x^3 + x^2 + x + 1$

Table 7.4: $\mathbb{Z}_2$-double cyclic codes from $\mathbb{Z}_2\mathbb{Z}_4$-additive cyclic codes

Example 7.66. Let $\mathcal{C} = \langle (1 + x + x^2 \mid 0), (1 \mid 3 + x + x^2) \rangle \subset \mathbb{Z}_2^3 \times \mathbb{Z}_4^3$ be a $\mathbb{Z}_2\mathbb{Z}_4$-additive cyclic code, where $f = 1$ and $h = 1 + x + x^2$. Then, we have that $\psi^{-1}(\tilde{f}^2\tilde{h}) = 3 + x + 3x^2$: $\tilde{f}^2\tilde{h} \longrightarrow (1,1,1,0,0,0) \xrightarrow{\tau^{-1}} (1,0,1,0,1,0) \xrightarrow{\phi^{-1}} (3,1,3) \longrightarrow 3 + x + 3x^2$. Therefore, the polynomial p such that $p(fh + 2f) = \psi^{-1}(\tilde{f}^2\tilde{h})$ is $p = 3x$ and then

$$\Psi(\mathcal{C}) = \langle (1 + x + x^2 \mid 0), (x \mid 1 + x + x^2) \rangle \subset \mathbb{Z}_2^3 \times \mathbb{Z}_2^6.$$

$\triangle$

Example 7.67. Table 7.4 gives some $\mathbb{Z}_2\mathbb{Z}_4$-additive cyclic codes $\mathcal{C} = \langle (b \mid 0), (\ell \mid fh + 2f) \rangle$, satisfying $\gcd(\frac{\tilde{f}b}{\gcd(b, \ell\tilde{g})}, (\tilde{g} \otimes \tilde{g})) = 1$. By Theorem 7.65, we have that $\Psi(\mathcal{C})$ are $\mathbb{Z}_2$-double cyclic codes of length $\alpha + 2\beta$. $\triangle$

7.5 Rank and Kernel of the Image of $\mathbb{Z}_2\mathbb{Z}_4$-Additive Cyclic Codes

In Section 7.4.1, it is completely characterized when a $\mathbb{Z}_2\mathbb{Z}_4$-additive cyclic code $\mathcal{C}$ has a binary linear image under the Gray map Φ, just by considering its generator polynomials. In this section, the rank and dimension of the kernel for those codes whose Gray map image $C = \Phi(\mathcal{C})$ is not linear is given. Results on the rank and kernel of a general $\mathbb{Z}_2\mathbb{Z}_4$-linear code C are shown in Chapter 5.

Let $\mathcal{C}$ be a $\mathbb{Z}_2\mathbb{Z}_4$-additive code, and $C = \Phi(\mathcal{C})$. Recall that

$$\mathcal{R}(\mathcal{C}) = \Phi^{-1}(\langle C \rangle), \quad \text{and}$$
$$\mathcal{K}(\mathcal{C}) = \Phi^{-1}(K(C)) = \Phi^{-1}(\{x \in \mathbb{Z}_2^{\alpha+2\beta} : C + x = C\}).$$

Since $K(C) \subseteq C \subseteq \langle C \rangle$, we have that $\mathcal{K}(\mathcal{C}) \subseteq \mathcal{C} \subseteq \mathcal{R}(\mathcal{C})$.

In Section 5.4, we have seen that if C is a $\mathbb{Z}_2\mathbb{Z}_4$-linear code, then $\langle C \rangle$ and $K(C)$ are both $\mathbb{Z}_2\mathbb{Z}_4$-linear codes. Therefore, $\mathcal{R}(\mathcal{C})$ and $\mathcal{K}(\mathcal{C})$ are both $\mathbb{Z}_2\mathbb{Z}_4$-additive codes. In the next subsections, we show that if the code $\mathcal{C}$ is a $\mathbb{Z}_2\mathbb{Z}_4$-additive cyclic code, then $\mathcal{R}(\mathcal{C})$ and $\mathcal{K}(\mathcal{C})$ are also $\mathbb{Z}_2\mathbb{Z}_4$-additive cyclic.

The following proposition is useful to relate the generator polynomials of $\mathcal{C}$ to the generator polynomials of $\mathcal{R}(\mathcal{C})$ and $\mathcal{K}(\mathcal{C})$.

Proposition 7.68 ([33]). *Let* $\mathcal{C}_1 = \langle (b \mid 0), (\ell \mid fh + 2f) \rangle$ *and* $\mathcal{C}_2 = \langle (b' \mid 0), (\ell' \mid f'h' + 2f') \rangle$ *be* $\mathbb{Z}_2\mathbb{Z}_4$-*additive cyclic codes with* $\mathcal{C}_1 \subseteq \mathcal{C}_2$. *Then,*

i) f' divides f,

ii) $\gcd(b', \ell')$ divides $\gcd(b, \ell)$.

Proof. Since $\mathcal{C}_1 \subseteq \mathcal{C}_2$, we have that $(\mathcal{C}_1)_Y = \langle fh + 2f \rangle \subseteq (\mathcal{C}_2)_Y = \langle f'h' + 2f' \rangle$. Therefore, by [72, Theorem 3], f' divides f.

Since $\mathcal{C}_1$ and $\mathcal{C}_2$ are $\mathbb{Z}_2\mathbb{Z}_4$-additive cyclic codes, clearly $(\mathcal{C}_1)_X = \langle \gcd(b, \ell) \rangle$ and $(\mathcal{C}_2)_X = \langle \gcd(b', \ell') \rangle$. Finally, $(\mathcal{C}_1)_X \subseteq (\mathcal{C}_2)_X$ implies that $\gcd(b', \ell')$ divides $\gcd(b, \ell)$. □

In order to study the rank and kernel of the Gray map image of a $\mathbb{Z}_2\mathbb{Z}_4$-additive code $\mathcal{C}$, it is necessary to consider the code $\mathcal{C}_b$. By Proposition 7.24, we have that $\mathcal{C}_b = \langle (b \mid 0), (\tilde{\mu}\ell\tilde{g} \mid 2f) \rangle$. Note that, if $\mathcal{C} = \mathcal{C}_b$, then $\Phi(\mathcal{C})$ is linear. In this case, $\delta = 0$ and, by Theorem 7.26, $g = 1$. Therefore $\mathcal{C}_b = \langle (b \mid 0), (\ell \mid 2f) \rangle$.

Example 7.69. By using MAGMA, we construct a $\mathbb{Z}_2\mathbb{Z}_4$-additive cyclic code $\mathcal{C}$, compute the $\mathbb{Z}_2\mathbb{Z}_4$-additive codes $\mathcal{R}(\mathcal{C})$ and $\mathcal{K}(\mathcal{C})$, and check that they are cyclic. We also compute the values of $\text{rank}(\Phi(\mathcal{C}))$ and $\ker(\Phi(\mathcal{C}))$. This code appears later in Examples 7.74 and 7.85.

```
> PR2<x> := PolynomialRing(Integers(2));
> PR4<y> := PolynomialRing(Integers(4));
> alpha :=  2;
> beta := 7;
> b := x - 1;
> l := PR2!1;
> f := y^3 + 2*y^2 + y + 3;
> h := y - 1;
> C := Z2Z4CyclicCode(alpha, beta, b, l, f, h);
> Z2Z4Type(C);
[ 2, 7, 2, 3, 2 ]
> SC, SCbin := SpanZ2Code(C);
> KC, KCbin := KernelZ2Code(C);
> IsCyclic(SC) and IsCyclic(KC);
true
> DimensionOfSpanZ2(C);
11
> DimensionOfSpanZ2(C) eq Dimension(SCbin);
true
> DimensionOfKernelZ2(C);
5
> DimensionOfKernelZ2(C) eq Dimension(KCbin);
true                                                                    △
```

7.5.1   Rank of the Image of $\mathbb{Z}_2\mathbb{Z}_4$-Additive Cyclic Codes

In this section, the rank of the Gray map image of a $\mathbb{Z}_2\mathbb{Z}_4$-additive cyclic code $\mathcal{C}$ is given. We prove that $\mathcal{R}(\mathcal{C})$ is also cyclic and we establish some properties of its generator polynomials. We also show that there does not exist a $\mathbb{Z}_2\mathbb{Z}_4$-additive cyclic code for all possible values of the rank, in contrast to what is exhibited in Proposition 5.23 and Theorem 5.28 for a $\mathbb{Z}_2\mathbb{Z}_4$-additive code.

The following theorem shows that if $\mathcal{C}$ is a $\mathbb{Z}_2\mathbb{Z}_4$-additive cyclic code, then $\mathcal{R}(\mathcal{C})$ is also cyclic. This theorem is a generalization of the case when $\mathcal{C}$ is a quaternary cyclic code [72].

Theorem 7.70 ([33]). *Let $\mathcal{C}$ be a $\mathbb{Z}_2\mathbb{Z}_4$-additive cyclic code. Then, $\mathcal{R}(\mathcal{C})$ is a $\mathbb{Z}_2\mathbb{Z}_4$-additive cyclic code.*

Proof. Let $\mathbf{x} \in \mathcal{R}(\mathcal{C})$. By Corollary 5.19, $\mathcal{R}(\mathcal{C})$ is a $\mathbb{Z}_2\mathbb{Z}_4$-additive code generated by $\mathcal{C}$ and $\{2\mathbf{v} * \mathbf{w} : \mathbf{v}, \mathbf{w} \in \mathcal{C}\}$. Then, $\mathbf{x} = \mathbf{u} + 2\mathbf{v} * \mathbf{w}$ for some $\mathbf{u}, \mathbf{v}, \mathbf{w} \in \mathcal{C}$. Since $\mathcal{C}$ is a $\mathbb{Z}_2\mathbb{Z}_4$-additive cyclic code, $\pi(\mathbf{u}), \pi(\mathbf{v}), \pi(\mathbf{w}) \in \mathcal{C}$ and

$2\pi(\mathbf{v}) * \pi(\mathbf{w}) \in \mathcal{R}(\mathcal{C})$. Thus, $\pi(\mathbf{x}) = \pi(\mathbf{u}) + 2\pi(\mathbf{v}) * \pi(\mathbf{w}) \in \mathcal{R}(\mathcal{C})$ and $\mathcal{R}(\mathcal{C})$ is a $\mathbb{Z}_2\mathbb{Z}_4$-additive cyclic code.   □

Proposition 7.71 ([33]). *Let* $\mathcal{C} = \langle (b \mid 0), (\ell \mid fh + 2f) \rangle$ *be a* $\mathbb{Z}_2\mathbb{Z}_4$-*additive cyclic code, where* $fgh = x^\beta - 1$. *Then,*

$$\alpha - \deg(b) + \deg(h) + 2\deg(g) \leq \operatorname{rank}(\Phi(\mathcal{C})) \leq$$
$$\min\left(\alpha + \beta + \deg(g) - \deg(\gcd(b, \ell\tilde{g})),\right.$$
$$\left.\alpha - \deg(b) + \deg(h) + 2\deg(g) + \binom{\deg(g)}{2}\right).$$

Proof. Straightforward from Theorems 5.28 and 7.26.   □

Theorem 7.72 ([33]). *Let* $\mathcal{C} = \langle (b \mid 0), (\ell \mid fh + 2f) \rangle$ *be a* $\mathbb{Z}_2\mathbb{Z}_4$-*additive cyclic code. Then,*

$$\mathcal{R}(\mathcal{C}) = \langle (b_r \mid 0), (\ell_r \mid fh + 2\frac{f}{r}) \rangle,$$

where r *divides* f *and* b_r *divides* b.

Proof. By Theorem 7.70, $\mathcal{R}(\mathcal{C})$ is a $\mathbb{Z}_2\mathbb{Z}_4$-additive cyclic code, and therefore $\mathcal{R}(\mathcal{C}) = \langle (b_r \mid 0), (\ell_r \mid f_r h_r + 2f_r) \rangle$. Since $(b \mid 0) \in \mathcal{R}(\mathcal{C})$, it is clear that b_r divides b. By Corollary 5.19, the δ generators of order four of $\mathcal{C}$ and $\mathcal{R}(\mathcal{C})$ are the same. Since $\mathcal{C} \subseteq \mathcal{R}(\mathcal{C})$, we have that $f_r h_r = fh$. Then, $g_r = g$. By Proposition 7.68, we know that f_r divides f and hence there exists $r \in \mathbb{Z}_4[x]$ such that $f_r = \frac{f}{r}$ and $h_r = hr$. Therefore, $\mathcal{R}(\mathcal{C}) = \langle (b_r \mid 0), (\ell_r \mid fh + 2\frac{f}{r}) \rangle$.   □

Let $\mathcal{C} = \langle (b \mid 0), (\ell \mid fh + 2f) \rangle$ be a $\mathbb{Z}_2\mathbb{Z}_4$-additive cyclic code and $\mathcal{R}(\mathcal{C}) = \langle (b_r \mid 0), (\ell_r \mid f_r h_r + 2f_r) \rangle$. The following example shows a code $\mathcal{C}$ with $b_r \neq b$.

Example 7.73. We have that $x^7 - 1 = (x - 1)p_1 p_2$ over $\mathbb{Z}_4$, where $p_1 = 3 + x + 2x^2 + x^3$ and $p_2 = 3 + 2x + 3x^2 + x^3$. Let $\mathcal{C} = \langle (x-1 \mid 0), (1 \mid (x-1)+2) \rangle$ be a $\mathbb{Z}_2\mathbb{Z}_4$-additive cyclic code, with $\beta = 7$, $f = 1$, $h = x - 1$ and $g = p_1 p_2$. If we compute $\mathcal{R}(\mathcal{C})$, then we obtain that $\mathcal{R}(\mathcal{C}) = \langle (1 \mid 0), (0 \mid (x-1)+2) \rangle$.   △

As it is shown in Theorem 5.28, there exists a $\mathbb{Z}_2\mathbb{Z}_4$-additive code of type $(\alpha, \beta; \gamma, \delta; \kappa)$ for any possible value of the rank. Nevertheless, the following example gives that, for a particular type $(\alpha, \beta; \gamma, \delta; \kappa)$, it is not possible to construct a $\mathbb{Z}_2\mathbb{Z}_4$-additive cyclic code with a specific, and valid, value of the rank.

Example 7.74. Let $\mathcal{C} = \langle (b \mid 0), (\ell \mid fh+2f) \rangle$ be a $\mathbb{Z}_2\mathbb{Z}_4$-additive cyclic code of type $(2, 7; 2, 3; \kappa)$, where $fgh = x^7 - 1$. By Theorem 5.28, rank$(\Phi(\mathcal{C})) \in \{8, 9, 10, 11\}$. In this example, we show that there does not exist any $\mathbb{Z}_2\mathbb{Z}_4$-additive cyclic code $\mathcal{C}$ of type $(2, 7; 2, 3; \kappa)$ with rank$(\Phi(\mathcal{C})) \in \{8, 9, 10\}$. We have that $x^7 - 1 = (x - 1)(3 + x + 2x^2 + x^3)(3 + 2x + 3x^2 + x^3)$. Let $\{p_1, p_2\} = \{3 + x + 2x^2 + x^3, 3 + 2x + 3x^2 + x^3\}$.

By Theorem 7.26, $\deg(g) = 3$ and $\deg(b) = \deg(h) \leq 2$. Without loss of generality, we assume that $g = p_1$ and, since $\deg(h) \leq 2$, we have that p_2 divides f. It is easy to see that $\gcd(\tilde{p}_2, (\tilde{p}_1 \otimes \tilde{p}_1)) \neq 1$ and therefore $\gcd(\frac{\tilde{f}b}{\gcd(b, \ell\tilde{g})}, (\tilde{g} \otimes \tilde{g})) \neq 1$. Hence, by Theorem 7.59, there does not exist a $\mathbb{Z}_2\mathbb{Z}_4$-additive cyclic code of type $(2, 7; 2, 3; \kappa)$ such that $\Phi(\mathcal{C})$ is linear. Thus, rank$(\Phi(\mathcal{C})) \neq 8$.

By Theorem 7.72, $\mathcal{R}(\mathcal{C}) = \langle (b_r \mid 0), (\ell_r \mid f_r h_r + 2f_r) \rangle$, where r divides f and $h_r = hr$. Since rank$(\Phi(\mathcal{C})) \in \{9, 10, 11\}$ and $|\mathcal{R}(\mathcal{C})| = 4^3 2^{2+\deg(r)} \leq 2^{11}$, we have that $\deg(r) \leq 3$. Moreover, since $\gcd(\tilde{p}_2, (\tilde{p}_1 \otimes \tilde{p}_1)) \neq 1$, we have that p_2 must divide r. Therefore, $\deg(r) \geq 3$ and, by the previous argument, we know that $r = p_2$. So, rank$(\Phi(\mathcal{C})) \notin \{9, 10\}$.

Finally, we give $\mathbb{Z}_2\mathbb{Z}_4$-additive cyclic codes of type $(2, 7; 2, 3; \kappa)$ such that rank$(\Phi(\mathcal{C})) = 11$, for different values of κ. Recall that $\kappa \leq \min\{\alpha, \gamma\} = 2$ and $\kappa = \alpha - \deg(\gcd(b, \ell\tilde{g}))$. Then,

   i) $\kappa = 2$: $\mathcal{C} = \langle (x - 1 \mid 0), (1 \mid p(x - 1) + 2p) \rangle$, for $p \in \{p_1, p_2\}$.

   ii) $\kappa = 1$: $\mathcal{C} = \langle (x - 1 \mid 0), (0 \mid p(x - 1) + 2p) \rangle$, for $p \in \{p_1, p_2\}$.

   iii) $\kappa = 0$: In this case, $\deg(\gcd(b, \ell\tilde{g})) = 2$ and therefore, $\gcd(b, \ell\tilde{g}) = x^2 - 1$. Note that $\tilde{p}_1$ and $\tilde{p}_2$ are not divisors of $x^2 - 1$ over $\mathbb{Z}_2$, thus there does not exist ℓ with $\deg(\ell) < 2$ such that $\ell\tilde{g} = x^2 - 1$. Thus, there does not exist a $\mathbb{Z}_2\mathbb{Z}_4$-additive cyclic code of type $(2, 7; 2, 3; 0)$.

$\triangle$

Let $\mathcal{C} = \langle fh + 2f \rangle$ be a quaternary cyclic code. We have that $\mathcal{R}(\mathcal{C}) = \langle fh + 2\frac{f}{r} \rangle$, where r divides f. By the definition of $\mathcal{R}(\mathcal{C})$, we have that $\mathcal{R}(\mathcal{C})$ is the minimum quaternary cyclic code containing $\mathcal{C}$ whose image under the Gray map is linear. Thus, r is the polynomial of minimum degree dividing f and satisfying that $\langle fh + 2\frac{f}{r} \rangle$ has linear image. This is equivalent, by Theorem 7.54, to the condition $\gcd\left(\frac{\tilde{f}}{r}, (\tilde{g} \otimes \tilde{g})\right) = 1$. Therefore, r is the Hensel lift of $\gcd(\tilde{f}, (\tilde{g} \otimes \tilde{g}))$. As a result, we obtain the following statement.

Proposition 7.75 ([33]). *Let $\mathcal{C} = \langle fh + 2f \rangle$ be a quaternary cyclic code of length n, where $fgh = x^n - 1$. Then,*

$$\mathcal{R}(\mathcal{C}) = \langle fh + 2\frac{f}{r} \rangle,$$

where r is the Hensel lift of $\gcd(\tilde{f},(\tilde{g}\otimes\tilde{g}))$.

Now, we generalize the previous statement to a $\mathbb{Z}_2\mathbb{Z}_4$-additive cyclic code $\mathcal{C}$. First, we establish the generator polynomials in the case that $\phi(\mathcal{C}_Y)$ is linear.

Proposition 7.76 ([33]). *Let $\mathcal{C} = \langle(b\mid0),(\ell\mid fh+2f)\rangle$ be a $\mathbb{Z}_2\mathbb{Z}_4$-additive cyclic code, where $fgh = x^\beta - 1$, such that $\Phi(\mathcal{C})$ is not linear and $\phi(\mathcal{C}_Y)$ is linear. Then,*

$$\mathcal{R}(\mathcal{C}) = \langle(b_r\mid0),(\ell_r\mid fh+2f)\rangle,$$

where $b_r = \gcd(b,\tilde{\mu}\tilde{g}\ell)$, μ is as in (7.3), and $\ell_r = \ell - \tilde{\mu}\tilde{g}\ell \pmod{b_r}$.

Proof. Let $\mathcal{C} = \langle(b\mid0),(\ell\mid fh+2f)\rangle$ be a $\mathbb{Z}_2\mathbb{Z}_4$-additive cyclic code. We have that $\mathcal{C}_Y = \langle fh+2f\rangle$. Let $\mathcal{G}$ be a generator matrix of $\mathcal{C}$ as in (2.7) and let $\{\mathbf{u}_i = (u_i\mid u_i')\}_{i=1}^\gamma$ be the rows of order two and $\{\mathbf{v}_j = (v_j\mid v_j')\}_{j=1}^\delta$ the rows of order four of $\mathcal{G}$. By Corollary 5.19, $\mathcal{R}(\mathcal{C}) = \langle\mathcal{C},\{2\mathbf{v}_j*\mathbf{v}_k\}_{1\le j<k\le\delta}\rangle$. Since $\Phi(\mathcal{C})$ is not linear, there exist $j,k\in\{1,\ldots,\delta\}$ such that $2\mathbf{v}_j*\mathbf{v}_k = (\mathbf{0}\mid2v_j'*v_k')\notin\mathcal{C}$. Since $\phi(\mathcal{C}_Y)$ is linear, $2v_j'*v_k'\in\langle2fh,2f\rangle$. Moreover, $\langle(0\mid2fh)\rangle\subseteq\mathcal{C}$ and therefore $\mathcal{R}(\mathcal{C}) = \langle\mathcal{C},(0\mid2f)\rangle$. Considering the generator polynomials of $\mathcal{R}(\mathcal{C})$ and μ as in (7.3), we have that

$$\begin{aligned}\mathcal{R}(\mathcal{C}) &= \langle(b\mid0),(\ell\mid fh+2f),(0\mid2f)\rangle\\ &= \langle(b\mid0),(\ell-\tilde{\mu}\tilde{g}\ell\mid fh),(\tilde{\mu}\tilde{g}\ell\mid2f),(0\mid2f)\rangle\\ &= \langle(b\mid0),(\tilde{\mu}\tilde{g}\ell\mid0),(\ell-\tilde{\mu}\tilde{g}\ell\mid fh+2f)\rangle\\ &= \langle(\gcd(b,\tilde{\mu}\tilde{g}\ell)\mid0),(\ell-\tilde{\mu}\tilde{g}\ell\mid fh+2f)\rangle.\end{aligned}$$

Therefore, considering the generator polynomials of $\mathcal{R}(\mathcal{C})$ in standard form, we have that $b_r = \gcd(b,\tilde{\mu}\tilde{g}\ell)$ and $\ell_r = \ell - \tilde{\mu}\tilde{g}\ell \pmod{b_r}$. $\qquad\square$

Theorem 7.77 ([33]). *Let $\mathcal{C} = \langle(b\mid0),(\ell\mid fh+2f)\rangle$ be a $\mathbb{Z}_2\mathbb{Z}_4$-additive cyclic code, where $fgh = x^\beta - 1$. Then,*

$$\mathcal{R}(\mathcal{C}) = \langle(b_r\mid0),(\ell_r\mid fh+2\frac{f}{r})\rangle,$$

where r is the Hensel lift of $\gcd(\tilde{f},(\tilde{g}\otimes\tilde{g}))$, $b_r = \gcd(b,\tilde{\mu}\tilde{g}\ell)$, μ is as in (7.3), and $\ell_r = \ell - \tilde{\mu}\tilde{g}\ell \pmod{b_r}$.

Proof. Let $\mathcal{C} = \langle(b\mid0),(\ell\mid fh+2f)\rangle$ be a $\mathbb{Z}_2\mathbb{Z}_4$-additive cyclic code. By Theorem 7.72, $\mathcal{R}(\mathcal{C}) = \langle(b_r\mid0),(\ell_r\mid fh+2\frac{f}{r})\rangle$, where r is a divisor of f and b_r divides b.

Consider the quaternary cyclic code $\mathcal{C}_Y = \langle fh + 2f \rangle$. By Lemma 5.32, we have that $\mathcal{R}(\mathcal{C})_Y = \mathcal{R}(\mathcal{C}_Y)$. Therefore, by Proposition 7.75, $\mathcal{R}(\mathcal{C}_Y) = \langle fh + 2\frac{\ell}{r} \rangle$, where r is the Hensel lift of $\gcd(\tilde{f}, (\tilde{g} \otimes \tilde{g}))$. Note that $\mathcal{R}(\mathcal{C}_Y) = \langle \mathcal{C}_Y, 2\frac{\ell}{r} \rangle$.

From Corollary 5.19, $\mathcal{R}(\mathcal{C}) = \langle \mathcal{C}, \{2\mathbf{v}_j * \mathbf{v}_k\}_{1 \leq j < k \leq \delta} \rangle$, where $\{\mathbf{v}_i = (v_i \mid v_i')\}_{i=1}^{\delta}$ are the rows of order four of a generator matrix of $\mathcal{C}$ as in (2.7). Note that, for all $\mathbf{v}_j, \mathbf{v}_k$, $1 \leq j \leq k \leq \delta$, $2\mathbf{v}_j * \mathbf{v}_k = (0 \mid 2v_j' * v_k')$, where $2v_j' * v_k' \in \mathcal{R}(\mathcal{C}_Y) = \langle \mathcal{C}_Y, 2\frac{\ell}{r} \rangle$ and, therefore, $(0 \mid 2v_j' * v_k') \in \langle \mathcal{C}, (0 \mid 2\frac{\ell}{r}) \rangle$. Hence, we have that $\mathcal{R}(\mathcal{C}) = \langle \mathcal{C}, (0 \mid 2\frac{\ell}{r}) \rangle$. Therefore, for μ as in (7.3),

$$
\begin{aligned}
\mathcal{R}(\mathcal{C}) &= \langle (b \mid 0), (\ell \mid fh + 2f), (0 \mid 2f) \rangle \\
&= \langle (b \mid 0), (\ell - \tilde{\mu}\tilde{g}\ell \mid fh), (\tilde{\mu}\tilde{g}\ell \mid 2f), (0 \mid 2\frac{\ell}{r}) \rangle \\
&= \langle (b \mid 0), (\tilde{\mu}\tilde{g}\ell \mid 0), (\ell - \tilde{\mu}\tilde{g}\ell \mid fh + 2\frac{\ell}{r}) \rangle \\
&= \langle (\gcd(b, \tilde{\mu}\tilde{g}\ell) \mid 0), (\ell - \tilde{\mu}\tilde{g}\ell \mid fh + 2\frac{\ell}{r}) \rangle.
\end{aligned}
$$

From the last equation, and considering the generator polynomials of $\mathcal{R}(\mathcal{C})$ in standard form, we have that $b_r = \gcd(b, \tilde{\mu}\tilde{g}\ell)$ and $\ell_r = \ell - \tilde{\mu}\tilde{g}\ell$ (mod b_r). □

Example 7.78. Let $\alpha = 3$ and $\beta = 7$. Consider, as in Example 7.57, $\mathcal{C} = \langle (x - 1 \mid 0), (0 \mid x - 1) \rangle$, where $f = x - 1$ and $h = 1$. As we have seen, $\Phi(\mathcal{C})$ is not linear. Then, by Theorem 7.77, we have that $\mathcal{R}(\mathcal{C}) = \langle (1 \mid 0), (0 \mid (x - 1) + 2) \rangle$, where $r = f = x - 1$. △

Example 7.79. Let $\alpha = 3$ and $\beta = 15$. Consider $\mathcal{C} = \langle (x - 1 \mid 0), (1 \mid fh + 2f) \rangle$, where $f = 1 + 3x + 2x^2 + x^4$ and $h = (x - 1)(1 + x + x^2 + x^3 + x^4)$. Then, by Theorem 7.77, we have that $\mathcal{R}(\mathcal{C}) = \langle (1 \mid 0), (0 \mid fh + 2\frac{\ell}{r}) \rangle$, where $r = f = 1 + 3x + 2x^2 + x^4$. △

7.5.2 Kernel of the Image of $\mathbb{Z}_2\mathbb{Z}_4$-Additive Cyclic Codes

In this section, the kernel of the Gray map image of a $\mathbb{Z}_2\mathbb{Z}_4$-additive cyclic code $\mathcal{C}$ is given. We prove that the code $\mathcal{K}(\mathcal{C})$ is also cyclic, and we establish some properties of its generator polynomials. We also show that there does not exist a $\mathbb{Z}_2\mathbb{Z}_4$-additive cyclic code for all possible values of the dimension of the kernel as for general $\mathbb{Z}_2\mathbb{Z}_4$-additive codes.

The following theorem shows that if $\mathcal{C}$ is a $\mathbb{Z}_2\mathbb{Z}_4$-additive cyclic code, then $\mathcal{K}(\mathcal{C})$ is also cyclic.

Theorem 7.80 ([33]). *Let $\mathcal{C}$ be a $\mathbb{Z}_2\mathbb{Z}_4$-additive cyclic code. Then, $\mathcal{K}(\mathcal{C})$ is a $\mathbb{Z}_2\mathbb{Z}_4$-additive cyclic code.*

Proof. By Corollary 5.38, $\mathcal{K}(\mathcal{C})$ is a $\mathbb{Z}_2\mathbb{Z}_4$-additive code. Let $\mathbf{u} = (u \mid u') \in \mathcal{K}(\mathcal{C})$. We just have to show that $\pi(\mathbf{u}) \in \mathcal{K}(\mathcal{C})$, that is, $2\pi(\mathbf{u}) * \mathbf{w} \in \mathcal{C}$ for all $\mathbf{w} \in \mathcal{C}$. Note that $2\pi(\mathbf{u}) * \mathbf{w} = \pi(2\mathbf{u} * \pi^{-1}(\mathbf{w}))$. We have that $\mathbf{u} \in \mathcal{K}(\mathcal{C})$ and $\pi^{-1}(\mathbf{w}) \in \mathcal{C}$, therefore $2\mathbf{u} * \pi^{-1}(\mathbf{w}) \in \mathcal{C}$. Since the code $\mathcal{C}$ is cyclic, $\pi(2\mathbf{u} * \pi^{-1}(\mathbf{w})) \in \mathcal{C}$, which gives that $2\pi(\mathbf{u}) * \mathbf{w} \in \mathcal{C}$, and $\pi(\mathbf{u}) \in \mathcal{K}(\mathcal{C})$. □

Proposition 7.81 ([33]). *Let* $\mathcal{C} = \langle (b \mid 0), (\ell \mid fh + 2f) \rangle$ *be a* $\mathbb{Z}_2\mathbb{Z}_4$-*additive cyclic code, where* $fgh = x^\beta - 1$. *Then,*

$$\alpha - \deg(b) + \deg(h) + \deg(g) \leq \ker(\Phi(\mathcal{C}))$$
$$\leq \alpha - \deg(b) + \deg(h) + 2\deg(g).$$

Proof. Straightforward from Corollary 2.13 and Theorem 7.26. □

Note that the upper bound of $\ker(\Phi(\mathcal{C}))$ is sharp when $\Phi(\mathcal{C})$ is linear, i.e., when $\mathcal{C} = \mathcal{K}(\mathcal{C})$. Moreover, the lower bound is tight when $\mathcal{K}(\mathcal{C})$ has only the all-zero vector and all order two codewords; that is, when $\mathcal{C} = \mathcal{C}_b$.

Now, we establish the kernel of the image of a $\mathbb{Z}_2\mathbb{Z}_4$-additive cyclic code taking into account its generator polynomials.

Corollary 7.82 ([33]). *Let* $\mathcal{C} = \langle (b \mid 0), (\ell \mid fh + 2f) \rangle$ *be a* $\mathbb{Z}_2\mathbb{Z}_4$-*additive cyclic code, where* $fhg = x^\beta - 1$. *Then,*

$$\mathcal{K}(\mathcal{C}) = \langle (b_k \mid 0), (\ell_k \mid f_k h_k + 2f_k) \rangle,$$

where $f_k h_k g_k = x^\beta - 1$, f *divides* f_k *and* $\gcd(b, \ell)$ *divides* $\gcd(b_k, \ell_k)$.

Proof. By Theorem 7.80, $\mathcal{K}(\mathcal{C})$ is a $\mathbb{Z}_2\mathbb{Z}_4$-additive cyclic code, and therefore $\mathcal{K}(\mathcal{C}) = \langle (b_k \mid 0), (\ell_k \mid f_k h_k + 2f_k) \rangle$, where $f_k h_k g_k = x^\beta - 1$. Since $\mathcal{K}(\mathcal{C}) \subseteq \mathcal{C}$, the result follows from Proposition 7.68. □

Lemma 7.83 ([33]). *Let* $\mathcal{C} = \langle (b \mid 0), (\ell \mid fh+2f) \rangle$ *be a* $\mathbb{Z}_2\mathbb{Z}_4$-*additive cyclic code, where* $fgh = x^\beta - 1$. *Let* $\langle (b \mid 0), (\ell_k \mid fhk + 2f) \rangle \subset \mathcal{C}$, *where* k *divides* g. *Then,* $\ell_k = \tilde{k}\ell + (1 - \tilde{k})\tilde{\mu}\ell\tilde{g} \pmod{b}$, *where* μ *is as in* (7.3).

Proof. In Lemma 7.58, it is proved that $\mathcal{C} = \langle (b \mid 0), (\ell' \mid fh), (\tilde{\mu}\ell\tilde{g} \mid 2f) \rangle$, where $\ell' = \ell - \tilde{\mu}\ell\tilde{g}$, and μ as in (7.3). Since $(\ell_k \mid fhk + 2f) \in \mathcal{C}$ and $(\ell_k \mid fhk + 2f) = c_1 \star (b \mid 0) + c_2 \star (\ell' \mid fh) + c_3 \star (\tilde{\mu}\ell\tilde{g} \mid 2f)$, we obtain $c_2 = k$, $c_3 = 1$ and $\ell_k = \tilde{k}\ell' + \tilde{\mu}\ell\tilde{g} \pmod{b} = \tilde{k}\ell + (1 - \tilde{k})\tilde{\mu}\ell\tilde{g} \pmod{b}$. □

Theorem 7.84 ([33]). *Let* $\mathcal{C} = \langle (b \mid 0), (\ell \mid fh + 2f) \rangle$ *be a* $\mathbb{Z}_2\mathbb{Z}_4$-*additive cyclic code, where* $fgh = x^\beta - 1$. *Then,*

$$\mathcal{K}(\mathcal{C}) = \langle (b \mid 0), (\ell_k \mid fhk + 2f) \rangle,$$

where k *divides* g, $\ell_k = \tilde{k}\ell + (1 - \tilde{k})\tilde{\mu}\ell\tilde{g} \pmod{b}$, *and* μ *is as in* (7.3).

Proof. By Theorem 7.80, $\mathcal{K}(\mathcal{C})$ is a $\mathbb{Z}_2\mathbb{Z}_4$-additive cyclic code, and then $\mathcal{K}(\mathcal{C}) = \langle (b_k \mid 0), (\ell_k \mid f_k h_k + 2f_k) \rangle$. Recall that $\mathcal{C}_0 = \{(v \mid 0) \in \mathcal{C}\}$. First, note that $\mathcal{C}_0 = \mathcal{K}(\mathcal{C})_0$, and hence $b_k = b$. Second, $\mathcal{C}_b \subseteq \mathcal{K}(\mathcal{C}) \subseteq \mathcal{C}$ and, by Propositions 7.24 and 7.68, we conclude that $f_k = f$. Next, since $\mathcal{K}(\mathcal{C})_Y = \langle f h_k + 2f \rangle \subseteq \mathcal{C}_Y$, with an argument analogous to that of [72, Theorem 9], we obtain that $h_k = hk$ with k a divisor of g. Finally, by Lemma 7.83, $\ell_k = \tilde{k}\ell + (1 - \tilde{k})\tilde{\mu}\ell\tilde{g} \pmod{b}$. $\qquad\square$

Theorem 5.50 shows that there exists a $\mathbb{Z}_2\mathbb{Z}_4$-additive code for all possible values of the dimension of the kernel for a given type $(\alpha, \beta; \gamma, \delta; \kappa)$. Considering the last theorem, the next example illustrates that this result is not true for $\mathbb{Z}_2\mathbb{Z}_4$-additive cyclic codes; i.e., for a given type $(\alpha, \beta; \gamma, \delta; \kappa)$ there does not always exist a $\mathbb{Z}_2\mathbb{Z}_4$-additive cyclic code for all possible values of the kernel. Furthermore, it shows that there does not always exist a $\mathbb{Z}_2\mathbb{Z}_4$-additive cyclic code for a given type $(\alpha, \beta; \gamma, \delta; \kappa)$.

Example 7.85. By Theorem 5.50, there exists a $\mathbb{Z}_2\mathbb{Z}_4$-additive code $\mathcal{C}$ of type $(2, 7; 2, 3; \kappa)$ with $\ker(\Phi(\mathcal{C})) = k_d$, for all $k_d \in \{5, 6, 8\}$. In this example, we show that there does not exist any $\mathbb{Z}_2\mathbb{Z}_4$-additive cyclic code $\mathcal{C}$ of type $(2, 7; 2, 3; \kappa)$ with $\ker(\Phi(\mathcal{C})) \in \{6, 8\}$. Let $\mathcal{C} = \langle (b \mid 0), (\ell \mid fh + 2f) \rangle$ be a $\mathbb{Z}_2\mathbb{Z}_4$-additive cyclic code of type $(2, 7; 2, 3; \kappa)$, where $fgh = x^7 - 1 = (x - 1)p_1 p_2$ over $\mathbb{Z}_4$, with p_1 and p_2 as in Example 7.74.

By Theorem 7.26, $\deg(g) = 3$ and $\deg(b) = \deg(h) \leq 2$. Without loss of generality, we assume that $g = p_1$ and, since $\deg(h) \leq 2$, we have that p_2 divides f. We have already proved, in Example 7.74, that there does not exist a $\mathbb{Z}_2\mathbb{Z}_4$-additive cyclic code of type $(2, 7; 2, 3; \kappa)$ such that $\Phi(\mathcal{C})$ is linear. Thus, $\ker(\Phi(\mathcal{C})) \neq 8$.

By Theorem 7.84, $\mathcal{K}(\mathcal{C}) = \langle (b \mid 0), (\ell_k \mid fhk + 2f) \rangle$, where k divides g. By the previous argument, $k \neq 1$ and then we have that $k = g = p_1$ and $\mathcal{K}(\mathcal{C}) = \langle (b \mid 0), (\ell_k \mid 2f) \rangle$. Therefore, $\mathcal{K}(\mathcal{C})$ does not contain codewords of order 4, thus $\ker(\Phi(\mathcal{C})) = \gamma + \delta = 5$.

Finally, we give $\mathbb{Z}_2\mathbb{Z}_4$-additive cyclic codes of type $(2, 7; 2, 3; \kappa)$ such that $\ker(\Phi(\mathcal{C})) = 5$, for different values of κ. Recall that $\kappa \leq \min\{\alpha, \gamma\} = 2$ and $\kappa = \alpha - \deg(\gcd(b, \ell\tilde{g}))$. Then,

i) $\kappa = 2$: $\mathcal{C} = \langle (x - 1 \mid 0), (1 \mid p(x - 1) + 2p) \rangle$, for $p \in \{p_1, p_2\}$.

ii) $\kappa = 1$: $\mathcal{C} = \langle (x - 1 \mid 0), (0 \mid p(x - 1) + 2p) \rangle$, for $p \in \{p_1, p_2\}$.

iii) $\kappa = 0$: As in Example 7.74, there does not exist any $\mathbb{Z}_2\mathbb{Z}_4$-additive cyclic code of type $(2, 7; 2, 3; 0)$.

$\triangle$

The statement in Theorem 7.84 is also true for any maximal $\mathbb{Z}_2\mathbb{Z}_4$-additive cyclic subcode of a $\mathbb{Z}_2\mathbb{Z}_4$-additive cyclic code $\mathcal{C}$ whose Gray map image is a linear subcode of $\Phi(\mathcal{C})$.

Corollary 7.86 ([33]). *Let* $\mathcal{C} = \langle (b \mid 0), (\ell \mid fh + 2f) \rangle$ *be a* $\mathbb{Z}_2\mathbb{Z}_4$-*additive cyclic code, where* $fgh = x^\beta - 1$. *If* $\mathcal{C}_1$ *is a maximal* $\mathbb{Z}_2\mathbb{Z}_4$-*additive cyclic subcode of* $\mathcal{C}$ *such that* $\Phi(\mathcal{C}_1)$ *is linear, then* $\mathcal{C}_1 = \langle (b \mid 0), (\ell_k \mid fhk + 2f) \rangle$, *where* k *divides* g, $\ell_k = \tilde{k}\ell + (1 - \tilde{k})\tilde{\mu}\ell\tilde{g}$ (mod b), *and* μ *is as in* (7.3).

The kernel of a binary code is the intersection of all its maximal linear subspaces [121]. Therefore, if $\mathcal{C}_1, \mathcal{C}_2, \dots, \mathcal{C}_s$ are all the maximal subcodes of a $\mathbb{Z}_2\mathbb{Z}_4$-additive code $\mathcal{C}$ such that $\Phi(\mathcal{C}_i)$ is a linear subcode of $\Phi(\mathcal{C})$, for $1 \leq i \leq s$, then

$$\mathcal{K}(\mathcal{C}) = \bigcap_{i=1}^{s} \mathcal{C}_i. \tag{7.15}$$

In [72], it is proved that if $\mathcal{C}_1 = \langle fh_1 + 2f \rangle$ and $\mathcal{C}_2 = \langle fh_2 + 2f \rangle$ are quaternary cyclic codes of odd length n, then $\mathcal{C}_1 \cap \mathcal{C}_2 = \langle f \operatorname{lcm}(h_1, h_2) + 2f \rangle$. We give a similar result for $\mathbb{Z}_2\mathbb{Z}_4$-additive cyclic codes.

Proposition 7.87 ([33]). *Let* $\mathcal{C} = \langle (b \mid 0), (\ell \mid fh + 2f) \rangle$ *be a* $\mathbb{Z}_2\mathbb{Z}_4$-*additive cyclic code, where* $fgh = x^\beta - 1$. *Let* $\mathcal{C}_1 = \langle (b \mid 0), (\ell_{k_1} \mid fhk_1 + 2f) \rangle$ *and* $\mathcal{C}_2 = \langle (b \mid 0), (\ell_{k_2} \mid fhk_2 + 2f) \rangle$ *be maximal* $\mathbb{Z}_2\mathbb{Z}_4$-*additive cyclic subcodes of* $\mathcal{C}$ *whose images under the Gray map are linear subcodes of* $\Phi(\mathcal{C})$. *Then,*

$$\mathcal{C}_1 \cap \mathcal{C}_2 = \langle (b \mid 0), (\ell_{k'} \mid fhk' + 2f) \rangle,$$

where $k' = \operatorname{lcm}(k_1, k_2)$, $\ell_{k'} = \tilde{k}'\ell + (1 - \tilde{k}')\tilde{\mu}\ell\tilde{g}$ (mod b), *and* μ *is as in* (7.3).

Proof. Let $\mathcal{C} = \langle (b \mid 0), (\ell \mid fh + 2f) \rangle$ be a $\mathbb{Z}_2\mathbb{Z}_4$-additive cyclic code and let $\mathcal{C}_i = \langle (b \mid 0), (\ell_{k_i} \mid fhk_i + 2f) \rangle$ be a maximal $\mathbb{Z}_2\mathbb{Z}_4$-additive subcode of $\mathcal{C}$ whose image under the Gray map is a linear subcode of $\Phi(\mathcal{C})$, $i \in \{1, 2\}$.

We first consider $(\mathcal{C}_1)_Y \cap (\mathcal{C}_2)_Y$. By [72], $(\mathcal{C}_1)_Y \cap (\mathcal{C}_2)_Y = \langle f \operatorname{lcm}(hk_1, hk_2) + 2f \rangle = \langle fhk' + 2f \rangle$, where $k' = \operatorname{lcm}(k_1, k_2)$. Since $\langle (b \mid 0) \rangle \in \mathcal{C}_1 \cap \mathcal{C}_2$, we have $\mathcal{C}_1 \cap \mathcal{C}_2 = \langle (b \mid 0), (\ell_{k'} \mid fhk' + 2f) \rangle$, where $\ell_{k'} = \tilde{k}'\ell + (1 - \tilde{k}')\tilde{\mu}\ell\tilde{g}$ (mod b) for μ as in (7.3), by Lemma 7.83. $\qquad\square$

Lemma 7.88 ([33]). *Let* $\mathcal{C}$ *be a* $\mathbb{Z}_2\mathbb{Z}_4$-*additive cyclic code and let* $\mathcal{D}$ *be a maximal* $\mathbb{Z}_2\mathbb{Z}_4$-*additive cyclic subcode of* $\mathcal{C}$ *such that* $\Phi(\mathcal{D})$ *is linear. Then,* $\mathcal{K}(\mathcal{C}) \subseteq \mathcal{D}$.

Proof. If $\mathcal{K}(\mathcal{C}) \not\subseteq \mathcal{D}$, then we consider the $\mathbb{Z}_2\mathbb{Z}_4$-additive code $\mathcal{D}'$ generated by $\mathcal{K}(\mathcal{C}) \cup \mathcal{D} \cup \{2\mathbf{u} * \mathbf{v} : \mathbf{u}, \mathbf{v} \in \mathcal{K}(\mathcal{C}) \cup \mathcal{D}\}$. Since the binary Gray map image of $\mathcal{K}(\mathcal{C}) \cup \mathcal{D}$ is cyclic, $\mathcal{D}'$ is a cyclic subcode of $\mathcal{C}$. Moreover, since $2\mathbf{u} * \mathbf{v} \in \mathcal{D}'$ for all $\mathbf{u}, \mathbf{v} \in \mathcal{D}'$, we have that $\Phi(\mathcal{D}')$ is linear, leading to a contradiction since we are assuming that $\mathcal{D}$ is maximal. $\qquad\square$

Theorem 7.89 ([33]). *Let $\mathcal{C} = \langle (b \mid 0), (\ell \mid fh + 2f) \rangle$ be a $\mathbb{Z}_2\mathbb{Z}_4$-additive cyclic code, where $fgh = x^\beta - 1$. Assume that $k_1, \ldots, k_s$ are the divisors of g of minimum degree such that*

$$\gcd\left(\frac{\tilde{f}b}{\gcd(b, \ell\frac{\tilde{g}}{k_i})}, \left(\frac{\tilde{g}}{\tilde{k}_i} \otimes \frac{\tilde{g}}{\tilde{k}_i}\right)\right) = 1$$

for $i \in \{1, \ldots, s\}$. Then,

$$\mathcal{K}(\mathcal{C}) = \langle (b \mid 0), (\ell'_k \mid fhk' + 2f) \rangle,$$

where $k' = \mathrm{lcm}(k_1, \ldots, k_s)$ and $\ell_{k'} = \tilde{k}'\ell + (1 - \tilde{k}')\tilde{\mu}\ell\tilde{g} \pmod{b}$ for μ as in (7.3).

Proof. Assume that $k_1, \ldots, k_s$ are the divisors of g of minimum degree such that

$$\gcd\left(\frac{\tilde{f}b}{\gcd(b, \ell\frac{\tilde{g}}{k_i})}, \left(\frac{\tilde{g}}{\tilde{k}_i} \otimes \frac{\tilde{g}}{\tilde{k}_i}\right)\right) = 1$$

for $i \in \{1, \ldots, s\}$. Let $\mathcal{D}_i$ be a cyclic subcode of $\mathcal{C}$, where $\mathcal{D}_i = \langle (b \mid 0), (\ell_{k_i} \mid fhk_i + 2f) \rangle$ for some ℓ_{k_i}. Note that $\Phi(\mathcal{D}_i)$ is linear by Theorem 7.59. Since k_i is a polynomial of minimum degree dividing g, we have that $\mathcal{D}_i$ is a maximal cyclic subcode of $\mathcal{C}$ with binary linear image. Then, each $\mathcal{D}_i$ extends to $\mathcal{C}_i$ which is a maximal subcode of $\mathcal{C}$, not necessarily cyclic, with binary linear image. Note that every maximal code with linear image must contain a cyclic code with linear image, e.g., every maximal code contains $\mathcal{K}(\mathcal{C})$ that is cyclic with linear image. By Lemma 7.88, we know $\mathcal{K}(\mathcal{C}) \subseteq \mathcal{D}_i$ and, therefore, $\mathcal{K}(\mathcal{C}) \subseteq \cap_{i=1}^s \mathcal{D}_i$. Since $\cap_{i=1}^s \mathcal{C}_i = \mathcal{K}(\mathcal{C}) \subseteq \cap_{i=1}^s \mathcal{D}_i \subseteq \cap_{i=1}^s \mathcal{C}_i$, we have that $\mathcal{K}(\mathcal{C}) = \cap_{i=1}^s \mathcal{D}_i$. Finally, by Corollary 7.86 and Proposition 7.87, the result follows. $\square$

Example 7.90. Let $x^7 - 1 = (x - 1)p_1 p_2$ over $\mathbb{Z}_4$, with p_1 and p_2 as in Example 7.74. Let $\mathcal{C} = \langle (1 \mid 0), (0 \mid f) \rangle$ of type $(1, 7; 1, 6; 1)$ with $f = x - 1$, $h = 1$ and $g = p_1 p_2$.

Note that $\gcd(\frac{\tilde{f}b}{\gcd(b, \ell\tilde{g})}, (\tilde{g} \otimes \tilde{g})) = x - 1 \neq 1$. We have that all maximal cyclic subcodes of $\mathcal{C}$ with binary linear image are $\mathcal{C}_1 = \langle (1 \mid 0), (0 \mid fp_1 + 2f) \rangle$, and $\mathcal{C}_2 = \langle (1 \mid 0), (0 \mid fp_2 + 2f) \rangle$. Clearly, $k' = \mathrm{lcm}(p_1, p_2) = p_1 p_2$ and then $\mathcal{K}(\mathcal{C}) = \langle (1 \mid 0), (0 \mid fp_1 p_2 + 2f) \rangle = \langle (1 \mid 0), (0 \mid 2f) \rangle$. $\triangle$

Chapter 8

Encoding and Decoding $\mathbb{Z}_2\mathbb{Z}_4$-Linear Codes

In this section, we describe two encoding and two decoding methods for $\mathbb{Z}_2\mathbb{Z}_4$-linear codes. The first encoding process is the usual one that consists of multiplying by a generator matrix, and the second one is always a systematic encoding, so the coordinates of the information are included in its corresponding codeword. The decoding methods are syndrome decoding and permutation decoding. Encoding and decoding $\mathbb{Z}_2\mathbb{Z}_4$-linear codes were first studied in [134]. Subsequently, in [45], a systematic encoding for 1-perfect $\mathbb{Z}_2\mathbb{Z}_4$-linear codes was presented. The systematic encoding and permutation decoding for $\mathbb{Z}_2\mathbb{Z}_4$-linear codes included in this chapter are described in [22].

8.1 Encoding and Decoding

Figure 8.1 shows a general scheme of the process of information transmission through a noisy channel, from an emitter to a receiver, by using an encoder and decoder in order to detect and correct errors.

Let C be a binary linear code of length n and dimension k. Let G be a generator matrix of C. Recall that the process of encoding a binary vector i of length k can be done simply multiplying by G, $v = iG$, which gives a codeword of length n. The vector i is the information we want to send through the noisy channel, and v is the codeword that is sent; that is, the information i with some redundancy that will help to detect and correct the

There are parts of this chapter that have been previously published. Reprinted by permission from Springer Nature Customer Service Centre GmbH: J. J. Bernal, J. Borges, C. Fernández-Córdoba, and M. Villanueva. "Permutation decoding of $\mathbb{Z}_2\mathbb{Z}_4$-linear codes". *Designs, Codes andCryptography*, v. 76, 2, pp. 269–277. ©(2015).

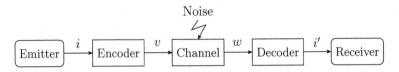

Figure 8.1: Communication scheme through a noisy channel using an error correcting code

errors in the decoding process.

Example 8.1. Let C be the binary Hamming code (linear 1-perfect code) of length 7 and dimension 4, or equivalently, a $\mathbb{Z}_2\mathbb{Z}_4$-additive code of type $(7, 0; 4, 0; 4)$, generated by

$$G = \begin{pmatrix} 1 & 0 & 0 & 0 & 0 & 1 & 1 \\ 0 & 1 & 0 & 0 & 1 & 0 & 1 \\ 0 & 0 & 1 & 0 & 1 & 1 & 0 \\ 0 & 0 & 0 & 1 & 1 & 1 & 1 \end{pmatrix}.$$

Let 10111110 be the sequence of information that we want to encode to be sent through the channel. Since the dimension is $k = 4$, we consider the following information vectors: $i_1 = (1, 0, 1, 1)$, $i_2 = (1, 1, 1, 0)$. By the encoding process, $v_j = i_j G$, for $j \in \{1, 2\}$, we obtain the codewords $v_1 = (1, 0, 1, 1, 0, 1, 0)$ and $v_2 = (1, 1, 1, 0, 0, 0, 0)$ and hence the encoded sequence is 10110101110000.   △

Let C be a binary, not necessarily linear, code of length n and size $|C| = 2^k$. For a vector $u \in \mathbb{Z}_2^n$ and a set $I \subseteq \{1, \dots, n\}$, we denote the projection of u onto the coordinates of I by u_I. We say that the code C is *systematic* if there is a set $I \subseteq \{1, \dots, n\}$ of size k such that $C_I = \{u_I : u \in C\} = \mathbb{Z}_2^k$. The set I is called an *information set*. Given a systematic code C and an information set I, a *systematic encoding* for I is a one-to-one map $f : \mathbb{Z}_2^k \longrightarrow \mathbb{Z}_2^n$ such that for any information vector $i \in \mathbb{Z}_2^k$, the corresponding codeword $v = f(i)$ satisfies $v_I = i$.

Let C be a binary linear code of length n and dimension k. It is well known that C is permutation equivalent to a code C' with generator matrix and parity check matrix of the form

$$G' = (I_k \; P) \quad \text{and} \quad H' = (P^T \; I_{n-k}), \tag{8.1}$$

respectively. Let $\pi \in \mathcal{S}_n$ such that $C = \pi(C')$. Note that C' is systematic with information set $I' = \{1, \dots, k\}$ and, therefore, C is systematic with information set $I = \{\pi(1), \ldots, \pi(k)\}$. A systematic encoding for I is

$$f(i) = i\pi(G'),\qquad(8.2)$$

where $\pi(G')$ is obtained from G' by permuting its columns using π.

Example 8.2. Let C be the binary linear code given in Example 8.1. We have that C is linear and therefore systematic. Moreover, since the generator matrix G is of the form $(I_4 \; P)$, $I = \{1, 2, 3, 4\}$ is an information set. As we have seen in Example 8.1 the associated codewords to the information vectors $i_1 = (1, 0, 1, 1)$ and $i_2 = (1, 1, 1, 0)$ are $v_1 = (1, 0, 1, 1, 0, 1, 0)$ and $v_2 = (1, 1, 1, 0, 0, 0, 0)$, respectively. Note that $(v_1)_I = i_1$ and $(v_2)_I = i_2$. $\triangle$

Let C be a binary, not necessarily linear, code with minimum distance d. The amount of errors that C is able to detect or correct depends on the value of d. Specifically,

i) C detects up to $d - 1$ errors, and

ii) C corrects up to $\lfloor \frac{d-1}{2} \rfloor$ errors.

The value $\tau = \lfloor \frac{d-1}{2} \rfloor$ is the *error correcting capability* of the code.

Let C be a binary code with error correcting capability τ. Assume that a codeword v is sent, and the vector $w = v + e$ is received, where e is the error vector. Note that if there are at most τ errors in the transmission of v, that is, $\text{wt}_H(e) \leq \tau$, then there is a unique codeword at a minimum distance from the received vector w, which is exactly $v' = v$, the sent codeword, and hence the errors are corrected. The decoding method for unrestricted codes is called *minimum distance decoding* and it is based on looking for the closest codeword v' to the received vector w, so it is necessary to search through a list of $|C|$ codewords. Such method is not efficient in general, and it is necessary to search alternative methods that improve the decoding process.

The general decoding method for linear codes is based on multiplying by a parity check matrix of the code, and is called *syndrome decoding*. Another general decoding method is the *permutation decoding*, which is based on finding a set of automorphisms of the code that satisfies certain conditions. This second decoding method is suitable for codes with a large automorphism group.

Let v be the codeword sent through the channel and v' the codeword obtained after the process of correcting errors in the receive vector w, as shown in Figure 8.1. Note that if the encoding method is systematic with information set I, then the information vector i' can be obtained easily by considering the coordinates of v' in the set I, that is, $i' = v'_I$. If the number of errors produced in the transmission is at most τ, then we have that $v' = v$ and $i' = i$. In other words, the receiver obtains the exact information sent by the emitter after correcting the errors produced in the channel.

8.2   Encoding $\mathbb{Z}_2\mathbb{Z}_4$-Linear Codes

In this section, we describe two encoding processes for $\mathbb{Z}_2\mathbb{Z}_4$-linear codes, a general method and another one which is systematic.

Let $\mathcal{C}$ be a $\mathbb{Z}_2\mathbb{Z}_4$-additive code of type $(\alpha, \beta; \gamma, \delta; \kappa)$. Let $\mathcal{G}$ be a generator matrix of $\mathcal{C}$. Let $C = \Phi(\mathcal{C})$ be the corresponding $\mathbb{Z}_2\mathbb{Z}_4$-linear code of length $n = \alpha + 2\beta$. Note that, in general, C is not linear and hence we cannot use the encoding process described in Section 8.1 for linear codes. A general encoding process for C, shown in Figure 8.2, can be given by using the generator matrix $\mathcal{G}$ as it is described in Algorithm 8.1. In this algorithm, the Gray map Φ is applied to vectors in $\mathbb{Z}_2^\alpha \times \mathbb{Z}_4^\beta$ as usual or, similarly, in $\mathbb{Z}_2^\gamma \times \mathbb{Z}_4^\delta$. By the context, it is clear the domain which is considered.

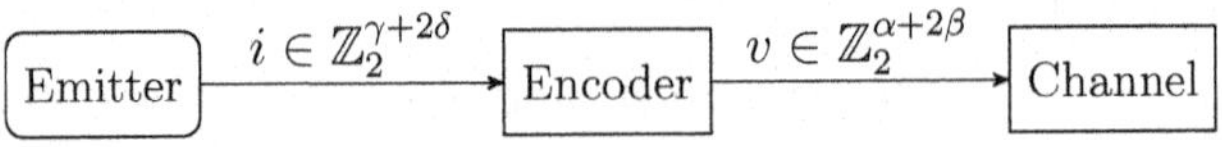

Figure 8.2: Encoding scheme for a $\mathbb{Z}_2\mathbb{Z}_4$-linear code of type $(\alpha, \beta; \gamma, \delta; \kappa)$

Recall that the product of a vector $\mathbf{i} \in \mathbb{Z}_2^\gamma \times \mathbb{Z}_4^\delta$ by a generator matrix $\mathcal{G}$ of $\mathcal{C}$ is $\mathbf{i}\mathcal{G} = \chi^{-1}(\iota(\mathbf{i})\chi(\mathcal{G}))$ as given in (2.12).

Algorithm 8.1 Encoding for a $\mathbb{Z}_2\mathbb{Z}_4$-linear code C

Require: A generator matrix $\mathcal{G}_S$ of $\mathcal{C} = \Phi^{-1}(C)$ in standard form (2.7) and an information vector $i \in \mathbb{Z}_2^{\gamma+2\delta}$.
1: $\mathbf{i} \leftarrow \Phi^{-1}(i)$
2: $\mathbf{v} \leftarrow \mathbf{i}\mathcal{G}_S$
3: **return** $v = \Phi(\mathbf{v})$

Note that the generator matrix used in Algorithm 8.1 does not need to be in standard form. However, it has to have $\gamma + \delta$ rows, the first γ rows of order two and the last δ rows of order four.

Example 8.3. Let $\mathcal{C}$ be a $\mathbb{Z}_2\mathbb{Z}_4$-additive code of type $(2, 3; 2, 1; 1)$ with generator matrix in standard form

$$\mathcal{G}_S = \left(\begin{array}{cc|ccc} 1 & 1 & 2 & 0 & 0 \\ 0 & 0 & 2 & 2 & 0 \\ 0 & 1 & 1 & 1 & 1 \end{array} \right).$$

To encode information using the code $\mathcal{C}$, we consider binary information vectors of length $\gamma + 2\delta = 4$. Let $i = (0, 1, 1, 1)$ be the binary information vector that we want to encode. Applying Algorithm 8.1, we obtain:

i) The information over $\mathbb{Z}_2^2 \times \mathbb{Z}_4$ is $\mathbf{i} = \Phi^{-1}(i) = (0, 1 \mid 2)$.

ii) The matrix $\chi(\mathcal{G}_S)$ is

$$\begin{pmatrix} 2 & 2 & 2 & 0 & 0 \\ 0 & 0 & 2 & 2 & 0 \\ 0 & 2 & 1 & 1 & 1 \end{pmatrix}.$$

The codeword over $\mathbb{Z}_2^2 \times \mathbb{Z}_4^3$ is

$$\begin{aligned} \mathbf{v} = \mathbf{i}\mathcal{G} &= \chi^{-1}(\iota(\mathbf{i})\chi(\mathcal{G}_S)) = \chi^{-1}((0, 1, 2)\chi(\mathcal{G}_S)) \\ &= \chi^{-1}(0, 0, 0, 0, 2) = (0, 0 \mid 0, 0, 2). \end{aligned}$$

iii) The encoded binary codeword is

$$v = \Phi(\mathbf{v}) = (0, 0, 0, 0, 0, 0, 1, 1).$$

Note that this encoding is not systematic since there is not any set $I \subseteq \{1, \ldots, 8\}$ such that $v_I = (0, 1, 1, 1) = i$. $\triangle$

Example 8.4. Let $\mathcal{C}$ be the $\mathbb{Z}_2\mathbb{Z}_4$-additive code of type $(7, 4; 5, 3; 5)$ with error correcting capability $\tau = 1$ and generator matrix in standard form

$$\mathcal{G}_S = \begin{pmatrix} 1 & 0 & 0 & 0 & 0 & 0 & 0 & 2 & 0 & 0 & 0 \\ 0 & 1 & 0 & 0 & 0 & 1 & 1 & 0 & 0 & 0 & 0 \\ 0 & 0 & 1 & 0 & 0 & 1 & 1 & 2 & 0 & 0 & 0 \\ 0 & 0 & 0 & 1 & 0 & 1 & 0 & 2 & 0 & 0 & 0 \\ 0 & 0 & 0 & 0 & 1 & 0 & 1 & 2 & 0 & 0 & 0 \\ 0 & 0 & 0 & 0 & 0 & 1 & 1 & 3 & 1 & 0 & 0 \\ 0 & 0 & 0 & 0 & 0 & 1 & 0 & 1 & 0 & 1 & 0 \\ 0 & 0 & 0 & 0 & 0 & 0 & 1 & 1 & 0 & 0 & 1 \end{pmatrix}.$$

Let $i = (1, 0, 1, 1, 1, 1, 1, 0, 1, 1, 0)$ be the binary information vector that we want to encode. Applying Algorithm 8.1, we obtain the following results:

i) The information over $\mathbb{Z}_2^5 \times \mathbb{Z}_4^3$ is $\mathbf{i} = \Phi^{-1}(i) = (1, 0, 1, 1, 1 \mid 2, 1, 3)$.

ii) The matrix $\chi(\mathcal{G}_S)$ is

$$\begin{pmatrix}
2 & 0 & 0 & 0 & 0 & 0 & 0 & 2 & 0 & 0 & 0 \\
0 & 2 & 0 & 0 & 0 & 2 & 2 & 0 & 0 & 0 & 0 \\
0 & 0 & 2 & 0 & 0 & 2 & 2 & 2 & 0 & 0 & 0 \\
0 & 0 & 0 & 2 & 0 & 2 & 0 & 2 & 0 & 0 & 0 \\
0 & 0 & 0 & 0 & 2 & 0 & 2 & 2 & 0 & 0 & 0 \\
0 & 0 & 0 & 0 & 0 & 2 & 2 & 3 & 1 & 0 & 0 \\
0 & 0 & 0 & 0 & 0 & 2 & 0 & 1 & 0 & 1 & 0 \\
0 & 0 & 0 & 0 & 0 & 0 & 2 & 1 & 0 & 0 & 1
\end{pmatrix}.$$

The codeword over $\mathbb{Z}_2^\alpha \times \mathbb{Z}_4^\beta$ is

$$\mathbf{v} = \mathbf{i}\mathcal{G}_S = \chi^{-1}(\iota(\mathbf{i})\chi(\mathcal{G}_S)) = \chi^{-1}((1,0,1,1,1,2,1,3)\chi(\mathcal{G}_S))$$
$$= \chi^{-1}(2,0,2,2,2,2,2,2,1,3) = (1,0,1,1,1,1,1 \mid 2,2,1,3).$$

iii) The encoded binary codeword is

$$v = \Phi(\mathbf{v}) = (1,0,1,1,1,1,1,1,1,1,1,0,1,1,0).$$

$\triangle$

Note that the described encoding method given by Algorithm 8.1 is not always systematic, as shown in Example 8.3. However, $\mathbb{Z}_2\mathbb{Z}_4$-linear codes are systematic, and we show this by giving a systematic encoding method for a specific information set.

Let $\mathcal{C}$ be a $\mathbb{Z}_2\mathbb{Z}_4$-additive code of type $(\alpha, \beta; \gamma, \delta; \kappa)$ with a generator matrix $\mathcal{G}_S$ in standard form (2.7) and $C = \Phi(\mathcal{C})$. For each quaternary coordinate position $\alpha + i$, with $i \in \{1, \ldots, \beta\}$, we denote the corresponding pair of binary coordinates in $\{1, \ldots, \alpha + 2\beta\}$ by $\varphi_1(i)$ and $\varphi_2(i)$, respectively. That is, $\varphi_1(i) = \alpha + 2i - 1$ and $\varphi_2(i) = \alpha + 2i$. We define the following sets of coordinate positions in $\{1, \ldots, \alpha + 2\beta\}$:

i) $J_1 = \{1, \ldots, \kappa\}$, $|J_1| = \kappa$,

ii) $J_2 = \{j_1, \ldots, j_{\gamma-\kappa}\}$, where $j_i = \varphi_1(\beta + \kappa - \gamma - \delta + i)$, $|J_2| = \gamma - \kappa$,

iii) $J_3 = \{\varphi_1(\beta - \delta + 1), \varphi_2(\beta - \delta + 1), \ldots, \varphi_1(\beta), \varphi_2(\beta)\}$, $|J_3| = 2\delta$.

Theorem 8.5 shows that $J = J_1 \cup J_2 \cup J_3$ is an information set for the $\mathbb{Z}_2\mathbb{Z}_4$-linear code C. We shall refer to J as the *standard information set*.

Theorem 8.5 ([22]). *Let C be a $\mathbb{Z}_2\mathbb{Z}_4$-linear code of type $(\alpha, \beta; \gamma, \delta; \kappa)$ and let $\mathcal{G}_S$ be a generator matrix of $\mathcal{C} = \Phi^{-1}(C)$ in standard form. Then, C is a systematic code and J is a standard information set for C.*

Proof. Let $i = (i_1, \ldots, i_{\gamma+2\delta}) \in \mathbb{Z}_2^{\gamma+2\delta}$ be an information vector. Consider the representation $i = (x, y, z)$, where $x = (i_1, \ldots, i_\kappa)$, $y = (y_1, \ldots, y_{\gamma-\kappa}) = (i_{\kappa+1}, \ldots, i_\gamma)$ and $z = (i_{\gamma+1}, \ldots, i_{\gamma+2\delta})$. Let $v = \Phi\left(\Phi^{-1}(i)\mathcal{G}_S\right) \in \mathbb{Z}_2^{\alpha+2\beta}$ and consider the set of coordinate positions $J_2 = \{j_1, \ldots, j_{\gamma-\kappa}\}$. Let $\eta = v_{J_2} + y$ and let $\sigma : \mathbb{Z}_2^\gamma \times \mathbb{Z}_4^\delta \longrightarrow \mathbb{Z}_2^\gamma \times \mathbb{Z}_4^\delta$ be the one-to-one map given by $\sigma(\Phi^{-1}(i)) = \sigma(\Phi^{-1}(x, y, z)) = \Phi^{-1}(x, y+\eta, z)$. Since

$$\eta = v_{J_2} + y = \left(\Phi(\Phi^{-1}(i)\mathcal{G}_S)\right)_{J_2} + y \ .$$

as a binary vector, it follows that

$$\sigma(\Phi^{-1}(i)) + \sigma(\Phi^{-1}(i')) = \sigma(\Phi^{-1}(i) + \Phi^{-1}(i')),$$

for all $i, i' \in \mathbb{Z}_2^{\gamma+2\delta}$, and therefore σ is a group automorphism of $\mathbb{Z}_2^\gamma \times \mathbb{Z}_4^\delta$. Finally, it is straightforward that the codeword $\sigma(\Phi^{-1}(i))\mathcal{G}_S$ verifies

$$\left(\Phi(\sigma(\Phi^{-1}(i))\mathcal{G}_S)\right)_J = (x, y, z) = i,$$

and hence C is systematic. Since $J = J_1 \cup J_2 \cup J_3$ and $|J| = \kappa + \gamma - \kappa + 2\delta = \gamma + 2\delta$, we have that J is a set of systematic coordinates. $\square$

Corollary 8.6 ([22]). *Let C be a $\mathbb{Z}_2\mathbb{Z}_4$-linear code of length n and size $|C| = 2^k$. Let $\mathcal{G}_S$ be a generator matrix of $\mathcal{C} = \Phi^{-1}(C)$ in standard form. Then, the function $f : \mathbb{Z}_2^k \longrightarrow \mathbb{Z}_2^n$ defined as*

$$f(i) = \Phi\left(\sigma\left(\Phi^{-1}(i)\right)\mathcal{G}_S\right), \quad \forall i \in \mathbb{Z}_2^k \tag{8.3}$$

is a systematic encoding for C with respect to the standard information set J.

Algorithm 8.2 Systematic encoding for a $\mathbb{Z}_2\mathbb{Z}_4$-linear code C

Require: A generator matrix $\mathcal{G}_S$ of $\mathcal{C} = \Phi^{-1}(C)$ in standard form (2.7), the standard information set $J_1 \cup J_2 \cup J_3$, and an information vector $i \in \mathbb{Z}_2^{\gamma+2\delta}$.

1: $x \leftarrow i[1, \ldots, \kappa]$
2: $y \leftarrow i[\kappa+1, \ldots, \gamma]$
3: $z \leftarrow i[\gamma+1, \ldots, \gamma+2\delta]$
4: $\mathbf{i} \leftarrow \Phi^{-1}(x, y, z)$
5: $v \leftarrow \Phi(\mathbf{i}\mathcal{G}_S)$
6: $\eta \leftarrow v[J_2] + y$
7: $\mathbf{i}' \leftarrow \Phi^{-1}(x, y+\eta, z)$
8: **return** $\Phi(\mathbf{i}'\mathcal{G}_S)$

The following example shows the systematic encoding process given by Corollary 8.6, and also described in Algorithm 8.2.

Example 8.7. Let $\mathcal{C}$ be the $\mathbb{Z}_2\mathbb{Z}_4$-additive code given in Example 8.3. Let $C = \Phi(\mathcal{C})$ be the corresponding $\mathbb{Z}_2\mathbb{Z}_4$-linear code of length 8. Note that $J_1 = \{1\}$, $J_2 = \{5\}$, and $J_3 = \{7,8\}$, so the standard information set is $J = \{1,5,7,8\}$. Note that there are other information sets. For example, it can be checked that $\{2,4,6,8\}$ is also an information set.

In order to encode the information vector $i = (0,1,0,1)$, by using the systematic encoding (8.3) with respect to the standard information set J, we apply Algorithm 8.2. We have that $x = (0)$, $y = (1)$ and $z = (0,1)$. Then, $\mathbf{i} = \Phi^{-1}(i) = (0,1 \mid 1)$ and $v = \Phi(\mathbf{i}\mathcal{G}_S) = \Phi(0,1 \mid 3,3,1) = (0,1,1,0,1,0,0,1)$. Since $y = v_{J_2} = (1)$, $\eta = (0)$ and σ is the identity map. Therefore, $f(i) = \Phi(0,1 \mid 3,3,1) = (0,1,1,0,1,0,0,1)$ and $f(i)_J = (0,1,0,1) = i$.

Now, for $i = (0,1,1,1)$, we have $x = (0)$, $y = (1)$ and $z = (1,1)$. Then, $\mathbf{i} = \Phi^{-1}(i) = (0,1 \mid 2)$ and $v = \Phi(\mathbf{i}\mathcal{G}_S) = \Phi(0,0 \mid 0,0,2) = (0,0,0,0,0,0,1,1)$. Since $v_{J_2} = (0)$ and $y = (1)$, $\eta = (1)$ and $\sigma(0,1 \mid 2) = (0,0 \mid 2)$. Then, $f(i) = \Phi((0,0 \mid 2)\mathcal{G}_S) = \Phi(0,0 \mid 2,2,2) = (0,0,1,1,1,1,1,1)$ and $f(i)_J = (0,1,1,1) = i$.

Finally, for $i = (0,0,1,1)$, we have $x = (0)$, $y = (0)$, $z = (1,1)$. Since $v_{J_2} = \Phi(\Phi^{-1}(i)\mathcal{G}_S)_{J_2} = (1)$ and $y = (0)$, $\eta = (1)$ and $\sigma(0,0 \mid 2) = (0,1 \mid 2)$. Then, $f(i) = \Phi((0,1 \mid 2)\mathcal{G}_S) = \Phi(0,1 \mid 0,0,2) = (0,1,0,0,0,0,1,1)$ and $f(i)_J = (0,0,1,1) = i$. $\triangle$

Note that in the case $\gamma = \kappa$, we have $i = (x,z)$, so η is the all-zero vector. Hence σ is the identity map and the above systematic encoding (8.3) coincides with the one given by Algorithm 8.1, that is,

$$f(i) = \Phi(\Phi^{-1}(i)\mathcal{G}_S), \quad \forall i \in \mathbb{Z}_2^k. \tag{8.4}$$

Example 8.8. Let $\mathcal{C}$ be the $\mathbb{Z}_2\mathbb{Z}_4$-additive code given in Example 8.4. Let $C = \Phi(\mathcal{C})$ be the $\mathbb{Z}_2\mathbb{Z}_4$-linear code of length 15 with $|C| = 2^{5+2\cdot3} = 2^{11}$. In this case, we have that $\gamma = \kappa$, so σ is the identity map and the encoding method given in (8.4) is systematic. We have that $J_1 = \{1,2,3,4,5\}$, $J_2 = \emptyset$, and $J_3 = \{10,11,12,13,14,15\}$, so $J = \{1,2,3,4,5,10,11,12,13,14,15\}$. From Example 8.4, we have that the corresponding codeword relative to the information vector $i = (1,0,1,1,1,1,1,0,1,1,0)$ is $v = (1,0,1,1,1,1,1,1,1,1,1,0,1,1,0)$. Note that we have that $v_J = i$. $\triangle$

Note that the systematic encoding method given by Corollary 8.6 or Algorithm 8.2 requires, in some cases, two products by the generator matrix. However, there is not any change in the computation complexity order.

8.3 Decoding $\mathbb{Z}_2\mathbb{Z}_4$-Linear Codes

In this section, we describe two general decoding methods for $\mathbb{Z}_2\mathbb{Z}_4$-linear codes, called syndrome and permutation decoding. The first one was first briefly described in [134], and the second one was introduced in [22].

Let $\mathcal{C}$ be a $\mathbb{Z}_2\mathbb{Z}_4$-additive code with minimum distance $d(\mathcal{C}) = d$, and let $C = \Phi(\mathcal{C})$. In Chapter 2, it is shown that $d_H(\Phi(\mathcal{C})) = d(\mathcal{C}) = d$. Therefore, the code C detects up to $d - 1$ errors and corrects up to $\tau = \lfloor \frac{d-1}{2} \rfloor$ errors. A scheme for the decoding process by using a $\mathbb{Z}_2\mathbb{Z}_4$-linear code C of type $(\alpha, \beta; \gamma, \delta; \kappa)$ is given in Figure 8.3.

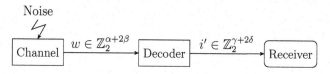

Figure 8.3: Decoding scheme for a $\mathbb{Z}_2\mathbb{Z}_4$-linear code of type $(\alpha, \beta; \gamma, \delta; \kappa)$

8.3.1 Syndrome Decoding

Let C be a binary linear $[n, k, d]$ code with parity check matrix H and error correcting capability τ. Consider $\{e_i\}_{i=1}^{r}$ the set of all error vectors with $\mathrm{wt}_H(e_i) \leq \tau$. For a vector $w \in \mathbb{Z}_2^n$, we define the *syndrome of w* to be $s = wH^T$. We consider Table 8.1 in order to compare the syndrome of the received vector w with the syndrome of all the possible error vectors of weight at most τ. We decode the vector w by v' considering the following cases:

i) If $s = \mathbf{0}$, then we have that w is a codeword, $w \in C$, and we assume no errors have occurred in the transmission. Hence, $v' = w$.

ii) If $s = s_i$ for some $i \in \{1, \dots, r\}$, then we assume that the errors produced in the transmission are given by e_i. Hence, we decode w by $v' = w - e_i$.

iii) If $s \neq s_i$ for all $i \in \{0, 1, \dots, r\}$, then more than τ errors have occurred in the transmission. In incomplete decoding, no codeword is assigned to w.

Now, we describe a syndrome decoding method for $\mathbb{Z}_2\mathbb{Z}_4$-linear codes. Since $\mathbb{Z}_2\mathbb{Z}_4$-linear codes may not be linear, we can not apply the syndrome decoding for linear codes.

error vector in $\mathbb{Z}_2^n$	syndrome in $\mathbb{Z}_2^{n-k}$
$e_0 = \mathbf{0}$	$s_0 = \mathbf{0}$
e_1	$s_1 = e_1 H^T$
$\vdots$	$\vdots$
e_r	$s_r = e_r H^T$

Table 8.1: Syndrome table for a linear code with parity check matrix H

Let $C = \Phi(\mathcal{C})$ be a $\mathbb{Z}_2\mathbb{Z}_4$-linear code of type $(\alpha, \beta; \gamma, \delta; \kappa)$. Let $\mathcal{H}$ be a parity check matrix of $\mathcal{C}$. From Section 3.1, we have that $\mathcal{H}$ is a generator matrix of $\mathcal{C}^\perp$, which is of type $(\alpha, \beta; \bar{\gamma}, \bar{\delta}; \bar{\kappa})$. Recall from (3.9) that the product of a vector $\mathbf{w} \in \mathbb{Z}_2^\alpha \times \mathbb{Z}_4^\beta$ by a matrix $\mathcal{H}^T$ is $\mathbf{w}\mathcal{H}^T = \chi^{-1}(\iota(\mathbf{w})\chi(\mathcal{H})^T)$. For a vector $w \in \mathbb{Z}_2^{\alpha+2\beta}$, we define the *syndrome of w* to be

$$\mathbf{s} = \Phi^{-1}(w)\mathcal{H}^T \in \mathbb{Z}_2^{\bar{\gamma}} \times \mathbb{Z}_4^{\bar{\delta}}.$$

Let τ be the error correcting capability of C. Consider $\{e_i\}_{i=1}^r$ the set of all error vectors with $\mathrm{wt}_H(e_i) \leq \tau$. Using the syndrome table, given in Table 8.2, we can describe the syndrome decoding algorithm for C, as shown in Algorithm 8.3.

error vector in $\mathbb{Z}_2^{\alpha+2\beta}$	syndrome in $\mathbb{Z}_2^{\bar{\gamma}} \times \mathbb{Z}_4^{\bar{\delta}}$
$e_0 = \mathbf{0}$	$\mathbf{s}_0 = \mathbf{0}$
e_1	$\mathbf{s}_1 = \Phi^{-1}(e_1)\mathcal{H}^T$
$\vdots$	$\vdots$
e_r	$\mathbf{s}_r = \Phi^{-1}(e_r)\mathcal{H}^T$

Table 8.2: Syndrome table for a $\mathbb{Z}_2\mathbb{Z}_4$-additive code with parity check matrix $\mathcal{H}$

Example 8.9. Let $\mathcal{C}$ be the $\mathbb{Z}_2\mathbb{Z}_4$-additive code given in Example 8.4. Let

$$\mathcal{H}_S = \begin{pmatrix} 0 & 1 & 1 & 1 & 0 & 1 & 0 & | & 0 & 2 & 2 & 0 \\ 0 & 1 & 1 & 0 & 1 & 0 & 1 & | & 0 & 2 & 0 & 2 \\ 1 & 0 & 1 & 1 & 1 & 0 & 0 & | & 1 & 1 & 3 & 3 \end{pmatrix}$$

be the parity check matrix obtained from the matrix $\mathcal{G}_S$, given in Example 8.4, and Theorem 3.23. Since $\tau = 1$, the syndrome table given in Table 8.3 contains the syndrome of all error vectors of weight at most 1. Moreover,

---

Algorithm 8.3 Syndrome decoding for a $\mathbb{Z}_2\mathbb{Z}_4$-linear code C

Require: A parity check matrix $\mathcal{H}$ of $\mathcal{C} = \Phi^{-1}(C)$, the syndrome table of size $r + 1$ given in Table 8.2, and a received vector $w \in \mathbb{Z}_2^{\alpha+2\beta}$.

1: $\mathbf{w} \leftarrow \Phi^{-1}(w)$
2: $\mathbf{s} \leftarrow \mathbf{w}\mathcal{H}^T$
3: found $\leftarrow$ False
4: $j \leftarrow 0$
5: **while** $j \leq r$ **and not** found **do**
6:    **if** $\mathbf{s} = \mathbf{s}_j$ **then**
7:       $e \leftarrow e_j$
8:       found $\leftarrow$ True
9:    **end if**
10:   $j \leftarrow j + 1$
11: **end while**
12: **if** found **then**
13:    **return** $v' = w - e$
14: **else**
15:    **return** "The errors can not be corrected"
16: **end if**

---

from Example 8.8, the encoding method given in (8.4) is systematic with respect to the information set $J = \{1, 2, 3, 4, 5, 10, 11, 12, 13, 14, 15\}$.

If we receive the sequence 100111111110110001001111110001, then we consider the received vectors $w_1 = (1, 0, 0, 1, 1, 1, 1, 1, 1, 1, 1, 0, 1, 1, 0)$ and $w_2 = (0, 0, 1, 0, 0, 1, 1, 1, 1, 1, 1, 0, 0, 0, 1) \in \mathbb{Z}_2^{15}$.

  i) For w_1, we have $\mathbf{s} = \Phi^{-1}(w_1)\mathcal{H}^T = (1, 1 \mid 2) = \mathbf{s}_3$. Therefore, the error vector is e_3 and $v_1' = w_1 - e_3 = (1, 0, 1, 1, 1, 1, 1, 1, 1, 1, 1, 0, 1, 1, 0)$.

  ii) For w_2, we have $\mathbf{s} = \Phi^{-1}(w_2)\mathcal{H}^T = (0, 1 \mid 1) = \mathbf{s}_{14}$, and $v_2' = w_2 - e_{14} = (0, 0, 1, 0, 0, 1, 1, 1, 1, 1, 1, 0, 0, 1, 1)$.

Since the encoding is systematic with information set J, the information vector relative to v_1' is $i_1' = (v_1')_J = (1, 0, 1, 1, 1, 1, 1, 0, 1, 1, 0)$ and the information vector relative to v_2' is $i_2' = (v_2')_J = (0, 0, 1, 0, 0, 1, 1, 0, 0, 1, 1)$. Hence, the decoded sequence is 10111110110 00100110011. $\triangle$

Now, we focus on perfect $\mathbb{Z}_2\mathbb{Z}_4$-linear codes, which are defined in Chapter 6. Let $\mathcal{C}$ be a $\mathbb{Z}_2\mathbb{Z}_4$-additive perfect code of type $(\alpha, \beta; \gamma, \delta; \kappa)$ with parity check matrix $\mathcal{H}$ and let $C = \Phi(\mathcal{C})$ of length $n = \alpha + 2\beta$. Since C is perfect, the syndrome decoding can be simplified and the syndrome table is not needed. In this case, every vector of length n is at a distance at most 1 of a codeword.

error vector in $\mathbb{Z}_2^{15}$	syndrome in $\mathbb{Z}_2^2 \times \mathbb{Z}_4^1$
$e_0 = \mathbf{0}$	$s_0 = \mathbf{0}$
$e_1 = (1,0,0,0,0,0,0,0,0,0,0,0,0,0,0)$	$s_1 = (0,0 \mid 2)$
$e_2 = (0,1,0,0,0,0,0,0,0,0,0,0,0,0,0)$	$s_2 = (1,1 \mid 0)$
$e_3 = (0,0,1,0,0,0,0,0,0,0,0,0,0,0,0)$	$s_3 = (1,1 \mid 2)$
$e_4 = (0,0,0,1,0,0,0,0,0,0,0,0,0,0,0)$	$s_4 = (1,0 \mid 2)$
$e_5 = (0,0,0,0,1,0,0,0,0,0,0,0,0,0,0)$	$s_5 = (0,1 \mid 2)$
$e_6 = (0,0,0,0,0,1,0,0,0,0,0,0,0,0,0)$	$s_6 = (1,0 \mid 0)$
$e_7 = (0,0,0,0,0,0,1,0,0,0,0,0,0,0,0)$	$s_7 = (0,1 \mid 0)$
$e_8 = (0,0,0,0,0,0,0,1,0,0,0,0,0,0,0)$	$s_8 = (0,0 \mid 3)$
$e_9 = (0,0,0,0,0,0,0,0,1,0,0,0,0,0,0)$	$s_9 = (0,0 \mid 1)$
$e_{10} = (0,0,0,0,0,0,0,0,0,1,0,0,0,0,0)$	$s_{10} = (1,1 \mid 3)$
$e_{11} = (0,0,0,0,0,0,0,0,0,0,1,0,0,0,0)$	$s_{11} = (1,1 \mid 1)$
$e_{12} = (0,0,0,0,0,0,0,0,0,0,0,1,0,0,0)$	$s_{12} = (1,0 \mid 1)$
$e_{13} = (0,0,0,0,0,0,0,0,0,0,0,0,1,0,0)$	$s_{13} = (1,0 \mid 3)$
$e_{14} = (0,0,0,0,0,0,0,0,0,0,0,0,0,1,0)$	$s_{14} = (0,1 \mid 1)$
$e_{15} = (0,0,0,0,0,0,0,0,0,0,0,0,0,0,1)$	$s_{15} = (0,1, \mid 3)$

Table 8.3: Syndrome table for the code $\mathcal{C}$ given in Example 8.9

Then, the set of all error vectors that can be corrected are $\{e_1, \ldots, e_n\}$, where e_j have 1 in the jth coordinate and zero elsewhere for all $j \in \{1, \ldots, n\}$. The set of the corresponding vectors in $\mathbb{Z}_2^\alpha \times \mathbb{Z}_4^\beta$ that are the preimage of the error vectors under the Gray map are $\{e_1, \ldots, e_\alpha, \lambda e_{\alpha+1}, \ldots, \lambda e_{\alpha+\beta}\}_{\lambda \in \{1,3\}}$. Therefore, for a syndrome $s_i = \Phi^{-1}(e_i)\mathcal{H}^T$ the value of $\chi(s_i)$ corresponds to a column of $\chi(\mathcal{H})$ when $i \in \{1, \ldots, \alpha\}$, and to a column of $\chi(\mathcal{H})$, up to change of sign, when $i \in \{\alpha+1, \ldots, n\}$. More specifically, Algorithm 8.4 describes the syndrome decoding algorithm for C.

Example 8.10. Let $\mathcal{C}$ be the $\mathbb{Z}_2\mathbb{Z}_4$-additive code given in Example 8.4 and let $\mathcal{H}_S$ be the parity check matrix given in Example 8.9. Then,

$$\chi(\mathcal{H}_S) = \begin{pmatrix} 0 & 2 & 2 & 2 & 0 & 2 & 0 & 0 & 2 & 2 & 0 \\ 0 & 2 & 2 & 0 & 2 & 0 & 2 & 0 & 2 & 0 & 2 \\ 2 & 0 & 2 & 2 & 2 & 0 & 0 & 1 & 1 & 3 & 3 \end{pmatrix}.$$

Note that $C = \Phi(\mathcal{C})$ is a perfect $\mathbb{Z}_2\mathbb{Z}_4$-linear code, since $\mathcal{H}_S$ can be obtained from the matrix $\mathcal{H}$, given in Example 6.9, by permuting some columns and changing the sign of the last two columns. As in Example 8.9, if we receive the sequence 100111111110110001001111110001, we

---

Algorithm 8.4 Syndrome decoding for a perfect $\mathbb{Z}_2\mathbb{Z}_4$-linear code C

Require: A parity check matrix $\mathcal{H}$ of $C = \Phi^{-1}(C)$, and a received vector $w \in \mathbb{Z}_2^{\alpha+2\beta}$.

1: $M \leftarrow \chi(\mathcal{H})^T$
2: $\mathbf{w} \leftarrow \Phi^{-1}(w)$
3: $\mathbf{v}' \leftarrow \mathbf{w}$
4: $s \leftarrow \iota(\mathbf{w})M$
5: found $\leftarrow$ False
6: $j \leftarrow 1$
7: **while** $j \le \alpha + \beta$ **and not** found **do**
8:    **if** $s = M[j]$ **then**
9:       $\mathbf{v}'[j] \leftarrow \mathbf{v}'[j] - 1$
10:      found $\leftarrow$ True
11:    **else**
12:      **if** $s = -M[j]$ **then**
13:        $\mathbf{v}'[j] \leftarrow \mathbf{v}'[j] - 3$
14:        found $\leftarrow$ True
15:      **end if**
16:    **end if**
17:    $j \leftarrow j + 1$
18: **end while**
19: **return** $\Phi(\mathbf{v}')$

---

consider the vectors $w_1 = (1,0,0,1,1,1,1,1,1,1,1,0,1,1,0)$ and $w_2 = (0,0,1,0,0,1,1,1,1,1,1,0,0,0,1)$. We decode these vectors by using Algorithm 8.4.

i) For w_1, we have $\mathbf{w}_1 = (1,0,0,1,1,1 \mid 2,2,1,3)$ and $\mathbf{s} = \mathbf{w}_1 \mathcal{H}^T = (1,1 \mid 2)$. Since $\chi(\mathbf{s}) = (2,2,2)$ is the third column of $\chi(\mathcal{H}_S)$, the third coordinate of $\mathbf{v}_1'$ is $0 - 1 = 1$, so $\mathbf{v}_1' = (1,0,1,1,1,1 \mid 2,2,1,3)$ and $v_1' = \Phi(\mathbf{v}_1') = (1,0,1,1,1,1,1,1,1,1,1,0,1,1,0)$.

ii) For w_2, we have $\mathbf{w}_2 = (0,0,1,0,0,1,1 \mid 2,2,0,1)$ and $\mathbf{s} = \mathbf{w}_2 \mathcal{H}^T = (0,1 \mid 1)$. We have that $\chi(\mathbf{s}) = (0,2,1)$ is the 11th column of $\chi(\mathcal{H}_S)$, after a change of sign, and thus the 11th coordinate of $\mathbf{v}_2'$ is $1 - 3 = 2$, so $\mathbf{v}_2' = (0,0,1,0,0,1,1 \mid 2,2,0,2)$ and $v_2' = \Phi(\mathbf{v}_2') = (0,0,1,0,0,1,1,1,1,1,1,0,0,1,1)$.

Moreover, as it is said in Examples 8.8 and 8.9, the encoding is systematic and the information set is $J = \{1,2,3,4,5,10,11,12,13,14,15\}$. Hence, $i_1' = (1,0,1,1,1,1,1,0,1,1,0)$, $i_2' = (0,0,1,0,0,1,1,0,0,1,1)$ and the decoded sequence is 10111110110001001101011. $\triangle$

8.3.2   Permutation Decoding

Recall that $\mathcal{S}_n$ is the group of permutations on n symbols acting on $\mathbb{Z}_2^n$ by permuting the coordinates of each vector, and $\mathrm{PAut}(C) = \{\pi \in \mathcal{S}_n : \pi(C) = C\}$ is the permutation automorphism group of a binary code C of length n.

Permutation decoding is a technique introduced by Prange [128] and developed by MacWilliams [111] for linear codes. A description of the standard method for linear codes can be found in [112, p. 513]. Let C be a binary linear code of length n, with error correction capability τ and information set I. Let $w = v + e$ be the received vector, where $v \in C$ and e is the error vector. Assume that w has less than $\tau + 1$ errors, that is, $\mathrm{wt}_H(e) \leq \tau$. The idea of permutation decoding is to use the elements of $\mathrm{PAut}(C)$ in order to move the non-zero coordinates of e out of I. Therefore, on the one hand, the method is based on the existence of some special subsets $S \subseteq \mathrm{PAut}(C)$, called PD-sets, verifying that for any vector $e \in \mathbb{Z}_2^n$ with $\mathrm{wt}_H(e) \leq \tau$, there is an element $\pi \in S$ such that $\mathrm{wt}_H(\pi(e)_I) = 0$. On the other hand, the main tool of this decoding algorithm is the following theorem which gives a necessary and sufficient condition for a received vector $w \in \mathbb{Z}_2^n$ to have its systematic coordinates corrected.

Theorem 8.11 ([112]). *Let C be a binary linear code with error correcting capability τ, information set I and parity check matrix H in standard form. Let $w = v + e$, where $v \in C$ and e verifies that $\mathrm{wt}_H(e) \leq \tau$. Then*

$$\mathrm{wt}_H(wH^T) = \mathrm{wt}_H(eH^T) \leq \tau \iff \mathrm{wt}_H(e_I) = 0. \qquad (8.5)$$

Let C be a binary linear code with error correcting capability τ, information set I and parity check matrix H in standard form. Assume that we have found a PD-set, for the information set I, and denote by $w = v + e$ the received vector, where $v \in C$ and e is the error vector. Algorithm 8.5 describes the permutation decoding algorithm for C.

Let C be a $\mathbb{Z}_2\mathbb{Z}_4$-linear code with error correcting capability τ and standard information set J. Let $\mathcal{C} = \Phi^{-1}(C)$ be the corresponding $\mathbb{Z}_2\mathbb{Z}_4$-additive code of type $(\alpha, \beta; \gamma, \delta; \kappa)$. If $\mathcal{C}$ is linear, then the usual systematic encoding can be applied considering the matrices as in (8.1), so Theorem 8.11 is satisfied. On the other hand, if $\gamma = \kappa$, considering the generator and parity check matrix as in (2.7) and (3.11), respectively, Theorem 8.11 is easily adapted so that, for $w = e + v$ with $v \in C$ and $\mathrm{wt}_H(e) \leq \tau$, we have

$$\mathrm{wt}(\Phi^{-1}(w)\mathcal{H}^T) = \mathrm{wt}(\Phi^{-1}(e)\mathcal{H}^T) \leq \tau \iff \mathrm{wt}_H(e_J) = 0. \qquad (8.6)$$

We say that C satisfies (8.6) when this equation is satisfied for all error vectors e such that $\mathrm{wt}_H(e) \leq \tau$. The following result shows that in the non-linear case, (8.6) holds if and only if $\gamma = \kappa$.

Algorithm 8.5 Permutation decoding for a binary linear code

Require: A binary linear code C of length n, the error correcting capability τ of C, a parity check matrix H of C, a PD-set S for the information set I, and a received vector $w \in \mathbb{Z}_2^n$.

1: **if** $\mathrm{wt}_H(wH^T) \leq \tau$ **then**
2:     **return** w_I
3: **else**
4:     **for** $\pi \in S$ **do**
5:         **if** $\mathrm{wt}_H(\pi(w)H^T) \leq \tau$ **then**
6:             **return** $\pi^{-1}(\pi(w)_I G)_I$
7:         **end if**
8:     **end for**
9: **end if**
10: **return** "The errors can not be corrected"

Proposition 8.12 ([22]). *Let $\mathcal{C}$ be a $\mathbb{Z}_2\mathbb{Z}_4$-additive code of type $(\alpha, \beta; \gamma, \delta; \kappa)$, error correcting capability τ and parity check matrix $\mathcal{H}$, such that $C = \Phi(\mathcal{C})$ is a binary non-linear code. Then, C satisfies (8.6) if and only if $\gamma = \kappa$.*

Proof. The case $\gamma = \kappa$ has been discussed above, so assume $\gamma > \kappa$.

Denote by e_i the binary vector of length $n = \alpha + 2\beta$ which has 1 in the ith coordinate and zero elsewhere, for $i \in \{1, \ldots, n\}$. Define the three sets of coordinate positions:

i) $L_1 = \{\kappa + 1, \ldots, \alpha\}$,

ii) $L_2 = \{\varphi_1(1), \varphi_2(1), \ldots, \varphi_1(\beta + \kappa - \gamma - \delta), \varphi_2(\beta + \kappa - \gamma - \delta)\}$,

iii) $L_3 = \{j_1, \ldots, j_{\gamma-\kappa}\}$, where $j_i = \varphi_2(\beta + \kappa - \gamma - \delta + i)$.

We have that $L = L_1 \cup L_2 \cup L_3 = \{1, \ldots, n\} \setminus J$, where J is the standard information set. First, note that, by the definition of $\mathcal{H}$ as in (3.11), it is easy to check that for $k_1, \ldots, k_r \in L_3$ we have that $\mathrm{wt}(\Phi^{-1}(e_{k_1} + \cdots + e_{k_r})\mathcal{H}^T) \geq 2r$. Second, $\mathrm{wt}(\Phi^{-1}(e)\mathcal{H}^T) = \mathrm{wt}(\Phi^{-1}(\varepsilon_1)\mathcal{H}^T) + \mathrm{wt}(\Phi^{-1}(\varepsilon_2)\mathcal{H}^T)$, where $\varepsilon_1 = (\varepsilon_1^1, \ldots, \varepsilon_1^n) \in \mathbb{Z}_2^n$ is given by $(\varepsilon_1)_{L_1} = e_{L_1}$, $\varepsilon_1^i = 0$ if $i \notin L_1$, and $\varepsilon_2 = (\varepsilon_2^1, \ldots, \varepsilon_2^n) \in \mathbb{Z}_2^n$ is given by $(\varepsilon_2)_{L\setminus L_1} = e_{L\setminus L_1}$, $\varepsilon_2^i = 0$ if $i \in L_1$. By using these properties, we see that there exists an error vector of weight at most τ not satisfying (8.6).

Consider an error vector $e \in \mathbb{Z}_2^n$ such that $\mathrm{wt}_H(e) = \tau$, $\mathrm{wt}_H(e_J) = 0$ and $\mathrm{wt}_H(e_{L_3}) \neq 0$. On the one hand, if $\mathrm{wt}_H(e_{L_2}) = 0$, we obtain $\mathrm{wt}(\Phi^{-1}(e)\mathcal{H}^T) = \mathrm{wt}(\Phi^{-1}(\varepsilon_1)\mathcal{H}^T) + \mathrm{wt}(\Phi^{-1}(\varepsilon_2)\mathcal{H}^T) \geq \mathrm{wt}_H(\varepsilon_1) + 2\,\mathrm{wt}_H(\varepsilon_2) > \mathrm{wt}_H(e) = \tau$ and e does not satisfy (8.6). On the other hand, assume $\mathrm{wt}_H(e_{L_2}) \neq 0$. If

$\mathrm{wt}(\Phi^{-1}(e)\mathcal{H}^T) > \tau$, then we have finished. Otherwise, the minimum value of $\mathrm{wt}(\Phi^{-1}(e)\mathcal{H}^T)$ is τ. This case occurs when $\mathrm{wt}(\Phi^{-1}(e_{L_2})\mathcal{H}^T + \Phi^{-1}(e_{L_3})\mathcal{H}^T) = 2\,\mathrm{wt}_H(e_{L_3})$. However, we can then consider $i \in L_2$ such that $\mathrm{wt}_H(e + e_i) = \tau - 1$ and we obtain $\mathrm{wt}(\Phi^{-1}(e + e_i)\mathcal{H}^T) > \tau$. In both cases, we have that C does not satisfy (8.6). $\qquad\square$

As we have seen, when the $\mathbb{Z}_2\mathbb{Z}_4$-linear code is linear or $\gamma = \kappa$, we can use the standard permutation decoding method given by Algorithm 8.5. However, in the other cases, we need a different method. Now, we describe a permutation decoding method which only assumes that we have a systematic encoding, so it works for any $\mathbb{Z}_2\mathbb{Z}_4$-linear code since they have a systematic encoding by Corollary 8.6.

Theorem 8.13 ([22]). *Let C be a binary systematic code of length n and error correcting capability τ. Let I be an information set and let f be a systematic encoding with respect to I. Let $w = v + e$ be a received vector, where $v \in C$ and e verifies that $\mathrm{wt}_H(e) \leq \tau$. Then, the systematic coordinates of w are correct, i.e. $w_I = v_I$, if and only if $\mathrm{wt}_H(w + f(w_I)) \leq \tau$.*

Proof. If $\mathrm{wt}_H(w + f(w_I)) \leq \tau$, then $f(w_I)$ is the closest codeword to w, that is, $f(w_I) = v$. Hence, the systematic coordinates are the same, which means that $w_I = v_I$.

If $v_I = w_I$, then $\mathrm{wt}_H(w + f(w_I)) = \mathrm{wt}_H(w + v) = \mathrm{wt}_H(e) \leq \tau$. $\qquad\square$

Let C be a $\mathbb{Z}_2\mathbb{Z}_4$-linear code with information set I. Assume that $S \subseteq \mathrm{PAut}(C)$ is a PD-set for I and w is a received vector. Algorithm 8.6 gives a permutation decoding algorithm for C. Note that for $w \in \mathbb{Z}_2^n$ if $\mathrm{wt}_H(w + f(w_I)) \leq \tau$, then $f(w_I)$ is a codeword of C at a minimum distance from w.

Finally, we show through the next two examples, how to apply the permutation decoding algorithm for $\mathbb{Z}_2\mathbb{Z}_4$-linear codes given by Algorithm 8.6. Both examples correspond to Hadamard $\mathbb{Z}_4$-linear codes ($\mathbb{Z}_2\mathbb{Z}_4$-linear codes with $\alpha = 0$). Note that, using a code in standard form, it is easy to see that the binary α coordinates are always systematic, so they do not affect on the application of the algorithm. Hadamard $\mathbb{Z}_2\mathbb{Z}_4$-linear codes have been completely classified [43, 102] and they include the linear ones. The $\mathbb{Z}_2\mathbb{Z}_4$-additive codes considered in these examples are linear, but we do not use their linearity to apply the permutation decoding algorithm.

As mentioned in Section 6.2, recall that a binary Hadamard code is a binary code of length n, $2n$ codewords and minimum distance $n/2$. They have a high error correcting capability $\tau = \lfloor (n-2)/4 \rfloor$. However, linear Hadamard codes are not suitable for a syndrome decoding since the number

---

Algorithm 8.6 Permutation decoding for a $\mathbb{Z}_2\mathbb{Z}_4$-linear code

---

Require: A $\mathbb{Z}_2\mathbb{Z}_4$-linear code C of length n, the error correcting capability τ of C, a systematic encoding f for the information set I, a PD-set S for I, and a received vector $w \in \mathbb{Z}_2^n$.

1: **if** $\mathrm{wt}_H(w + f(w_I)) \leq \tau$ **then**
2:   **return** w_I
3: **else**
4:   **for** $\pi \in S$ **do**
5:    **if** $\mathrm{wt}_H(\pi(w) + f(\pi(w)_I)) \leq \tau$ **then**
6:     **return** $\pi^{-1}(f(\pi(w)_I))_I$
7:    **end if**
8:   **end for**
9: **end if**
10: **return** "The errors can not be corrected"

---

of syndromes is also very high. In [85], for linear Hadamard and simplex codes, a partial permutation decoding, that is, a permutation decoding up to $s < \tau$ errors, was presented by giving s-PD-sets for these two families of codes. Later, in [19, 20], s-PD-sets for Hadamard $\mathbb{Z}_4$-linear codes and other related $\mathbb{Z}_4$-linear families of codes were described.

Example 8.14. Consider the $\mathbb{Z}_2\mathbb{Z}_4$-additive code $\mathcal{C}$ with generator and parity check matrices

$$\mathcal{G}_S = \begin{pmatrix} 3 & 2 & 1 & 0 \\ 2 & 3 & 0 & 1 \end{pmatrix} \text{ and } \mathcal{H}_S = \begin{pmatrix} 1 & 0 & 1 & 2 \\ 0 & 1 & 2 & 1 \end{pmatrix},$$

respectively. The corresponding $\mathbb{Z}_2\mathbb{Z}_4$-linear code $C = \Phi(\mathcal{C})$ is a 1-error-correcting code of type $(0, 4; 0, 2; 0)$ (i.e., C is a $\mathbb{Z}_4$-linear code). Indeed, C is a Hadamard $\mathbb{Z}_4$-linear code [102].

Let $\vartheta = (1, 3, 5, 7)(2, 4, 6, 8)$. It is straightforward to check that $\vartheta \in \mathrm{PAut}(C)$ since $\mathcal{C}$ is a quaternary cyclic code [120]. Moreover, $S = \{id, \vartheta, \vartheta^2\}$ is a PD-set for the standard information set $I = \{5, 6, 7, 8\}$. Since $\gamma = \kappa$, we can use the systematic encoding f defined in (8.4). For the permutation decoding, we can use Algorithm 8.5 adapted to $\mathbb{Z}_2\mathbb{Z}_4$-linear codes by using $\mathcal{H}_S$ instead of H.

For example, let $i = (0, 1, 0, 1) \in \mathbb{Z}_2^4$ be an information vector. Then, the corresponding codeword is

$$v = f(i) = \Phi\left(\Phi^{-1}(i)\mathcal{G}_S\right) = \Phi\left((1, 1)\mathcal{G}_S\right) = \Phi(1, 1, 1, 1) = (0, 1, 0, 1, 0, 1, 0, 1).$$

Suppose that the received vector is $w = v + e$, where $e = (0, 0, 0, 0, 0, 0, 0, 1)$.

The syndrome of w is

$$\Phi\left(\Phi^{-1}(w)\mathcal{H}_S^T\right) = \Phi\left((1,1,1,0)\mathcal{H}_S^T\right) = \Phi(2,3) = (1,1,1,0),$$

which has weight $3 > t = 1$. However, considering the vector $u = \vartheta(w) = (0,0,0,1,0,1,0,1)$, we have that the syndrome of u is

$$\Phi\left(\Phi^{-1}(u)\mathcal{H}_S^T\right) = \Phi(3,0) = (1,0,0,0),$$

which has weight $1 \le \tau = 1$. Therefore, the systematic coordinates of u have no errors. Hence, we decode w as

$$\begin{aligned} v' &= \vartheta^{-1}\left(\Phi(\Phi^{-1}(u_I)\mathcal{G}_S)\right) = \vartheta^{-1}\left(\Phi((1,1)\mathcal{G}_S)\right) \\ &= \vartheta^{-1}(\Phi(1,1,1,1)) = (0,1,0,1,0,1,0,1), \end{aligned}$$

and the information vector is $i' = v_I' = (0,1,0,1)$. Since $\mathrm{wt}_H(e) \le \tau$, we have $v' = v$ and $i' = i$. $\triangle$

Example 8.15. Consider the $\mathbb{Z}_2\mathbb{Z}_4$-additive code $\mathcal{C}$ with generator matrix

$$\mathcal{G}_S = \begin{pmatrix} 2 & 2 & 2 & 0 & 0 & 2 & 0 & 0 \\ 3 & 2 & 1 & 2 & 3 & 0 & 1 & 0 \\ 2 & 3 & 0 & 3 & 2 & 1 & 0 & 1 \end{pmatrix}.$$

The corresponding $\mathbb{Z}_2\mathbb{Z}_4$-linear code $C = \Phi(\mathcal{C})$ is a 3-error-correcting code of type $(0,8;1,2;0)$ (i.e., C is a $\mathbb{Z}_4$-linear code). In fact, C is also a Hadamard $\mathbb{Z}_4$-linear code [102].
We know that $\langle \vartheta_1, \vartheta_2, \vartheta_3, \vartheta_4 \rangle \subseteq \mathrm{PAut}(C)$ [120], where

$$\begin{aligned} \vartheta_1 &= (1,5)(2,6)(3,11)(4,12)(9,13)(10,14)(7,15)(8,16), \\ \vartheta_2 &= (1,3,5,11)(2,4,6,12)(9,7,13,15)(10,8,14,16), \\ \vartheta_3 &= (9,13)(10,14)(7,15)(8,16), \\ \vartheta_4 &= (1,9)(2,10)(5,13)(6,14). \end{aligned} \qquad (8.7)$$

Moreover, it is easy to check using the MAGMA software package [18] that we can take the elements in the subgroup $S = \langle \vartheta_1, \vartheta_2, \vartheta_4 \rangle$ as a PD-set for the standard information set $I = \{11,13,14,15,16\}$. In this case, we can not use the standard permutation decoding, since $\gamma \neq \kappa$. However, we can still perform a permutation decoding using the alternative method given by Algorithm 8.6.
For example, let $i = (1,1,1,1,1) \in \mathbb{Z}_2^5$ be an information vector. Using the systematic encoding f defined in (8.3), the corresponding codeword is

$$\begin{aligned} v &= f(i) = \Phi\left(\sigma(\Phi^{-1}(i))\mathcal{G}_S\right) = \Phi\left((1+\eta_1, 2, 2)\mathcal{G}_S\right) \\ &= \Phi(2,2,2,2,2,2,2,2) = (1,1,1,1,1,1,1,1,1,1,1,1,1,1,1,1), \end{aligned}$$

where $\eta = (\eta_1) = (1)$. Suppose that the received vector is $w = v + e$, where $e = (0,0,0,0,0,0,0,0,0,0,0,0,1,0,1,1)$. By considering the standard information set I, the information coordinates of w are $w_I = (1,0,1,0,0)$ and

$$f(w_I) = \Phi(\sigma(\Phi^{-1}(w_I))\mathcal{G}_S) = (0,1,0,0,1,0,1,1,1,0,1,1,0,1,0,0),$$

so $\text{wt}_H(w + f(w_I)) = 5 > \tau = 3$. However, considering the vector $u = \vartheta_1(w) = (1,1,1,1,1,1,0,0,0,1,1,1,1,1,1,1)$, we have that $u_I = (1,1,1,1,1)$ and

$$f(u_I) = \Phi(\sigma(\Phi^{-1}(u_I))\mathcal{G}) = (1,1,1,1,1,1,1,1,1,1,1,1,1,1,1,1),$$

so $\text{wt}_H(u + f(u_I)) = 3 \leq \tau = 3$. Therefore, the systematic coordinates of u have no errors. Hence, we decode w as $v' = \vartheta_1^{-1}(f(u_I))$ and the information vector is $i' = v'_I = (1,1,1,1,1)$. Since $\text{wt}_H(e) \leq \tau$, we have $v' = v$ and $i' = i$. △

Chapter 9

Generalizations, Variants, and Applications of $\mathbb{Z}_2\mathbb{Z}_4$-Additive Codes

As mentioned in Section 1.1, $\mathbb{Z}_2\mathbb{Z}_4$-additive codes were first introduced in 1997 as abelian translation-invariant propelinear codes [134]. It was proved that they coincide with the additive codes in the sense of Delsarte [57], which are subgroups of the underlying abelian group in a translation association scheme when this is the binary Hamming scheme. In 2010, a general discussion about this type of codes, their generator matrix, rank, kernel and associated algebraic parameters was presented [33, 37]. In 2014, the concept of $\mathbb{Z}_2\mathbb{Z}_4$-additive cyclic codes was introduced [1]. A natural generalization of cyclic codes is the class of quasi-cyclic codes studied in [141]. In addition, in 2015 and 2018, $\mathbb{Z}_2\mathbb{Z}_4$-additive codes were generalized to $\mathbb{Z}_2\mathbb{Z}_{2^s}$-additive codes and $\mathbb{Z}_{p^r}\mathbb{Z}_{p^s}$-additive codes, where p is a prime, and r, s are positive integers [6, 13].

In this chapter, several generalizations and variants are reviewed, such as $\mathbb{Z}_{p^s}$-additive codes [70], $\mathbb{Z}_p\mathbb{Z}_{p^2}\cdots\mathbb{Z}_{p^s}$-additive codes [13], and more generally, additive codes in a product of commutative finite chain rings [11, 12, 13, 41, 50, 88, 145]. Also, there are several papers in the literature about additive cyclic codes over the product of different chain rings [1, 3, 9, 39, 40, 42, 59, 87, 98, 110, 116, 146, 165].

A new ring of order four, $\mathbb{Z}_2[u]$, where $u^2 = 0$, appears not as a generalization, but as a variant. In 2015 and 2016, the algebraic structure of

---

There are parts of this chapter that have been previously published. Reprinted by permission from Springer Nature Customer Service Centre GmbH: J. Borges and C. Fernández-Córdoba. "A characterization of $\mathbb{Z}_2\mathbb{Z}_2[u]$-linear codes". *Designs, Codes and Cryptography*, v. 86, 7, pp. 1377–1389 ©(2018).

$\mathbb{Z}_2\mathbb{Z}_2[u]$-linear cyclic codes and their duals was studied [7, 8]. In 2016, it was proved that the class of $\mathbb{Z}_2\mathbb{Z}_4$-linear codes which are also binary linear is strictly contained in the class of $\mathbb{Z}_2\mathbb{Z}_2[u]$-linear codes [35]. In 2018, the existence of asymptotically good additive cyclic codes was studied, and good long additive cyclic codes on any extension of a fixed degree of the base field were constructed [146]. In 2020, the concept of additive cyclic codes over $\mathbb{Z}_pR$, where $R = \mathbb{Z}_p[u]$, $u^2 = 0$ and p is a prime number, was considered [131].

Finally, an application of $\mathbb{Z}_2\mathbb{Z}_4$-additive codes to steganography is also described [140]. Data is embedded into a message by distorting each cover symbol by one unit at most (±1-steganography). The described method is optimal, and its performance is compared to that of other known methods.

9.1 Additive Codes over Mixed Alphabets and their Gray Map Images

In this section, we include different generalizations of $\mathbb{Z}_4$-additive and $\mathbb{Z}_2\mathbb{Z}_4$-additive codes, by considering linear codes over mixed alphabets. First, we focus on linear codes with some coordinates over $\mathbb{Z}_p$, some over $\mathbb{Z}_{p^2}$, and so on until $\mathbb{Z}_{p^s}$, where p is a prime number. Then, we consider codes over mixed alphabets which are product of commutative finite chain rings, in general. We also include here some of these generalizations when the codes are cyclic. Then, several generalizations of the usual Gray map ϕ from $\mathbb{Z}_4$ to $\mathbb{Z}_2^2$ are considered in order to be able to define not necessarily linear codes over $\mathbb{Z}_p$, as the generalized Gray map images of linear codes over mixed alphabets.

9.1.1 Additive Codes over Mixed Alphabets

Let $\mathbb{Z}_{p^s}$ be the ring of integers modulo p^s, where p is prime and $s \geq 1$. Let $\mathbb{Z}_{p^s}^n$ denote the set of all n-tuples over $\mathbb{Z}_{p^s}$. The elements of $\mathbb{Z}_{p^s}^n$ will also be called vectors of length n. The order of a vector $u \in \mathbb{Z}_{p^s}^n$, denoted by $o(u)$, is the smallest positive integer m such that $mu = \mathbf{0} \in \mathbb{Z}_{p^s}^n$.

A code over $\mathbb{Z}_p$ of length n is a non-empty subset of $\mathbb{Z}_p^n$, and it is linear if it is a subspace of $\mathbb{Z}_p^n$. Similarly, a non-empty subset of $\mathbb{Z}_{p^s}^n$ is a $\mathbb{Z}_{p^s}$-additive code if it is a subgroup of $\mathbb{Z}_{p^s}^n$. Let $\mathcal{C}$ be a $\mathbb{Z}_{p^s}$-additive code of length n. Since $\mathcal{C}$ is a subgroup of $\mathbb{Z}_{p^s}^n$, it is isomorphic to an abelian group $\mathbb{Z}_{p^s}^{t_1} \times \mathbb{Z}_{p^{s-1}}^{t_2} \times \cdots \times \mathbb{Z}_p^{t_s}$, and we say that $\mathcal{C}$ is of type $(p^s)^{t_1}(p^{s-1})^{t_2} \ldots p^{t_s}$ or $(n; t_1, \ldots, t_s)$ briefly. It is clear that a $\mathbb{Z}_{p^s}$-additive code of type $(n; t_1, \ldots, t_s)$ has $p^{st_1 + (s-1)t_2 + \cdots + t_s}$ codewords.

In general, $\mathbb{Z}_{p^s}$-additive codes are not free as submodules of $\mathbb{Z}_{p^s}^n$ (they are when $t_2 = \cdots = t_s = 0$), which means that they usually do not have a basis, that is, a generating set consisting of linearly independent elements. However, for any $\mathbb{Z}_{p^s}$-additive code $\mathcal{C}$, there exists a set of codewords $S = \{u_i^{(j)} : 1 \leq j \leq s, 1 \leq i \leq t_j\} \subseteq \mathcal{C}$, with $u_i^{(j)}$ of order p^{s-j+1} for all i and j, satisfying that any codeword in $\mathcal{C}$ can be expressed uniquely in the form

$$\sum_{j=1}^{s} \sum_{i=1}^{t_j} \lambda_i^{(j)} u_i^{(j)}, \qquad (9.1)$$

where $\lambda_i^{(j)} \in \mathbb{Z}_{p^{s-j+1}}$. The matrix whose rows are the codewords in S is a generator matrix of $\mathcal{C}$ having a minimum number of rows, that is, $t_1 + \cdots + t_s$ rows. Moreover, as for other families of codes, there also exists a standard form for the generator matrix of a $\mathbb{Z}_{p^s}$-additive code. Specifically, $\mathcal{C}$ is permutation equivalent to a $\mathbb{Z}_{p^s}$-additive code with a generator matrix of the form

$$\mathcal{G} = \begin{pmatrix} I_{t_1} & A_{0,1} & A_{0,2} & A_{0,3} & \cdots & \cdots & A_{0,s} \\ 0 & pI_{t_2} & pA_{1,2} & pA_{1,3} & \cdots & \cdots & pA_{1,s} \\ 0 & 0 & p^2 I_{t_3} & p^2 A_{2,3} & \cdots & \cdots & p^2 A_{2,s} \\ 0 & 0 & 0 & \ddots & \ddots & & \vdots \\ \vdots & \vdots & \vdots & \ddots & \ddots & \ddots & \vdots \\ 0 & 0 & 0 & \cdots & 0 & p^{s-1} I_{t_s} & p^{s-1} A_{s-1,s} \end{pmatrix}, \qquad (9.2)$$

where $A_{i,j}$ are matrices over $\mathbb{Z}_{p^s}$, for $0 \leq i \leq s-1$ and $1 \leq j \leq s$, and $\mathbf{0}$ is the all-zero matrix [49]. If the generator matrix of a $\mathbb{Z}_{p^s}$-additive code has this form, it is said to be in standard form.

In the same way, we can define additive codes over mixed alphabets, that is, with some coordinates over $\mathbb{Z}_p$, some over $\mathbb{Z}_{p^2}$, and so on until $\mathbb{Z}_{p^s}$. A $\mathbb{Z}_p \mathbb{Z}_{p^2} \ldots \mathbb{Z}_{p^s}$-additive code is a subgroup of $\mathbb{Z}_p^{\alpha_1} \times \mathbb{Z}_{p^2}^{\alpha_2} \times \cdots \times \mathbb{Z}_{p^s}^{\alpha_s}$ for some non-negative integers $\alpha_1, \alpha_2, \ldots, \alpha_s$. Again, a $\mathbb{Z}_p \mathbb{Z}_{p^2} \ldots \mathbb{Z}_{p^s}$-additive code $\mathcal{C}$ is isomorphic to an abelian group $\mathbb{Z}_{p^s}^{t_1} \times \mathbb{Z}_{p^{s-1}}^{t_2} \times \cdots \times \mathbb{Z}_p^{t_s}$, and we say that $\mathcal{C}$ is of type $(p^s)^{t_1}(p^{s-1})^{t_2} \ldots p^{t_s}$ or $(\alpha_1, \ldots, \alpha_s; t_1, \ldots, t_s)$ briefly. A generator matrix in standard form for these codes is given in [13].

As it is defined for vectors of $\mathbb{Z}_2^{\alpha_1} \times \mathbb{Z}_4^{\alpha_2}$ in Section 3.1, for vectors of $\mathbb{Z}_p^{\alpha_1} \times \mathbb{Z}_{p^2}^{\alpha_2} \times \cdots \times \mathbb{Z}_{p^s}^{\alpha_s}$, we can write the inner product given by (3.1) in the following way:

$$\mathbf{u} \cdot \mathbf{v} = p^{s-1}(u^{(1)} \cdot v^{(1)})_p + p^{s-2}(u^{(2)} \cdot v^{(2)})_{p^2} + \cdots + (u^{(s)} \cdot v^{(s)})_{p^s} \in \mathbb{Z}_{p^s}, \quad (9.3)$$

where $\mathbf{u} = (u^{(1)} \mid u^{(2)} \mid \cdots \mid u^{(s)}), \mathbf{v} = (v^{(1)} \mid v^{(2)} \mid \cdots \mid v^{(s)}) \in \mathbb{Z}_p^{\alpha_1} \times \mathbb{Z}_{p^2}^{\alpha_2} \times \cdots \times \mathbb{Z}_{p^s}^{\alpha_s}$. Note that the computations for vectors over $\mathbb{Z}_{p^i} \subseteq \mathbb{Z}_{p^s}$,

$i \in \{1, \ldots, s\}$, are always made as vectors over $\mathbb{Z}_{p^s}$. We also refer to this product as the *standard inner product*, and it can also be written as

$$
\mathbf{u} \cdot \mathbf{v} = \mathbf{u}
\begin{pmatrix}
p^{s-1} I_{\alpha_1} & 0 & \cdots & 0 & 0 \\
0 & p^{s-2} I_{\alpha_2} & \cdots & 0 & 0 \\
0 & 0 & \ddots & 0 & 0 \\
\vdots & \vdots & & \ddots & \vdots \\
0 & 0 & \cdots & 0 & I_{\alpha_s}
\end{pmatrix}
\mathbf{v}^t .
$$

Note that, when $\alpha_1 = \cdots = \alpha_{s-1} = 0$, the inner product is the usual one for vectors over $\mathbb{Z}_{p^s}$, and when $s = 2$ and $p = 2$ coincides with the one defined for vectors over $\mathbb{Z}_2^\alpha \times \mathbb{Z}_4^\beta$ in (3.2).

Let $\mathcal{C}$ be a $\mathbb{Z}_p \mathbb{Z}_{p^2} \ldots \mathbb{Z}_{p^s}$-additive code of type $(\alpha_1, \ldots, \alpha_s; t_1, \ldots, t_s)$. The *additive orthogonal code* of $\mathcal{C}$, denoted by $\mathcal{C}^\perp$, is defined in the standard way as

$$
\mathcal{C}^\perp = \{ \mathbf{v} \in \mathbb{Z}_p^{\alpha_1} \times \mathbb{Z}_{p^2}^{\alpha_2} \times \cdots \times \mathbb{Z}_{p^s}^{\alpha_s} : \mathbf{u} \cdot \mathbf{v} = 0 \text{ for all } \mathbf{u} \in \mathcal{C} \}.
$$

We also call $\mathcal{C}^\perp$ the *additive dual code* of $\mathcal{C}$. The additive dual code $\mathcal{C}^\perp$ is also a $\mathbb{Z}_p \mathbb{Z}_{p^2} \ldots \mathbb{Z}_{p^s}$-additive code, that is, a subgroup of $\mathbb{Z}_p^{\alpha_1} \times \mathbb{Z}_{p^2}^{\alpha_2} \times \cdots \times \mathbb{Z}_{p^s}^{\alpha_s}$.

Another approach to generalize $\mathbb{Z}_2 \mathbb{Z}_4$-additive codes is by using commutative finite chain rings R_1 and R_2 as studied in [41].

A chain ring R is a principal ideal ring such that the ideals are linearly ordered by set theoretic containment. That is, there exists an element $\gamma \in R$ and an integer $e \geq 1$ such that $0 \subseteq \langle \gamma^{e-1} \rangle \subseteq \langle \gamma^{e-2} \rangle \subseteq \cdots \subseteq \langle \gamma \rangle \subseteq R$. The value e is called the nilpotency index of R. Note that a chain ring R is also a Frobenious ring and, therefore, for any code C over R of length n, we have that $|C||C^\perp| = |R^n|$ [69, 162].

Let R_1 and R_2 be finite commutative chain rings with identity, where γ_1 and γ_2 are generators of the maximal ideals of R_1 and R_2 with nilpotency indices e_1 and e_2, respectively. Assume that $e_1 \leq e_2$. We suppose that R_1 and R_2 have the same residue field $K = R_1/\langle \gamma_1 \rangle = R_2/\langle \gamma_2 \rangle$, with $|K| = q = p^m$, p a prime number and m a positive integer. By $\bar{\ }: R_i \to K$, we denote the natural projection that maps $r \mapsto \bar{r} = r + \langle \gamma_i \rangle$, for $i \in \{1, 2\}$. For any two non-negative integers, α and β, the ring $R_1^\alpha \times R_2^\beta$ is an R_2-module [41]. A subset $\mathcal{C} \subseteq R_1^\alpha \times R_2^\beta$ is a code, which is called an $R_1 R_2$-additive code, if it is an R_2-submodule of $R_1^\alpha \times R_2^\beta$.

For $R_1 R_2$-additive codes, the concept of duality is defined in the following way. For $i \in \{1, 2\}$, the Teichmüller set of representatives of R_i is a subset of R_i, closed under the multiplication and isomorphic to the multiplicative semigroup of the finite field K under the natural projection. Let

$T_1 = \{r_0, \ldots, r_{q-1}\}$ and $T_2 = \{r'_0, \ldots, r'_{q-1}\}$ be the Teichmüller sets of representatives of R_1 and R_2, respectively. We can arrange the subscripts such that $\bar{r}_i = \bar{r}'_i$. Then, we consider the map

$$
\begin{aligned}
\iota : \quad R_1 &\rightarrow R_2 \\
\gamma_1 &\mapsto \gamma_2 \\
r_j &\mapsto r'_j,
\end{aligned}
$$

and we extend it linearly for all elements of R_1. Then, the inner product of $\mathbf{u} = (u_1, \ldots, u_\alpha \mid u'_1, \ldots, u'_\beta), \mathbf{v} = (v_1, \ldots, v_\alpha \mid v'_1, \ldots, v'_\beta) \in R_1^\alpha \times R_2^\beta$ is

$$
\mathbf{u} \cdot \mathbf{v} = \gamma_2^{e_2 - e_1} \sum_{i=1}^{\alpha} \iota(u_i v_i) + \sum_{j=1}^{\beta} u'_j v'_j \in R_2.
$$

We define the dual code of an $R_1 R_2$-additive code $\mathcal{C}$ in $R_1^\alpha \times R_2^\beta$ as

$$
\mathcal{C}^\perp = \{\mathbf{v} \in R_1^\alpha \times R_2^\beta : \mathbf{u} \cdot \mathbf{v} = 0 \text{ for all } \mathbf{u} \in \mathcal{C}\}.
$$

In the literature, there are papers about $\mathbb{Z}_2\mathbb{Z}_8$-additive codes [11], $\mathbb{Z}_2\mathbb{Z}_{2^s}$-additive codes [12] or, more general, about $\mathbb{Z}_{p^r}\mathbb{Z}_{p^s}$-additive codes [13], which can be considered as $\mathbb{Z}_p\mathbb{Z}_{p^2}\ldots\mathbb{Z}_{p^s}$-additive codes and also as $R_1 R_2$-additive codes. Other examples of $\mathbb{Z}_p\mathbb{Z}_{p^2}\ldots\mathbb{Z}_{p^s}$-additive codes are the $\mathbb{Z}_2\mathbb{Z}_4\mathbb{Z}_8$-additive codes studied in [50]. Considering $R_1 R_2$-additive codes, we have that one-weight $\mathbb{Z}_2\mathbb{Z}_4$-additive codes [73] are generalized to one-weight and two-weight $\mathbb{Z}_2 R_1$-additive codes [145] and $\mathbb{Z}_2 R_2$-additive codes [88], where $R_1 = \mathbb{Z}_2[u, v]/\langle u^2, v^2 \rangle$ and $R_2 = \mathbb{Z}_4[u]/\langle u^2 \rangle$. Furthermore, $\mathbb{Z}_p\mathbb{Z}_p[\theta]$-additive codes are studied for $p \in \{2, 3\}$ and $\theta \neq \theta^2 = 0$ in [16]. Other rings, such as $\mathbb{Z}_p[u]/\langle u^2 \rangle$, $\mathbb{Z}_p[v]/\langle v^2 - v \rangle$ or $\mathbb{Z}_2[u, v]/\langle u^2, v^2 \rangle$, are used for $R_1 R_2$-additive cyclic codes, as it is shown in Section 9.1.2.

9.1.2   Additive Cyclic Codes over Mixed Alphabets

Cyclic codes over fields and rings have been deeply studied. A code over a ring R is cyclic when a cyclic permutation of any codeword is also a codeword. Cyclic codes of length n are identified with ideals in the polynomial ring $R[x]/\langle x^n - 1 \rangle$. It was in 2014 that started a new consideration of cyclic codes over mixed alphabets, with $\mathbb{Z}_2\mathbb{Z}_4$-additive cyclic codes [1], which are studied in Chapter 7. In [1], the cyclicity of codes over mixed alphabets, their spanning sets, generator polynomials and duality are examined. Later, in [39], the duality of these codes is studied, and the generator polynomials of the dual code in terms of the generator polynomials of the code are also given. In this section, we review some approaches to cyclic codes over mixed alphabets.

Let R_1 and R_2 be finite commutative chain rings such that R_1 is an R_2-module. Let $\mathcal{C}$ be a code over the alphabets R_1 and R_2, that is, $\mathcal{C} \subseteq R_1^\alpha \times R_2^\beta$. We say that $\mathcal{C}$ is $R_1 R_2$-additive if $\mathcal{C}$ is an R_2-submodule of $R_1^\alpha \times R_2^\beta$. The code $\mathcal{C}$ is *cyclic* if

$$\mathbf{u} = (u_1, u_2, \ldots, u_{\alpha-1}, u_\alpha \mid u_1', u_2', \ldots, u_{\beta-1}', u_\beta') \in \mathcal{C}$$

implies that

$$\mathbf{u}^{(1)} = (u_\alpha, u_1, u_2, \ldots, u_{\alpha-1} \mid u_\beta', u_1', u_2', \ldots, u_{\beta-1}') \in \mathcal{C}.$$

Let $R_{\alpha,\beta} = R_1[x]/\langle x^\alpha - 1 \rangle \times R_2[x]/\langle x^\beta - 1 \rangle$. There is a bijective map between $R_1^\alpha \times R_2^\beta$ and $R_{\alpha,\beta}$ given by

$$(u_1, \ldots, u_\alpha \mid u_1', \ldots, u_\beta') \mapsto (u_1 + u_2 x + \cdots + u_\alpha x^{\alpha-1} \mid u_1' + u_2' x + \cdots + u_\beta' x^{\beta-1}).$$

Also, as for the case of $\mathbb{Z}_2 \mathbb{Z}_4$-additive cyclic codes, we have an operation

$$\star : R_2[x] \times R_{\alpha,\beta} \to R_{\alpha,\beta}$$

defined by $p(x) \star (f(x) \mid g(x)) = (\bar{p}(x) f(x) \mid p(x) g(x))$, where $p(x) \in R_2[x]$ and $(f(x) \mid g(x)) \in R_{\alpha,\beta}$. By using this operation, we can identify cyclic codes in $R_1^\alpha \times R_2^\beta$ as R_2-submodules of $R_{\alpha,\beta}$. Therefore, any R_2-submodule of $R_{\alpha,\beta}$ is an $R_1 R_2$-additive cyclic code.

There is the special case when $R_1 = R_2 = R$. An additive code $\mathcal{C} \subseteq R^\alpha \times R^\beta$ is in fact an additive code over R of length $\alpha + \beta$. However, a cyclic code in $R^\alpha \times R^\beta$ is not a cyclic code in $R^{\alpha+\beta}$ because the condition to be cyclic is different. A cyclic code in $R^\alpha \times R^\beta$ is also called a double cyclic code of length (α, β) over R, or an R-double cyclic code.

Example 9.1. Consider the binary code C of length 4,

$$C = \{(0,0,0,0), (0,0,1,1), (1,1,0,0), (1,1,1,1)\}.$$

Note that C is not cyclic. However, if we consider the code as a submodule of $\mathbb{Z}_2^2 \times \mathbb{Z}_2^2$, then we have that

$$C = \{(0,0 \mid 0,0), (0,0 \mid 1,1), (1,1 \mid 0,0), (1,1 \mid 1,1)\}$$

is a $\mathbb{Z}_2$-double cyclic code of length $(2,2)$. $\triangle$

Double cyclic codes were first studied over $\mathbb{Z}_4$ [87], and later over $\mathbb{Z}_2$ [40]. The techniques used are almost the same as for $\mathbb{Z}_2\mathbb{Z}_4$-additive cyclic codes [1]. In both cases, the authors determined the minimal generating sets of these codes, the generator polynomials for their duals, and obtained several optimal codes. It has also been studied the self-duality of double cyclic codes over $\mathbb{Z}_2$ [116]. Before considering double cyclic codes over $\mathbb{Z}_2$ and $\mathbb{Z}_4$, double cyclic codes were studied over other rings; for example, for q a power of a prime number p, over the ring $R = \mathbb{F}_q + u\mathbb{F}_q + u^2\mathbb{F}_q$ where $u^3 = 0$ [165], and $R = \mathbb{F}_q + v\mathbb{F}_q$ where $v^2 = v$ [59].

For the case of $\mathbb{Z}_2$-double cyclic codes, how $\mathbb{Z}_2$-double cyclic codes are related to $\mathbb{Z}_4$-cyclic codes and to $\mathbb{Z}_2\mathbb{Z}_4$-additive cyclic codes is shown in [40]. Specifically, it is proven that the Gray map image of a $\mathbb{Z}_4$-cyclic code is a $\mathbb{Z}_2$-double cyclic code under certain conditions. Also, the Nechaev-Gray image (see Section 7.4.2) of a $\mathbb{Z}_2\mathbb{Z}_4$-additive cyclic code is a $\mathbb{Z}_2$-double cyclic code if such image is a binary linear code.

In 2018, Borges et al. studied $\mathbb{Z}_{p^r}\mathbb{Z}_{p^s}$-additive cyclic codes and determined their structure [42]. They also discussed minimal generating sets for these codes and some properties of their duals. The existence of asymptotically good $\mathbb{Z}_{p^r}\mathbb{Z}_{p^s}$-additive cyclic codes is proved in [146]. Cyclic codes over the direct product of other finite chain rings have been studied, for example $\mathbb{Z}_2 R_1$-additive codes, where $R_1 = \mathbb{Z}_2[u]/\langle u^2 \rangle$ [9]; $\mathbb{Z}_2 R_2$-additive codes, where $R_2 = \mathbb{Z}_2[u]/\langle u^3 \rangle$ [14, 95]; $\mathbb{Z}_4 R_3$-additive codes, where $R_3 = \mathbb{Z}_4[u]/\langle u^2 \rangle$ [98]; $\mathbb{Z}_4 R_4$-additive codes, where $R_4 = \mathbb{Z}_4[u]/\langle u^3 \rangle$ [127]; $\mathbb{Z}_p R_5$-additive codes, where $R_5 = \mathbb{Z}_p[u]/\langle u^2 \rangle$ [110]; $R_1 R_6$-additive codes, where R_1 is as above and $R_6 = \mathbb{Z}_2[u,v]/\langle u^2, v^2 \rangle$, $uv = vu$ [3]; and $\mathbb{Z}_p R_7$-additive, where $R_7 = \mathbb{Z}_p[v]/\langle v^2 - v \rangle$ [60]. Also, some optimal codes over $\mathbb{Z}_p$ are given in [131] from $\mathbb{Z}_p R_4$-additive cyclic codes.

Several results given in the above mentioned papers have been generalized in [41] for cyclic codes over the direct product of any two finite commutative chain rings R_1 and R_2 with identity, where γ_1 and γ_2 are generators of the maximal ideals of R_1 and R_2 with nilpotency indices e_1 and e_2, respectively. We suppose that R_1 and R_2 have the same residue field $K = R_1/\langle \gamma_1 \rangle = R_2/\langle \gamma_2 \rangle$, with $|K| = q = p^m$, p a prime number and m a positive integer. We consider that α and β are coprime with p, so that we know that $R_1[x]/\langle x^\alpha - 1 \rangle$ and $R_2[x]/\langle x^\beta - 1 \rangle$ are principal ideal rings as shown in [61]. The generator polynomials of $R_1 R_2$-additive cyclic codes are given in the following theorem.

Theorem 9.2 ([41]). *Let $\mathcal{C}$ be a submodule of the $R_2[x]$-module $R_{\alpha,\beta}$. Then, there exist polynomials $\ell(x)$ and $b_{e_1-1}(x) \mid b_{e_1-2}(x) \mid \cdots \mid b_1(x) \mid b_0(x) \mid (x^\alpha - 1)$ over $R_1[x]$ and $a_{e_2-1}(x) \mid a_{e_2-2}(x) \mid \cdots \mid a_1(x) \mid a_0(x) \mid (x^\beta - 1)$ over*

$R_2[x]$ *such that*

$$\mathcal{C} = \langle (b_0(x) + \gamma_1 b_1(x) + \cdots + \gamma_1{}^{e_1-1} b_{e_1-1}(x) \mid 0),$$
$$(\ell(x) \mid a_0(x) + \gamma_2 a_1(x) + \cdots + \gamma_2{}^{e_2-1} a_{e_2-1}(x)) \rangle.$$

As a generalization of cyclic codes over the product of two different rings, there are some papers studying cyclic codes $\mathcal{C} \subseteq R_1^\alpha \times R_2^\beta \times R_3^\gamma$. First, when the three rings are the same $R_1 = R_2 = R_3 = R$, these codes are called triple cyclic codes over R. Structural properties and duality of triple cyclic codes are studied over $\mathbb{Z}_2$ [114, 151], over $\mathbb{Z}_4$ [163], and over $\mathbb{F}_q + u\mathbb{F}_q$ [166]. When there are two different rings, $R_1 = R_2 \neq R_3$, the specific case of $\mathbb{Z}_2\mathbb{Z}_2\mathbb{Z}_4$-additive cyclic codes, and the generator polynomials of these codes along with their duals, is studied in [164]. In [114, 163, 164], the authors studied additive cyclic codes over mixed alphabets considering the product of three not necessarily different rings, say $\mathbb{Z}_2$-additive triple cyclic codes, $\mathbb{Z}_4$-additive triple cyclic codes and $\mathbb{Z}_2\mathbb{Z}_2\mathbb{Z}_4$-additive cyclic codes, respectively. They defined these codes and gave their generator polynomials. However, as shown in [76], these generator polynomials do not allow to define the general case. In [76], $\mathbb{Z}_{p^r}\mathbb{Z}_{p^r}\mathbb{Z}_{p^s}$-additive cyclic codes are defined and the general form of their generator polynomials is given. The case when the three rings are different is studied in [10], where the authors study $\mathbb{Z}_2\mathbb{Z}_4\mathbb{Z}_8$-cyclic codes, and in [108], where the author studies $\mathbb{Z}_2(\mathbb{Z}_2+u\mathbb{Z}_2)(\mathbb{Z}_2+u\mathbb{Z}_2+u^2\mathbb{Z}_2)$-additive cyclic codes.

Additive cyclic codes over mixed alphabets have been raised to other constructions. For example, quantum codes from $\mathbb{Z}_p R$-additive codes, where $R = \mathbb{Z}_p[v]/\langle v^2 - v \rangle$ [60], and also from $\mathbb{F}_q RS$-additive codes, where $q = p^m$, $R = \mathbb{F}_q[u]/\langle u^2 - 1 \rangle$, and $S = \mathbb{F}_q[u, v]/\langle u^2 - 1, v^2 - 1, uv - vu \rangle$ are constructed in [62]. Furthermore, when $q = 4$, DNA codes from $\mathbb{F}_q RS$-additive cyclic codes are constructed in [63].

9.1.3   Generalizations of the Gray Map

In order to generalize the results about $\mathbb{Z}_4$-linear codes, that is, the Gray map image of quaternary linear or $\mathbb{Z}_4$-additive codes, as well as about $\mathbb{Z}_2\mathbb{Z}_4$-linear codes, to linear codes over mixed alphabets, we need to generalize the usual Gray map ϕ from $\mathbb{Z}_4$ to $\mathbb{Z}_2^2$.

Recall that the Gray map $\phi : \mathbb{Z}_4^n \longrightarrow \mathbb{Z}_2^{2n}$ is defined as $\phi(0) = (0, 0)$, $\phi(1) = (0, 1)$, $\phi(2) = (1, 1)$, and $\phi(3) = (1, 0)$ in each coordinate [92, 117]. In [92], it is mentioned that "In communication systems employing quadrature phase shift keying (QPSK), the preferred assignment of two information bits to the four possible phases is in which adjacent phases differ by only one binary digit. This mapping is called Gray encoding and has the advantage that, when a quaternary codeword is transmitted across an additive white Gaussian noise channel, the errors most likely to occur are those causing a single erroneously decoded information bit." Recall that the Lee weight of a vector over $\mathbb{Z}_4$ is defined as the addition of the Lee weights of each one of its coordinates over $\mathbb{Z}_4$, where $\mathrm{wt}_L(0) = 0$, $\mathrm{wt}_L(1) = \mathrm{wt}_L(3) = 1$, and $\mathrm{wt}_L(2) = 2$. Another essential property of the map ϕ is that it is an isometry which transforms Lee distances over $\mathbb{Z}_4^n$ into Hamming distances over $\mathbb{Z}_2^{2n}$.

The Hamming weight of a vector $u \in \mathbb{Z}_p^n$, denoted by $\mathrm{wt}_H(u)$, is the number of non-zero coordinates of u. The Hamming distance of two vectors $u, v \in \mathbb{Z}_p^n$, denoted by $d_H(u, v)$, is the number of coordinates in which they differ. Note that $d_H(u, v) = \mathrm{wt}_H(v - u)$. The minimum distance of a code C over $\mathbb{Z}_p$ is $d_H(C) = \min\{d_H(u, v) : u, v \in C, u \neq v\}$.

A natural generalization of the Lee weight to elements of $\mathbb{Z}_{2^s}$ is defined as

$$\bar{\mathrm{wt}}(x) = \begin{cases} x & \text{if } 0 \leq x \leq 2^{s-1}, \\ 2^s - x & \text{if } x > 2^{s-1}. \end{cases} \tag{9.4}$$

Related to this generalized Lee weight, a generalized Gray map $\bar{\phi}$ from $\mathbb{Z}_{2^s}$ to $\mathbb{Z}_2^{2^{s-1}}$ is defined in [38] as follows:

$$\bar{\phi}(x) = \begin{cases} 0^{2^{s-1}-x} 1^x & \text{if } 0 \leq x \leq 2^{s-1}, \\ 1^{2^{s-1}} + \bar{\phi}(x - 2^{s-1}) & \text{if } x > 2^{s-1}. \end{cases} \tag{9.5}$$

Recall that for any positive integer ℓ, $\mathbf{0}^\ell$ and $\mathbf{1}^\ell$ are the all-zero and the all-one vectors, respectively, of length ℓ. In Example 9.3, the map $\bar{\phi}$ for $\mathbb{Z}_8$ is shown. This generalized Gray map $\bar{\phi}$ respects, as the original one, that *adjacent phases*, i.e., the images of consecutive elements in $\mathbb{Z}_{2^s}$, differ just in one bit. In addition, it is an isometry, which transforms Lee weights of elements of $\mathbb{Z}_{2^s}$ into Hamming weights of elements of $\mathbb{Z}_2^{2^{s-1}}$ [167].

In Section 1.3, the Gray map $\bar{\phi}$ is used to study the propelinear structure of codes over $\mathbb{Z}_{2k}$. This map is also considered in [70], where the linearity and self-duality of the binary Gray map images of codes over $\mathbb{Z}_{2^s}$ is studied.

Another approach to generalize the usual Lee weight for elements of $\mathbb{Z}_4$ is given by the homogeneous weight metric defined in [53] for elements of $\mathbb{Z}_{p^s}$, and also used in [90, 142]:

$$\mathrm{wt}^*(x) = \begin{cases} 0 & \text{if } x = 0, \\ p^{s-1} & \text{if } x \in p^{s-1}\mathbb{Z}_{p^s} \setminus \{0\}, \\ (p-1)p^{s-2} & \text{otherwise.} \end{cases} \tag{9.6}$$

The weight of a vector $u = (u_1, u_2, \ldots, u_n) \in \mathbb{Z}_{p^s}^n$ is defined as $\mathrm{wt}^*(u) = \sum_{i=1}^n \mathrm{wt}^*(u_i) \in \mathbb{Z}_{p^s}$; and the distance between two vectors $u, v \in \mathbb{Z}_{p^s}^n$ as $d^*(u, v) = \mathrm{wt}^*(v - u)$. The minimum distance of a $\mathbb{Z}_{p^s}$-additive code $\mathcal{C}$ is $d^*(\mathcal{C}) = \min\{d^*(u, v) : u, v \in \mathcal{C}, u \neq v\}$. If C is a linear code over $\mathbb{Z}_p$, then $d_H(C)$ coincides with the minimum weight of C, that is, $\mathrm{wt}_H(C) = \min\{\mathrm{wt}_H(u) : u \in C, u \neq 0\}$ [112, p. 10]. Similarly, if $\mathcal{C}$ is a $\mathbb{Z}_{p^s}$-additive code, then $d^*(\mathcal{C})$ coincides with $\mathrm{wt}^*(\mathcal{C}) = \min\{\mathrm{wt}^*(u) : u \in \mathcal{C}, u \neq 0\}$.

In [51], Carlet presented another generalization ϕ^* of the Gray map that goes from $\mathbb{Z}_{2^s}$ to $\mathbb{Z}_2^{2^{s-1}}$. This map was generalized to any p prime in [90] as follows:

$$\phi^* : \quad \mathbb{Z}_{p^s} \longrightarrow \mathbb{Z}_p^{p^{s-1}} \tag{9.7}$$
$$x \longmapsto (x_0, x_1, \ldots, x_{s-1})M,$$

where $x \in \mathbb{Z}_{p^s}$; $[x_0, x_1, \ldots, x_{s-1}]_p$ is the p-ary expansion of x, that is, $x = \sum_{i=0}^{s-1} x_i p^i$ with $x_i \in \mathbb{Z}_p$; M is the following matrix of size $s \times p^{s-1}$ over $\mathbb{Z}_p$:

$$M = \begin{pmatrix} Y \\ 1 \end{pmatrix};$$

and Y is a matrix of size $(s-1) \times p^{s-1}$ whose columns are exactly all the elements of $\mathbb{Z}_p^{s-1}$. The elements of $\mathbb{Z}_p^{s-1}$ can be seen as the p-ary expansions of the elements of $\mathbb{Z}_{p^{s-1}}$. Therefore, we can always arrange the columns of Y in ascending order, as elements of $\mathbb{Z}_{p^{s-1}}$. In Example 9.3, the Gray map ϕ^* for elements of $\mathbb{Z}_8$ is given. In this case, the following matrix M is used:

$$M = \begin{pmatrix} 0 & 1 & 0 & 1 \\ 0 & 0 & 1 & 1 \\ 1 & 1 & 1 & 1 \end{pmatrix}$$

Note that ϕ^* is an isometry using the homogeneous weight metric defined in (9.6) for elements of $\mathbb{Z}_{p^s}$ and the Hamming weight for elements of $\mathbb{Z}_p^{p^{s-1}}$ [90].

When $p = 2$, the generalized Gray map ϕ^*, given by (9.7), holds that the Hamming weight of the image of any element $x \in \mathbb{Z}_{2^s}$ is half of the length, i.e., $\mathrm{wt}_H(\phi^*(x)) = 2^{s-2}$, except the images of 2^{s-1} and 0 that are $\mathrm{wt}_H(\phi^*(2^{s-1})) = 2^{s-1}$ and $\mathrm{wt}_H(\phi^*(0)) = 0$, respectively. This property is also held by the usual Gray map ϕ from $\mathbb{Z}_4$ to $\mathbb{Z}_2^2$.

In general, for any p prime and $s \geq 1$, note that the image of ϕ^* can be seen as a linear code H over $\mathbb{Z}_p$ generated by the matrix M. It is easy to check that H is a linear two-weight code of size p^s with non-zero weights $(p-1)p^{s-2}$ and p^{s-1}. Indeed, it is a linear generalized Hadamard code of length p^{s-1} over $\mathbb{Z}_p$, which comes from a generalized Hadamard matrix $H(p, p^{s-2})$ known as the Sylvester Hadamard matrix [154]. It can also be seen as the p-ary first-order Reed-Muller code as mentioned in [90].

As for vectors over $\mathbb{Z}_4$, we can define the coordinate-wise extended Gray map from $\mathbb{Z}_{p^s}^n$ to $\mathbb{Z}_p^{np^{s-1}}$, also denoted by ϕ^*. Specifically, if $u = (u_1, \ldots, u_n) \in \mathbb{Z}_{p^s}^n$, then $\phi^*(u) = (\phi^*(u_1), \ldots, \phi^*(u_n)) \in \mathbb{Z}_p^{np^{s-1}}$. Let $\mathcal{C}$ be a $\mathbb{Z}_{p^s}$-additive code, that is, a subgroup of $\mathbb{Z}_{p^s}^n$ for a positive integer n. The code $C = \phi^*(\mathcal{C})$ over $\mathbb{Z}_p$ (not necessarily linear) is called $\mathbb{Z}_{p^s}$-*linear code*. The $\mathbb{Z}_{p^s}$-linear codes obtained by using the Gray map given by (9.7) are distance invariant, that is, the Hamming weight distribution is invariant under translation by a codeword. This can be proven directly by using the fact that $d_H(\phi^*(x), \phi^*(y)) = \mathrm{wt}_H(\phi^*(x - y))$ for any $x, y \in \mathbb{Z}_{p^s}$ [51, 90].

The generalized Gray map (9.7) for $p = 2$ was studied in [51, 155]. Later, in [80, 81, 82, 83, 84, 91], this Gray map was considered to study and classify binary codes (not necessarily linear), from the families of Hadamard, simplex, and MacDonald, which are the Gray map image of $\mathbb{Z}_{2^s}$-additive codes, that is, $\mathbb{Z}_{2^s}$-linear codes. These results generalize some ones presented in [52, 91, 102, 122] for $\mathbb{Z}_4$-additive codes. More recently, for p prime, results on the classification of $\mathbb{Z}_{p^s}$-linear and $\mathbb{Z}_p\mathbb{Z}_{p^2}$-linear codes which are generalized Hadamard codes can be found in [23, 24, 25]. Also, $\mathbb{Z}_{p^s}$-linear generalized Hadamard codes are constructed in [4] as images under the Gray map of Butson Hadamard codes. More generally, the structure of two-weight codes over $\mathbb{Z}_{2^s}$ and $\mathbb{Z}_{p^m}$ is studied in [144, 147]. Finally, also mention that a systematic encoding and permutation decoding is described for $\mathbb{Z}_{p^s}$-linear codes in [156], generalizing the results given in Section 8.3.2 or [22].

In [51], it is pointed out that if $\mathcal{C}$ and $\mathcal{C}^\perp$ are additive dual codes, then the weight enumerators of $\phi^*(\mathcal{C})$ and $\phi^*(\mathcal{C}^\perp)$ are not in general related by the MacWilliams identity, contrarily to the case of $\mathbb{Z}_4$-linear codes. In this sense, [103] proposes a new generalized Gray map $\bar{\varphi}$ such that $\phi^*(\mathcal{C})$ and $\bar{\varphi}(\mathcal{C}^\perp)$ are formally dual, that is, the weight enumerators of these two codes satisfy the MacWilliams identity. In more detail, two generalized Gray maps from $\mathbb{Z}_{2m}$ are defined in [103]. The first one goes from $\mathbb{Z}_{2m}$ to $\mathbb{Z}_2^m$ and is defined, from a given Hadamard code $A = \{\mathbf{a}_0, \ldots, \mathbf{a}_{2m-1}\}$ of length m such that $\mathbf{a}_0 = \mathbf{0}$ and $\mathbf{a}_i + \mathbf{a}_{i+m} = \mathbf{1}$, as follows:

$$\varphi: \begin{array}{ccc} \mathbb{Z}_{2m} & \to & \mathbb{Z}_2^m \\ x & \mapsto & \mathbf{a}_x. \end{array} \tag{9.8}$$

The second one goes from $\mathbb{Z}_{2m}$ to $2^{\mathbb{Z}_2^m}$, where $2^{\mathbb{Z}_2^m}$ means the set of subsets of $\mathbb{Z}_2^m$, and requires a partition of $\mathbb{Z}_2^m$ into extended 1-perfect codes $\{H_0, \ldots, H_{2m-1}\}$ of length m:

$$\bar{\varphi}: \begin{array}{ccc} \mathbb{Z}_{2m} & \to & 2^{\mathbb{Z}_2^m} \\ x & \mapsto & H_x. \end{array} \tag{9.9}$$

In Example 9.3, Gray maps φ and $\bar{\varphi}$ for $\mathbb{Z}_8$, that is, for $m = 4$, are shown. Note that they depend on the Hadamard code or the partition of 1-perfect extended codes chosen. For $m = 4$, the map is unique. However, for greater values of m, several Gray maps can be defined. For example, for $m = 8$, we can define different Gray maps φ from $\mathbb{Z}_{16}$ to $\mathbb{Z}_2^8$, although there is just one non-equivalent Hadamard code of length 8, which is the linear one, as shown in Example 9.4.

The generalized Gray map ϕ^*, given by (9.7), is a particular case of the map φ given by (9.8), which satisfies $\sum \lambda_i \phi^*(2^i) = \phi^*(\sum \lambda_i 2^i)$ as it was shown in [81]. In fact, in [103], the author mentions that ϕ^* can be seen as a particular case of φ when A is the binary linear Hadamard code.

Example 9.3. Let $p = 2$ and $s = 3$. Let $\bar{\phi}$, ϕ^*, φ and $\bar{\varphi}$ be the generalized Gray maps defined in [38], [51], [103], and [103], given by (9.5), (9.7), (9.8), and (9.9), respectively. Let $A = \{\mathbf{a}_0, \ldots, \mathbf{a}_7\} = \{0000, 0101, 0011, 0110, 1111, 1010, 1100, 1001\}$, which is a Hadamard code of length 2^2. Note that A is linear. Let $H_i = \{\mathbf{a}_i, \mathbf{a}_{i+4}\}$ for all $0 \leq i \leq 3$, $H_4 = \{1000, 0111\}$, $H_5 = \{0100, 1011\}$, $H_6 = \{0010, 1101\}$, and $H_7 = \{0001, 1110\}$. Then, we have that the corresponding images for each generalized Gray map are the following:

$\mathbb{Z}_8$		$\bar{\phi}$	ϕ^*	φ	$\bar{\varphi}$
0	$\longmapsto$	0000	0000	0000	H_0
1	$\longmapsto$	0001	0101	0101	H_1
2	$\longmapsto$	0011	0011	0011	H_2
3	$\longmapsto$	0111	0110	0110	H_3
4	$\longmapsto$	1111	1111	1111	H_4
5	$\longmapsto$	1110	1010	1010	H_5
6	$\longmapsto$	1100	1100	1100	H_6
7	$\longmapsto$	1000	1001	1001	H_7.

Note that the images of ϕ^* and φ are the same, since there is a unique Hadamard code of length 4 that is linear. $\triangle$

Example 9.4. Let $p = 2$ and $s = 4$. Let $A = A' \cup (A' + \mathbf{1})$, where $A' = \{00000000, 01010101, 00110011, 01100110, 00001111, 01011010, 00111100, 01101001\}$ is the unique Hadamard code of length 2^3. Note that A' is linear. Depending on the order of the elements in A, we can define different Gray maps. For example, we can define the following ones:

$\mathbb{Z}_{16}$		φ_1	φ_2
0	$\longmapsto$	00000000	00000000
1	$\longmapsto$	01010101	01010101
2	$\longmapsto$	00110011	00110011
3	$\longmapsto$	01100110	00001111
4	$\longmapsto$	00001111	01100110
5	$\longmapsto$	01011010	01011010
6	$\longmapsto$	00111100	00111100
7	$\longmapsto$	01101001	01101001,

where $\varphi_1(x+8) = \varphi_1(x)+1$ and $\varphi_2(x+8) = \varphi_2(x)+1$ for all $x \in \{0,\ldots,7\}$. Note that φ_1 coincides with the Gray map ϕ^* defined in (9.7), by considering the following generator matrix M of the Hadamard code A:

$$M = \begin{pmatrix} 0 & 1 & 0 & 1 & 0 & 1 & 1 & 0 \\ 0 & 0 & 1 & 1 & 0 & 0 & 1 & 1 \\ 0 & 0 & 0 & 0 & 1 & 1 & 1 & 1 \\ 1 & 1 & 1 & 1 & 1 & 1 & 1 & 1 \end{pmatrix}.$$

Moreover, it satisfies $\sum_{i=0}^{3} \lambda_i \varphi_1(2^i) = \varphi_1(\sum_{i=0}^{3} \lambda_i 2^i)$, while the Gray map φ_2 does not satisfy this property, since $\varphi_2(1) + \varphi_2(2) = 01100110 \neq 00001111 = \varphi_2(3)$. $\triangle$

9.2 $\mathbb{Z}_2\mathbb{Z}_2[u]$-Linear Codes

In 2015, $\mathbb{Z}_2\mathbb{Z}_2[u]$-linear codes were introduced in [7]. They are binary images of $\mathbb{Z}_2\mathbb{Z}_2[u]$-additive codes, which are submodules of the ring $\mathbb{Z}_2^\alpha \times \mathbb{Z}_2[u]^\beta$, where $\mathbb{Z}_2[u]^\beta = (\mathbb{Z}_2 + u\mathbb{Z}_2)^\beta$ and $u^2 = 0$. These codes have some similarities with $\mathbb{Z}_2\mathbb{Z}_4$-linear codes. However, there is a key difference: every $\mathbb{Z}_2\mathbb{Z}_2[u]$-linear code is also binary linear after an appropriate Gray map, which is not true, in general, for $\mathbb{Z}_2\mathbb{Z}_4$-linear codes.

As can be seen in [35], a binary linear code is $\mathbb{Z}_2\mathbb{Z}_2[u]$-linear if and only if its automorphism group has even order. Also, in [35], it is shown that for a $\mathbb{Z}_2\mathbb{Z}_2[u]$-linear code, its $\mathbb{Z}_2\mathbb{Z}_2[u]$-dual is exactly its $\mathbb{Z}_2$-dual code, that is, its standard binary dual code. This, in turn, implies directly that the weight distribution of the code and the weight distribution of its dual are related by the MacWilliams identity. This fact was proved in [7]. Finally, following [35], it can be seen that the class of $\mathbb{Z}_2\mathbb{Z}_4$-linear codes which are also $\mathbb{Z}_2$-linear is strictly contained in the class of $\mathbb{Z}_2\mathbb{Z}_2[u]$-linear codes. All these results from [7] and [35] are included in this section.

We remark that cyclic $\mathbb{Z}_2\mathbb{Z}_2[u]$-linear codes have been studied in [2, 8, 152], where generator polynomials, independent generating sets, and optimal families of such codes are investigated.

9.2.1   The Ring $\mathbb{Z}_2[u]$ and $\mathbb{Z}_2\mathbb{Z}_2[u]$-Linear Codes

Consider the ring $\mathbb{Z}_2[u] = \mathbb{Z}_2 + u\mathbb{Z}_2 = \{0, 1, u, 1 + u\}$, where $u^2 = 0$. Note that $(\mathbb{Z}_2[u], +)$ is group isomorphic to the additive structure of GF(4). With the product operation, $(\mathbb{Z}_2[u], \cdot)$ is monoid isomorphic to $(\mathbb{Z}_4, \cdot)$. Define the map $\pi: \mathbb{Z}_2[u] \longrightarrow \mathbb{Z}_2$, such that $\pi(0) = \pi(u) = 0$ and $\pi(1) = \pi(1 + u) = 1$. Then, for $\lambda \in \mathbb{Z}_2[u]$ and $\mathbf{x} = (x_1, \ldots, x_\alpha \mid x_1', \ldots, x_\beta') \in \mathbb{Z}_2^\alpha \times \mathbb{Z}_2[u]^\beta$, we can consider the scalar product

$$\lambda\mathbf{x} = (\pi(\lambda)x_1, \ldots, \pi(\lambda)x_\alpha \mid \lambda x_1', \ldots, \lambda x_\beta') \in \mathbb{Z}_2^\alpha \times \mathbb{Z}_2[u]^\beta.$$

With this operation, $\mathbb{Z}_2^\alpha \times \mathbb{Z}_2[u]^\beta$ is a $\mathbb{Z}_2[u]$-module. Note that, a $\mathbb{Z}_2[u]$-submodule of $\mathbb{Z}_2^\alpha \times \mathbb{Z}_2[u]^\beta$ is not the same as a subgroup of $\mathbb{Z}_2^\alpha \times \mathbb{Z}_2[u]^\beta$. However, a $\mathbb{Z}_4$-submodule of $\mathbb{Z}_2^\alpha \times \mathbb{Z}_4^\beta$ is the same as an additive subgroup of $\mathbb{Z}_2^\alpha \times \mathbb{Z}_4^\beta$.

Definition 9.5 ([7]). *A $\mathbb{Z}_2\mathbb{Z}_2[u]$-additive code $\mathcal{C}$ with parameters (α, β) is a $\mathbb{Z}_2[u]$-submodule of $\mathbb{Z}_2^\alpha \times \mathbb{Z}_2[u]^\beta$.*

The following straightforward equivalence can be used as an alternative definition.

Lemma 9.6 ([35]). *A code $\mathcal{C} \subseteq \mathbb{Z}_2^\alpha \times \mathbb{Z}_2[u]^\beta$ is $\mathbb{Z}_2\mathbb{Z}_2[u]$-additive if and only if*

$$u\mathbf{z} \in \mathcal{C} \quad \text{for all} \quad \mathbf{z} \in \mathcal{C}, \text{ and}$$
$$\mathbf{x} + \mathbf{y} \in \mathcal{C} \quad \text{for all} \quad \mathbf{x}, \mathbf{y} \in \mathcal{C}.$$

As for $\mathbb{Z}_2^\alpha \times \mathbb{Z}_4^\beta$, we can also define a Gray map. Let $\psi: \mathbb{Z}_2[u] \longrightarrow \mathbb{Z}_2^2$ be defined as

$$\psi(0) = (0, 0), \quad \psi(1) = (0, 1), \quad \psi(u) = (1, 1), \quad \psi(1 + u) = (1, 0).$$

If $a = (a_1, \ldots, a_m) \in \mathbb{Z}_2[u]^m$, then the coordinate-wise extension of ψ is $\psi(a) = (\psi(a_1), \ldots, \psi(a_m))$. Now, we define the Gray map for elements $\mathbf{x} = (x \mid x') \in \mathbb{Z}_2^\alpha \times \mathbb{Z}_2[u]^\beta$ so that $\Psi(\mathbf{x}) = (x \mid \psi(x'))$.

The Lee weight of the elements $0, 1, u, 1 + u \in \mathbb{Z}_2[u]$ is $0, 1, 2, 1$, respectively. Denote by $\mathrm{wt}_L(x')$ the Lee weight of $x' \in \mathbb{Z}_2[u]^\beta$, which is the integer sum of the Lee weights of the coordinates of x'. For a vector

$\mathbf{x} = (x \mid x') \in \mathbb{Z}_2^\alpha \times \mathbb{Z}_2[u]^\beta$, define the weight of $\mathbf{x}$ as $\mathrm{wt}(\mathbf{x}) = \mathrm{wt}_H(x) + \mathrm{wt}_L(x')$. Clearly, $\mathrm{wt}(\mathbf{x}) = \mathrm{wt}_H(\Psi(\mathbf{x}))$.

If $\mathcal{C}$ is a $\mathbb{Z}_2\mathbb{Z}_2[u]$-additive code with parameters (α, β), then the binary image $C = \Psi(\mathcal{C})$ is called a $\mathbb{Z}_2\mathbb{Z}_2[u]$-linear code with parameters (α, β). Note that, unlike for $\mathbb{Z}_2\mathbb{Z}_4$-linear codes, C is a $\mathbb{Z}_2$-linear code of length $n = \alpha + 2\beta$. This fact is clear since for any pair of elements $\mathbf{x}, \mathbf{y} \in \mathbb{Z}_2^\alpha \times \mathbb{Z}_2[u]^\beta$, we have that $\Psi(\mathbf{x}) + \Psi(\mathbf{y}) = \Psi(\mathbf{x} + \mathbf{y})$.

The inner product of $\mathbf{u} = (u_1, \ldots, u_\alpha \mid u'_1, \ldots, u'_\beta)$, $\mathbf{v} = (v_1, \ldots, v_\alpha \mid v'_1, \ldots, v'_\beta) \in \mathbb{Z}_2^\alpha \times \mathbb{Z}_2[u]^\beta$, defined in [7], can be written as

$$\mathbf{u} \cdot \mathbf{v} = u \left(\sum_{i=1}^{\alpha} u_i v_i \right) + \sum_{j=1}^{\beta} u'_j v'_j \in \mathbb{Z}_2[u],$$

where the computations are made taking the zeros and ones in the α binary coordinates as zeros and ones in $\mathbb{Z}_2[u]$, respectively. The dual code of a $\mathbb{Z}_2\mathbb{Z}_2[u]$-additive code $\mathcal{C}$ is defined in the standard way as

$$\mathcal{C}^\perp = \{\mathbf{v} \in \mathbb{Z}_2^\alpha \times \mathbb{Z}_2[u]^\beta \; : \; \mathbf{u} \cdot \mathbf{v} = 0 \text{ for all } \mathbf{u} \in \mathcal{C}\}.$$

The $\mathbb{Z}_2\mathbb{Z}_2[u]$-dual code of $C = \Psi(\mathcal{C})$ is the code $\Psi(\mathcal{C}^\perp)$.

Recall that, as it is mentioned in Section 3.1, if $\mathcal{C}$ is a $\mathbb{Z}_2\mathbb{Z}_4$-additive code, then the codes $\Phi(\mathcal{C})$ and $\Phi(\mathcal{C}^\perp)$ are not necessarily linear, so they are not dual in the binary linear sense, in general. However, the weight enumerator polynomial of $\Phi(\mathcal{C}^\perp)$ is the MacWilliams transform of the weight enumerator polynomial of $\Phi(\mathcal{C})$ [57] (see also [58] and [134]) and therefore they are formally dual. We see in the next subsection that a $\mathbb{Z}_2\mathbb{Z}_2[u]$-linear code $\Psi(\mathcal{C})$ and its $\mathbb{Z}_2\mathbb{Z}_2[u]$-dual code $\Psi(\mathcal{C}^\perp)$ are not only formally dual, as it was proved in [7], but also dual in the binary usual sense, i.e. $\Psi(\mathcal{C})^\perp = \Psi(\mathcal{C}^\perp)$.

9.2.2 Duality of $\mathbb{Z}_2\mathbb{Z}_2[u]$-Linear Codes

It is readily verified that if $a, b \in \mathbb{Z}_2[u]$, then $\psi(a) \cdot \psi(b) = 1$ if and only if $ab \in \{1, u\}$. This property can be easily generalized for elements in $\mathbb{Z}_2[u]^\beta$.

Lemma 9.7 ([35]). *If $x', y' \in \mathbb{Z}_2[u]^\beta$, then $\psi(x') \cdot \psi(y') = 1$ if and only if $x' \cdot y' \in \{1, u\}$.*

Proposition 9.8 ([35]). *Let $\mathbf{x}, \mathbf{y} \in \mathbb{Z}_2^\alpha \times \mathbb{Z}_2[u]^\beta$.*

  i) *If $\mathbf{x} \cdot \mathbf{y} = 0$, then $\Psi(\mathbf{x}) \cdot \Psi(\mathbf{y}) = 0$.*

  ii) *If $\mathbf{x} \cdot \mathbf{y} \neq 0$ and $\Psi(\mathbf{x}) \cdot \Psi(\mathbf{y}) = 0$, then $\Psi(\mathbf{x}) \cdot \Psi((1 + u)\mathbf{y}) = 1$.*

Proof: Let $\mathbf{x} = (x \mid x')$ and $\mathbf{y} = (y \mid y')$ be elements in $\mathbb{Z}_2^\alpha \times \mathbb{Z}_2[u]^\beta$. We can write the inner product of $\mathbf{x}$ and $\mathbf{y}$ as $\mathbf{x} \cdot \mathbf{y} = u(x \cdot y) + (x' \cdot y')$.

i) If $\mathbf{x} \cdot \mathbf{y} = 0$, then either (a) $x \cdot y = x' \cdot y' = 0$, or (b) $x \cdot y = 1$ and $x' \cdot y' = u$.

   (a) By Lemma 9.7, we have that $\psi(x') \cdot \psi(y') = 0$ and hence $\Psi(\mathbf{x}) \cdot \Psi(\mathbf{y}) = 0$.

   (b) Again, by Lemma 9.7, we obtain $\psi(x') \cdot \psi(y') = 1$ and then $\Psi(\mathbf{x}) \cdot \Psi(\mathbf{y}) = 0$.

ii) If $\mathbf{x} \cdot \mathbf{y} \neq 0$, then either (a) $x \cdot y = 0$ and $x' \cdot y' \neq 0$, or (b) $x \cdot y = 1$ and $x' \cdot y' \neq u$.

   (a) In this case, $x' \cdot y' \in \{1, u, 1 + u\}$. Since $x \cdot y = 0$ and $\Psi(\mathbf{x}) \cdot \Psi(\mathbf{y}) = 0$, we have that $\psi(x') \cdot \psi(y') = 0$ and hence, by Lemma 9.7, the only possible case is that $x' \cdot y' = 1 + u$. Therefore, $x' \cdot ((1+u)y') = 1$ and $\psi(x') \cdot \psi((1+u)y') = 1$, again by Lemma 9.7. Thus, $\Psi(\mathbf{x}) \cdot \Psi((1+u)\mathbf{y}) = 1$.

   (b) We have $x' \cdot y' \in \{0, 1, 1+u\}$. Since $x \cdot y = 1$ and $\Psi(\mathbf{x}) \cdot \Psi(\mathbf{y}) = 0$, we obtain $\psi(x') \cdot \psi(y') = 1$. By Lemma 9.7, the only possibility is $x' \cdot y' = 1$. Hence, $x' \cdot ((1+u)y') = 1+u$ and $\psi(x') \cdot \psi((1+u)y') = 0$. We conclude that $\Psi(\mathbf{x}) \cdot \Psi((1+u)\mathbf{y}) = 1$.

□

Theorem 9.9 ([35]). *Let $\mathcal{C}$ be a $\mathbb{Z}_2\mathbb{Z}_2[u]$-additive code and let $C = \Psi(\mathcal{C})$ be the corresponding binary $\mathbb{Z}_2\mathbb{Z}_2[u]$-linear code. Then, $\Psi(\mathcal{C}^\perp) = C^\perp$.*

Proof: If $\mathbf{x} \in \mathcal{C}^\perp$, then $\mathbf{x} \cdot \mathbf{c} = 0$, for all $\mathbf{c} \in \mathcal{C}$. Hence, by Proposition 9.8-(i), we have that $\Psi(\mathbf{x}) \cdot \Psi(\mathbf{c}) = 0$ for all $\mathbf{c} \in \mathcal{C}$, implying that $\Psi(\mathbf{x}) \in C^\perp$. We have proved $\Psi(\mathcal{C}^\perp) \subseteq C^\perp$.

If $\mathbf{x} \notin \mathcal{C}^\perp$, then $\mathbf{x} \cdot \mathbf{c} \neq 0$ for some $\mathbf{c} \in \mathcal{C}$. Now, by Proposition 9.8-(ii), we have that $\Psi(\mathbf{x}) \cdot \Psi(\mathbf{c}) \neq 0$ or $\Psi(\mathbf{x}) \cdot \Psi((1 + u)\mathbf{c}) \neq 0$. It follows that $\Psi(\mathbf{x}) \notin C^\perp$ and therefore $C^\perp \subseteq \Psi(\mathcal{C}^\perp)$.   □

Let $\mathcal{C}$ be a $\mathbb{Z}_2\mathbb{Z}_2[u]$-additive code. Then, the following diagram commutes:

$$
\begin{array}{ccc}
\mathcal{C} & \xrightarrow{\Psi} & C \\
\perp \downarrow & & \perp \downarrow \\
\mathcal{C}^\perp & \xrightarrow{\Psi} & C^\perp
\end{array}
$$

Clearly, this immediately implies that the weight distributions of C and $C^\perp$ are related by the MacWilliams identity, as it is proved in [7]. To finish this subsection, we prove that the dual of a $\mathbb{Z}_2\mathbb{Z}_2[u]$-linear code is also $\mathbb{Z}_2\mathbb{Z}_2[u]$-linear with the same parameters (α, β).

Proposition 9.10 ([35]). *A binary code C is $\mathbb{Z}_2\mathbb{Z}_2[u]$-linear with parameters (α, β) if and only if $C^\perp$ is $\mathbb{Z}_2\mathbb{Z}_2[u]$-linear with the same parameters (α, β).*

Proof: Assume that C is a $\mathbb{Z}_2\mathbb{Z}_2[u]$-linear code with parameters (α, β). Let $\mathcal{C}^\perp = \Psi^{-1}(C^\perp)$. By linearity of $C^\perp$ and Lemma 9.6, we only need to prove that $u\Psi^{-1}(c) \in \mathcal{C}^\perp$ for all $c \in C^\perp$. For any codeword $\mathbf{x} \in \mathcal{C}$, we have $(u\Psi^{-1}(c)) \cdot \mathbf{x} = u\left(\Psi^{-1}(c) \cdot \mathbf{x}\right) = 0$, which implies that $u\Psi^{-1}(c) \in \mathcal{C}^\perp$. The converse follows from the fact that $(C^\perp)^\perp = C$. □

9.2.3 Characterization of $\mathbb{Z}_2\mathbb{Z}_2[u]$-Linear Codes

Given a binary linear code C of length n, a natural question is whether we can choose a set of β pairs of coordinates such that C is a $\mathbb{Z}_2\mathbb{Z}_2[u]$-linear code with parameters $(n - 2\beta, \beta)$. This is an open problem for $\mathbb{Z}_2\mathbb{Z}_4$-linear codes. For $\mathbb{Z}_2\mathbb{Z}_2[u]$-linear codes, the next lemma and corollary show us that it is enough to answer the question for a generator matrix of C.

Lemma 9.11 ([35]). *Let $S = \{\mathbf{x}_1, \ldots, \mathbf{x}_r\} \subset \mathbb{Z}_2^\alpha \times \mathbb{Z}_2[u]^\beta$ and let C be the linear code generated by the binary image vectors of S, $C = \langle \Psi(S) \rangle$. Then, C is a $\mathbb{Z}_2\mathbb{Z}_2[u]$-linear code with parameters (α, β) if and only if $\Psi(u\mathbf{x}_i) \in C$ for all $i \in \{1, \ldots, r\}$.*

Proof: Let $\mathcal{C} = \Psi^{-1}(C) \subset \mathbb{Z}_2^\alpha \times \mathbb{Z}_2[u]^\beta$. Then, C is $\mathbb{Z}_2\mathbb{Z}_2[u]$-linear if and only if $\mathcal{C}$ is $\mathbb{Z}_2\mathbb{Z}_2[u]$-additive. Clearly, for all $\mathbf{x}, \mathbf{y} \in \mathcal{C}$, $\Psi(\mathbf{x} + \mathbf{y}) = \Psi(\mathbf{x}) + \Psi(\mathbf{y}) \in C$ and hence $\mathbf{x} + \mathbf{y} \in \mathcal{C}$. Therefore, applying Lemma 9.6, we have that $\mathcal{C}$ is a $\mathbb{Z}_2\mathbb{Z}_2[u]$-additive code if and only $u\mathbf{x} \in \mathcal{C}$ for all $\mathbf{x} \in \mathcal{C}$. For $\mathbf{x} \in \mathcal{C}$, we have that $\Psi(\mathbf{x}) = \sum_{i=1}^r \lambda_i \Psi(\mathbf{x}_i) = \Psi(\sum_{i=1}^r \lambda_i \mathbf{x}_i)$ for some $\lambda_1, \ldots, \lambda_r \in \mathbb{Z}_2$. Thus $\mathbf{x} = \sum_{i=1}^r \lambda_i \mathbf{x}_i$ and $u\mathbf{x} = \sum_{i=1}^r \lambda_i u\mathbf{x}_i$. Hence, $u\mathbf{x} \in \mathcal{C}$ for all $\mathbf{x} \in \mathcal{C}$, if and only if $u\mathbf{x_i} \in \mathcal{C}$ for all $i \in \{1, \ldots, r\}$. □

Corollary 9.12 ([35]). *Let C be a binary linear code of length $n = \alpha + 2\beta$, for some $\alpha \geq 0$ and $\beta > 0$. Let*

$$G = \begin{pmatrix} v_1 \\ \vdots \\ v_r \end{pmatrix}$$

be a generator matrix of C. Then, C is a $\mathbb{Z}_2\mathbb{Z}_2[u]$-linear code with parameters (α, β) if and only if $\Psi\left(u\Psi^{-1}(v_i)\right) \in C$ for all $i \in \{1, \ldots, r\}$.

Now, we give a necessary and sufficient condition for a binary linear code to be $\mathbb{Z}_2\mathbb{Z}_2[u]$-linear, concerning its permutation automorphism group.

Proposition 9.13 ([35]). *Let C be a binary linear code. Then, C is permutation equivalent to a $\mathbb{Z}_2\mathbb{Z}_2[u]$-linear code with parameters (α, β), where $\beta > 0$, if and only if there exists an involution $\sigma \in \text{PAut}(C)$ fixing α coordinates.*

Proof: Assume that C is a $\mathbb{Z}_2\mathbb{Z}_2[u]$-linear code with $\beta > 0$ and let $\mathcal{C} = \Psi^{-1}(C)$. For any codeword $\mathbf{x} = (x_1, \ldots, x_\alpha \mid x'_1, \ldots, x'_\beta) \in \mathcal{C}$, we write its binary image as $x = (x_1, \ldots, x_\alpha \mid y_1, \ldots, y_{2\beta})$, where $\psi(x'_i) = (y_{2i-1}, y_{2i})$, for $i \in \{1, \ldots, \beta\}$. Let σ be the involution that transposes y_{2i-1} and y_{2i}, for all $i \in \{1, \ldots, \beta\}$. Clearly, $\Psi\left((1+u)\mathbf{x}\right) = \sigma(x)$. Since $(1+u)\mathbf{x} \in \mathcal{C}$, we have $\sigma \in \text{PAut}(C)$.

Conversely, if $\sigma \in \text{PAut}(C)$ has order two, then σ is a product of disjoint transpositions. We can assume that the α coordinates fixed by σ are the first ones, and the pairs of coordinates that σ transposes are consecutive. Then, considering the pairs of coordinates that σ transposes as the images of $\mathbb{Z}_2[u]$ coordinates, we obtain that $\sigma(x) = \Psi\left((1+u)\Psi^{-1}(x)\right)$, for any codeword $x \in C$. Since $\sigma(x) \in C$, we have that $(1+u)\mathbf{x} \in \mathcal{C} = \Psi^{-1}(C)$ for any $\mathbf{x} \in \mathcal{C}$. However, this condition implies that $\mathcal{C}$ is a $\mathbb{Z}_2\mathbb{Z}_2[u]$-additive code since $(1+u)\mathbf{x} = \mathbf{x} + u\mathbf{x}$ and thus $u\mathbf{x} \in \mathcal{C}$. For all $\mathbf{x}, \mathbf{y} \in \mathcal{C}$, $\Psi(\mathbf{x} + \mathbf{y}) = \Psi(\mathbf{x}) + \Psi(\mathbf{y}) \in C$ and hence $\mathbf{x} + \mathbf{y} \in \mathcal{C}$. Then, the result follows applying Lemma 9.6.   □

Theorem 9.14 ([35]). *A binary linear code C is permutation equivalent to a $\mathbb{Z}_2\mathbb{Z}_2[u]$-linear code with parameters (α, β), where $\beta > 0$, if and only if $\text{PAut}(C)$ has an even order.*

Proof: From Sylow first theorem, the group $\text{PAut}(C)$ has even order if and only if $\text{PAut}(C)$ contains an involution. The statement then follows by Proposition 9.13.   □

Remark 9.15. Note that for different involutions in $\text{PAut}(C)$, we have different $\mathbb{Z}_2\mathbb{Z}_2[u]$ structures and, possibly, with different parameters. Moreover, according to Proposition 9.13, for each $\sigma_i \in \text{PAut}(C)$ fixing α_i coordinates, we have a $\mathbb{Z}_2\mathbb{Z}_2[u]$ structure of C with parameters (α_i, β_i), where $\alpha_i + 2\beta_i$ is the length of C.

Example 9.16. Consider the code C with generator matrix

$$
\begin{pmatrix}
1 & 1 & 1 & 0 & 0 & 1 & 0 & 0 & 0 & 0 & 0 & 0 \\
0 & 0 & 1 & 1 & 1 & 0 & 0 & 0 & 0 & 0 & 0 & 0 \\
0 & 0 & 0 & 0 & 1 & 1 & 1 & 0 & 0 & 0 & 0 & 0 \\
0 & 0 & 0 & 0 & 0 & 0 & 1 & 1 & 1 & 0 & 0 & 0 \\
0 & 0 & 0 & 0 & 0 & 0 & 0 & 1 & 1 & 1 & 0 \\
1 & 0 & 0 & 0 & 0 & 0 & 0 & 0 & 0 & 1 & 1
\end{pmatrix}.
$$

As it is pointed out in [112, Problem (32), p. 230], $\mathrm{PAut}(C)$ is trivial, i.e. it only contains the identity permutation. Therefore, C is not $\mathbb{Z}_2\mathbb{Z}_2[u]$-linear for any $\beta > 0$. $\triangle$

It is natural to ask if there are interesting binary linear codes which are not $\mathbb{Z}_2\mathbb{Z}_2[u]$-linear codes and whether the automorphism group is always trivial or not in these cases. Denote by $C_B(n, k)$ the best-known linear code of length n and dimension k in the database of MAGMA [46]. The automorphism group of these codes in the database can be obtained using MAGMA software resulting that many linear codes have automorphism group of even order and, therefore, they are $\mathbb{Z}_2\mathbb{Z}_2[u]$-linear. However, several best-known linear codes have been found with automorphism group of odd order. We have obtained some linear codes with trivial automorphism group, for example, $C_B(20, 10)$, $C_B(32, 12)$, $C_B(135, 38)$, and also other codes with non-trivial and odd order automorphism group, for example, $C_B(78, 8)$, $C_B(81, 20)$, $C_B(128, 14)$ and $C_B(89, 11)$ with automorphism groups of order 3, 27, 889 and 979, respectively.

Some remarkable binary linear codes fall in the class of $\mathbb{Z}_2\mathbb{Z}_2[u]$-linear codes. For example, in [35], it is shown that all binary linear perfect codes are $\mathbb{Z}_2\mathbb{Z}_2[u]$-linear codes. In addition, the extended Hamming codes, the extended Golay code, as well as their duals, are $\mathbb{Z}_2\mathbb{Z}_2[u]$-linear codes. The possible parameters (α, β) for all such codes are determined in [35].

9.2.4 $\mathbb{Z}_2\mathbb{Z}_2[u]$-Linear and $\mathbb{Z}_2\mathbb{Z}_4$-Linear Codes

For a $\mathbb{Z}_2\mathbb{Z}_4$-linear code C, we say that it has parameters (α, β) if the corresponding $\mathbb{Z}_2\mathbb{Z}_4$-additive code $\mathcal{C} = \Phi^{-1}(C)$ has α coordinates in $\mathbb{Z}_2$ and β coordinates in $\mathbb{Z}_4$ so that $\mathcal{C} \subseteq \mathbb{Z}_2^\alpha \times \mathbb{Z}_4^\beta$.

In this subsection, we prove that any $\mathbb{Z}_2\mathbb{Z}_4$-linear code with parameters (α, β) which is also binary linear has a $\mathbb{Z}_2\mathbb{Z}_2[u]$ structure with the same parameters. It is not difficult to see this result using Theorem 9.14. However, we give here an independent proof in order to better clarify the relation between both classes of codes.

Recall that if $\mathcal{C}$ is a $\mathbb{Z}_2\mathbb{Z}_4$-additive code, then its binary image $C = \Phi(\mathcal{C})$ is linear if and only if $2(\mathbf{x} * \mathbf{y}) \in \mathcal{C}$ for all $\mathbf{x}, \mathbf{y} \in \mathcal{C}$ (see Corollary 5.2 or [79]). Define the map $\theta : \mathbb{Z}_2^\alpha \times \mathbb{Z}_4^\beta \longrightarrow \mathbb{Z}_2^\alpha \times \mathbb{Z}_2[u]^\beta$ such that, for every element $(x_1, \ldots, x_\alpha \mid y_1, \ldots, y_\beta) \in \mathbb{Z}_2^\alpha \times \mathbb{Z}_4^\beta$,

$$\theta(x_1, \ldots, x_\alpha \mid y_1, \ldots, y_\beta) = (x_1, \ldots, x_\alpha \mid \vartheta(y_1), \ldots, \vartheta(y_\beta)),$$

where $\vartheta(0) = 0, \vartheta(1) = 1, \vartheta(2) = u$ and $\vartheta(3) = 1 + u$. Note that $\theta = \Psi^{-1} \circ \Phi$.

Theorem 9.17 ([35]). *If $\mathcal{C} \subseteq \mathbb{Z}_2^\alpha \times \mathbb{Z}_4^\beta$ is a $\mathbb{Z}_2\mathbb{Z}_4$-additive code such that $\Phi(\mathcal{C})$ is linear, then $\mathcal{C}' = \theta(\mathcal{C}) \subseteq \mathbb{Z}_2^\alpha \times \mathbb{Z}_2[u]^\beta$ is a $\mathbb{Z}_2\mathbb{Z}_2[u]$-additive code.*

Proof: We use the characterization of Lemma 9.6 to prove the statement. First, given $\mathbf{x} \in \mathcal{C}'$, we need to prove that $u\mathbf{x} \in \mathcal{C}'$. Note that $u\mathbf{x} = \theta(2\theta^{-1}(\mathbf{x}))$, which is in $\mathcal{C}'$. Next, we want to prove that $\mathbf{x} + \mathbf{y} \in \mathcal{C}'$ for all $\mathbf{x}, \mathbf{y} \in \mathcal{C}'$. Clearly,

$$\mathbf{x} + \mathbf{y} = \Psi^{-1}\left(\Psi(\mathbf{x}) + \Psi(\mathbf{y})\right). \qquad (9.10)$$

By Proposition 5.1, we have that

$$\Psi(\mathbf{x}) + \Psi(\mathbf{y}) = \Phi\left(\Phi^{-1}\left(\Psi(\mathbf{x})\right) + \Phi^{-1}\left(\Psi(\mathbf{y})\right) + 2\left(\Phi^{-1}\left(\Psi(\mathbf{x})\right) * \Phi^{-1}\left(\Psi(\mathbf{y})\right)\right)\right). \qquad (9.11)$$

Combining (9.10) and (9.11), we obtain that

$$\mathbf{x} + \mathbf{y} = \theta\left(\theta^{-1}(\mathbf{x}) + \theta^{-1}(\mathbf{y}) + 2(\theta^{-1}(\mathbf{x}) * \theta^{-1}(\mathbf{y}))\right).$$

Since $\Phi(\mathcal{C})$ is linear, we have that $2\left(\theta^{-1}(\mathbf{x}) * \theta^{-1}(\mathbf{y})\right) \in \mathcal{C}$, by Corollary 5.2. It follows that $\mathbf{x} + \mathbf{y} \in \mathcal{C}'$. $\square$

The following corollary gives the minimum $\mathbb{Z}_2\mathbb{Z}_2[u]$-additive code containing the image under the map θ of a fixed $\mathbb{Z}_2\mathbb{Z}_4$-additive code.

Corollary 9.18 ([35]). *Let $\mathcal{C}$ be a $\mathbb{Z}_2\mathbb{Z}_4$-additive code and let $\mathcal{G}$ be a generator matrix of $\mathcal{C}$. Let $\{u_i\}_{i=1}^\gamma$ be the rows of order two and $\{v_j\}_{j=1}^\delta$ the rows of order four in $\mathcal{G}$. Then, the code $\mathcal{C}'$ generated by $\{\theta(u_i)\}_{i=1}^\gamma$, $\{\theta(v_j), \theta(2v_j)\}_{j=1}^\delta$ and $\{\theta(2v_j * v_k)\}_{1 \le j < k \le \delta}$ is the minimum $\mathbb{Z}_2\mathbb{Z}_2[u]$-additive code containing $\theta(\mathcal{C})$.*

Proof: By [79], we have that the minimum binary linear code containing $\Phi(\mathcal{C})$ is $\langle \Phi(\mathcal{C}) \rangle$, which is generated by $\{\Phi(u_i)\}_{i=1}^\gamma$, $\{\Phi(v_j), \Phi(2v_j)\}_{j=1}^\delta$ and $\{\Phi(2v_j * v_k)\}_{1 \le j < k \le \delta}$. Since $\theta = \Psi^{-1} \circ \Phi$, we have that

$$\mathcal{C}' = \langle \{\theta(u_i)\}_{i=1}^\gamma, \{\theta(v_j), \theta(2v_j)\}_{j=1}^\delta, \{\theta(2v_j * v_k)\}_{1 \le j < k \le \delta} \rangle$$

is the minimum $\mathbb{Z}_2\mathbb{Z}_2[u]$-additive code containing $\theta(\mathcal{C})$. $\square$

Let $\mathcal{C}$ be a $\mathbb{Z}_2\mathbb{Z}_4$-additive code and let $\mathcal{C}' = \theta(\mathcal{C})$. For all $\mathbf{x} \in \mathcal{C}'$, we have $u\mathbf{x} = \theta(\theta^{-1}(u\mathbf{x})) = \theta(2\theta^{-1}(\mathbf{x})) \in \mathcal{C}'$ because $2\theta^{-1}(\mathbf{x}) \in \mathcal{C}$. Therefore, for any $\mathbb{Z}_2\mathbb{Z}_4$-additive code $\mathcal{C}$ and $\mathcal{C}' = \theta(\mathcal{C})$, the first condition of Lemma 9.6 is always satisfied. From Theorem 9.17, if $\Phi(\mathcal{C})$ is linear then $\theta(\mathcal{C})$ is a $\mathbb{Z}_2\mathbb{Z}_2[u]$-additive code. However, it is not true when $\Phi(\mathcal{C})$ is not linear. Hence, the second condition of Lemma 9.6 is satisfied for $\mathcal{C}' = \theta(\mathcal{C})$ if and only if $\Phi(\mathcal{C})$ is linear. The next example will show that the second condition is not satisfied in $\mathcal{C}' = \theta(\mathcal{C})$ for a $\mathbb{Z}_2\mathbb{Z}_4$-additive code $\mathcal{C}$ whose image is not linear under the Gray map.

Example 9.19. Let $\mathcal{C}$ be the $\mathbb{Z}_2\mathbb{Z}_4$-additive code generated by

$$\begin{pmatrix} 1 & 0 & 1 & | & 0 & 0 & 0 & 0 \\ 0 & 1 & 1 & | & 0 & 2 & 0 & 0 \\ 0 & 0 & 0 & | & 1 & 0 & 1 & 0 \\ 0 & 0 & 1 & | & 1 & 0 & 0 & 1 \end{pmatrix}.$$

Note that $\Phi(\mathcal{C})$ is not linear by Corollary 5.2 due to the fact that $2(0,0,0 \mid 1,0,1,0) * (0,0,1 \mid 1,0,0,1) = (0,0,0 \mid 2,0,0,0)$ which is not in $\mathcal{C}$.

Now, $\theta(0,0,0 \mid 1,0,1,0) + \theta(0,0,1 \mid 1,0,0,1) = (0,0,1 \mid 0,0,1,1) \notin \mathcal{C}'$ because $\theta^{-1}(0,0,1 \mid 0,0,1,1) = (0,0,1 \mid 0,0,1,1) \notin \mathcal{C}$. And hence, the second condition of Lemma 9.6 is not satisfied. $\triangle$

There are $\mathbb{Z}_2\mathbb{Z}_2[u]$-linear codes which are not $\mathbb{Z}_2\mathbb{Z}_4$-linear, as we can see in the following example.

Example 9.20. Let $\mathcal{D} \subset \mathbb{Z}_2[u]^4$ be the code generated by $\mathbf{x} = (1,1,1,u)$ and $\mathbf{y} = (1,u,1,1)$. We can see that

$$\theta\left(\theta^{-1}(\mathbf{x}) + \theta^{-1}(\mathbf{y})\right) = \theta(2,3,2,3) = (u, 1+u, u, 1+u).$$

It is easy to check that the equation $\lambda\mathbf{x} + \mu\mathbf{y} = (u, 1+u, u, 1+u)$ has no solution for $\lambda, \mu \in \mathbb{Z}_2[u]$. Therefore, $\mathcal{C} = \theta^{-1}(\mathcal{D})$ is not a $\mathbb{Z}_2\mathbb{Z}_4$-additive code. $\triangle$

Remark 9.21. Note that Proposition 9.13 and Theorem 9.14 apply also to $\mathbb{Z}_2\mathbb{Z}_4$-linear codes but only in one direction. Indeed, if C is a $\mathbb{Z}_2\mathbb{Z}_4$-linear code, then $\mathrm{PAut}(C)$ has even order. However, the converse is not true, in general. In the previous example, the code $D = \Psi(\mathcal{D})$ is $\mathbb{Z}_2\mathbb{Z}_2[u]$-linear and hence $\mathrm{PAut}(D)$ is of even order, but D is not $\mathbb{Z}_2\mathbb{Z}_4$-linear.

It is worth noting that if $\mathcal{C}$ is a $\mathbb{Z}_2\mathbb{Z}_4$-additive code such that $\Phi(\mathcal{C})$ is linear, it is not yet true that $\Phi(\mathcal{C}^\perp) = \Phi(\mathcal{C})^\perp$ as we can see in the next example.

Example 9.22. Let $\mathcal{C} \subset \mathbb{Z}_4^3$ be the code generated by $\mathbf{x} = (1,1,1)$ and $\mathbf{y} = (0,2,3)$. It can be easily verified that $\Phi(\mathcal{C})$ is linear. However, we have that $(1,1,2) \in \mathcal{C}^\perp$, but $\Phi(1,1,2) = (0,1,0,1,1,1) \notin \Phi(\mathcal{C})^\perp$. $\triangle$

Therefore, we have that if $\mathcal{C}$ is a $\mathbb{Z}_2\mathbb{Z}_4$-additive code, whose binary image is linear, and $\mathcal{C}' = \theta(\mathcal{C})$, then $\Phi(\mathcal{C})$ and $\Phi(\mathcal{C}^\perp)$ are formally dual whereas $\Psi(\mathcal{C}')$ and $\Psi((\mathcal{C}')^\perp)$ are dual by Theorem 9.9. The relations among these codes are illustrated in Figure 9.1.

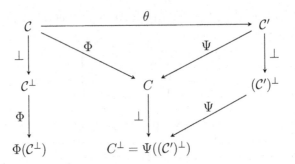

Figure 9.1: Relations among a $\mathbb{Z}_2\mathbb{Z}_4$-additive code (whose binary image is linear), its corresponding $\mathbb{Z}_2\mathbb{Z}_2[u]$-additive code, their binary images and their duals

9.3 $\mathbb{Z}_2\mathbb{Z}_4$-Additive Perfect Codes in Steganography

Steganography is a scientific discipline within *data hiding*, which hides information imperceptibly into innocuous media. In this section, we describe a coding method based on $\mathbb{Z}_2\mathbb{Z}_4$-additive codes, in which data is embedded by distorting each cover symbol by one unit at most (± 1-steganography). The proposed method is optimal and with a slight modification it can solve the problem of boundary values encountered by most of the known methods. The performance of this method is compared with that of the other known methods and with the well-known rate-distortion upper bound. The conclusion is that a higher payload can be obtained for a given distortion by using the proposed technique. A comprehensive overview of the core principles and mathematical methods used for data hiding can be found in [115].

We refer to the information hidden as the *secret message*. The innocuous media where the secret message is embedded, mainly image, video or sound, is called the *cover object*. Finally, the modified object where the secret message is embedded is called the *stego object*. The general scheme is shown in Figure 9.2. Let $\aleph$ be an alphabet, $x = (x_1, \ldots, x_N) \in \aleph^N$ the cover data and $s = (s_1, \ldots, s_M) \in \mathbb{Z}_2^M$ the secret message. A steganographic protocol consists of a pair of functions (e, r), where $e : \aleph^N \times \mathbb{Z}_2^M \longrightarrow \aleph^N$ and $r : \aleph^N \longrightarrow \mathbb{Z}_2^M$, such that $r(e(x, s)) = s$. The functions e and r are called *embedding* and *retrieval* functions, respectively.

An interesting steganographic method, known as *matrix encoding*, was

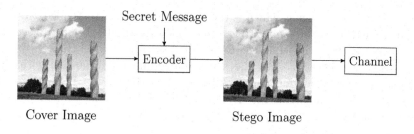

Figure 9.2: Hiding a secret message into a cover image

introduced by Crandall [55] and analysed by Bierbrauer et al. [27]. Matrix encoding requires the sender and the recipient to agree in advance on a parity check matrix H, and the secret message is then extracted by the recipient as the syndrome (with respect to H) of the stego object. This method was made popular by Westfeld [159], who incorporated a specific implementation using Hamming codes. The resulting method is known as the F5 algorithm, which can embed a secret message of M bits in $N = 2^M - 1$ cover symbols by changing one of them, at most. In this case, we have that $\aleph = \mathbb{Z}_2$ and H is a parity check matrix of size $M \times (2^M - 1)$ of a binary Hamming code. Given a cover vector $x \in \mathbb{Z}_2^{2^M-1}$ and a secret message $s \in \mathbb{Z}_2^M$, if $xH^T = s$, then the stego vector is $e(x, s) = x$. Otherwise, we consider the column vector h_j in H which is equal to $s - xH^T$, and then $e(x, s)$ consists of the vector x after changing its jth coordinate. The secret message is extracted by the recipient as the syndrome of the stego vector. Indeed, $e(x, s)H^T = xH^T + h_j = xH^T + s - xH^T = s$.

Example 9.23. Let C be the binary Hamming code of length 7 with parity check matrix

$$H = \begin{pmatrix} 1 & 1 & 1 & 0 & 1 & 0 & 0 \\ 1 & 1 & 0 & 1 & 0 & 1 & 0 \\ 0 & 1 & 1 & 1 & 0 & 0 & 1 \end{pmatrix}.$$

Let $x = (1, 1, 0, 0, 0, 1, 0)$ be the cover data and $s = (1, 0, 0)$ the secret message. We have that $xH^T = (0, 1, 1)$. The vector $s - (0, 1, 1) = (1, 1, 1)$ corresponds to the second column of H, hence $y = e(x, s) = (1, 0, 0, 0, 0, 1, 0)$. Note that $r(y) = yH^T = (1, 0, 0) = s$. △

There are several parameters which are used to evaluate the performance of a steganographic method over a cover message of N symbols: the *average distortion* $D = \frac{R_a}{N}$, where R_a is the expected number of changes over uniformly distributed messages; and the *embedding rate* $E = \frac{M}{N}$, which is the amount of bits that can be hidden in a cover message. Some authors use the *embedding efficiency*, which is the average number of embedded bits per change, instead of the embedding rate. Following the terminology used by Fridrich et al. [86], the tuple (D, E) is called the *CI-rate*.

Given two methods with the same embedding rate, the one with the smaller average distortion performs better than the other. A scheme with block length N, embedding rate E, and average distortion D is called *optimal* if all other schemes with the same block length N have an embedding rate $E' \leq E$ or average distortion $D' \geq D$.

As Willems et al. in [160], we also assume that a discrete source produces a sequence $x = (x_1, \ldots, x_N)$, where N is the block length, $x_i \in \aleph = \{0, 1, \ldots, 2^B - 1\}$, and $B \in \{8, 12, 16\}$ depends on the type of source. The secret message $s \in \mathbb{Z}_2^M$ produces a composite sequence $y = e(x, s)$ by distorting the source x, where $y = (y_1, \ldots, y_N)$ and each $y_i \in \aleph$. In Figure 9.2, the source x is the cover image and the distorted image y is the stego image. This distortion is assumed to be of the squared-error type [160]. In these conditions, we may deal with "binary steganography", in which hidden information is carried by the least significant bit (LSB) of each x_i and the appropriate solution comes from using binary Hamming codes [159]. This technique was improved by using product Hamming codes [139]. Instead of binary steganography, we may deal with "±1-steganography", where $y_i = x_i + c$ for $c \in \{0, +1, -1\}$, and the hidden information is carried by the two LSBs of x_i. The absolute value of c is called the *amplitude* of an embedding change.

There are some steganographic techniques in which messages carrying hidden information are statistically indistinguishable from those that do not carry hidden data. However, in general, the embedding becomes statistically detectable rather quickly with the increasing amplitude of embedding changes. One important goal is to avoid changes of amplitude greater than one. With this assumption, the embedding rate of the ±1-steganographic scheme presented in [140] and described below is compared with the upper bound $H(D) + D$ [160], where $H(D)$ is the binary entropy function $H(D) = -D\log_2(D) - (1 - D)\log_2(1 - D)$ and $0 \leq D \leq 2/3$ is the average distortion. One of the purposes of steganographers is to design schemes to approach this upper bound.

In most studies, ±1-steganography has been treated using ternary codes. Willems et al. [160] proposed a scheme based on ternary Golay and Hamming codes, which were proven to be optimal except for a remark that exposed a problem related to boundary values. Fridrich et al. [86] proposed a method based on rainbow colouring graphs using q-ary Hamming codes, where q is a prime power. This method performs better than the scheme from [160] when q is not a power of 3. However, the authors of both methods suggest making a change of magnitude greater than one to avoid applying the change $x_i - 1$

and $x_i + 1$ to a host sequence of value $x_i = 0$ and $x_i = 2^B - 1$, respectively. Note that this would introduce a larger distortion and, therefore, make the embedding more detectable. The treatment of boundary greyscale values in steganography is important and, to the best of our knowledge, not many papers have focused on this issue.

The method [140] described below is based on $\mathbb{Z}_2\mathbb{Z}_4$-additive perfect codes that, although they are not linear, they have a representation using a parity check matrix that makes them as computationally efficient as the q-ary Hamming codes. This method considers ± 1-steganography, is optimal, and performs better than the method obtained by the direct sum of ternary Hamming codes from [160] and the method based on rainbow colouring of graphs using q-ary Hamming codes [86] for the specific case $q = 3$. Furthermore, it also deals better with boundary greyscale values, because the magnitude of embedding changes is under no circumstances greater than one.

Let $\mathcal{C}$ be a $\mathbb{Z}_2\mathbb{Z}_4$-additive perfect code (described in Section 6.1) and consider its additive dual, which is of type $(\alpha, \beta; \gamma, \delta)$. The corresponding $\mathbb{Z}_2\mathbb{Z}_4$-linear code has length $\alpha + 2\beta = 2^M - 1$. Let $\mathcal{H}$ be a parity check matrix of $\mathcal{C}$ with γ rows of order two and δ rows of order four. Consider the matrix $\chi(\mathcal{H})$, where χ is the map defined on page 18. Let h_j be the jth column vector of $\chi(\mathcal{H})$, for $j \in \{1, \ldots, \alpha + \beta\}$. We consider the columns of $\chi(\mathcal{H})$ (and hence the columns of $\mathcal{H}$) in the following order. Note that $\mathbf{2}$, the vector with all coordinates equal to two, is always one of the columns of $\chi(\mathcal{H})$ and, for the sake of simplicity, it is written as column h_1. We place the remaining first α columns of $\chi(\mathcal{H})$ such that, for any $1 < i \leq \frac{\alpha+1}{2}$, vector h_{2i-2} is paired with its complementary vector $\bar{h}_{2i-2} = h_{2i-1}$. Note that $\bar{h}_{2i-2} = h_{2i-2} + \mathbf{2}$.

Example 9.24. By Theorem 6.11, for $M = 4$, there are three different $\mathbb{Z}_2\mathbb{Z}_4$-additive perfect codes such that their corresponding $\mathbb{Z}_2\mathbb{Z}_4$-linear codes have length $2^4 - 1 = 15$. They correspond to the possible values of $\delta \in \{0, \ldots, \lfloor \frac{M}{2} \rfloor\} = \{0, 1, 2\}$. For $\delta = 0$, the corresponding $\mathbb{Z}_2\mathbb{Z}_4$-additive perfect code is the usual binary Hamming code, whereas for $\delta = 2$ the $\mathbb{Z}_2\mathbb{Z}_4$-additive perfect code $\mathcal{C}$ has parameters $\alpha = 3$, $\beta = 6$, $\gamma = 0$, $\delta = 2$ and the following parity check matrix:

$$\mathcal{H} = \begin{pmatrix} 1 & 0 & 1 & 0 & 1 & 1 & 1 & 1 & 2 \\ 1 & 1 & 0 & 1 & 0 & 1 & 2 & 3 & 1 \end{pmatrix}. \tag{9.12}$$

Note that in $\chi(\mathcal{H})$, we have $\bar{h}_2 = h_3$, that is $h_2 + h_3 = \mathbf{2}$. $\triangle$

Take $N = 2^{M-1} = \frac{\alpha+1}{2} + \beta$, since $2^M - 1 = \alpha + 2\beta$. Let $x = (x_1, \ldots, x_N)$ be an N-length source of greyscale symbols such that $x_i \in \aleph = \{0, 1, \ldots, 2^B - 1\}$, where, for instance, $B = 8$ for greyscale images. Let $[u_{s-1}, \ldots, u_1, u_0]_4$ be the 4-ary expansion of $u \in \aleph$, that is $u = \sum_{i=0}^{s-1} u_i 4^i$ with $u_i \in \mathbb{Z}_4$. Define $\eta_k : \aleph \longrightarrow \mathbb{Z}_2^k$ the function that, for $u \in \aleph$, return the last k bits of $(\phi(u_{s-1}), \phi(u_{s-2}), \ldots, \phi(u_0))$. Finally, define $\eta : \aleph^N \longrightarrow \mathbb{Z}_2^\alpha \times \mathbb{Z}_4^\beta$ as $\eta(x_1, \ldots, x_N) = \Phi^{-1}(\eta_1(x_1), \eta_2(x_2), \eta_2(x_3), \ldots, \eta_2(x_N))$.

Each symbol x_i, for $1 \le i \le N$, is linked with one or two columns of $\chi(\mathcal{H})$. Symbol x_1 corresponds to column h_1, symbol x_2 corresponds to columns h_2 and $\bar{h}_2$, and so on. Specifically,

$$\begin{cases} x_1 & \text{corresponds to} \quad h_1 \\ x_i & \text{corresponds to} \quad h_{2i-2} \text{ and } h_{2i-1} \quad \text{for } 1 < i \le \frac{\alpha+1}{2} \\ x_i & \text{corresponds to} \quad h_{i+\frac{\alpha-1}{2}} \qquad\qquad \text{for } \frac{\alpha+1}{2} < i \le N = \frac{\alpha+1}{2} + \beta. \end{cases} \tag{9.13}$$

Example 9.25. Let $B = 8$ and $M = 4$. Then, we have that $N = 2^{M-1} = 8$. Consider the following N-length source of greyscale symbols: $x = (239, 251, 90, 224, 226, 187, 229, 180)$, where $x_i \in \aleph = \{0, \ldots, 255\}$. We have that $\eta(x) = \Phi^{-1}(\eta_1(239), \eta_2(251), \eta_2(90), \eta_2(224), \eta_2(226),\ \eta_2(187), \eta_2(229), \eta_2(180))$. Then, by considering the values in Table 9.1 and $\eta_1(239) = 0$, since $\eta_2(239) = (1, 0)$ we obtain that $\eta(x) = \Phi^{-1}(0, 1, 0, 1, 1, 0, 0, 1, 1, 1, 0, 0, 1, 0, 0)$. The source x is then translated into the vector $\eta(x) = (0, 1, 0 \mid 2, 0, 2, 3, 1, 0)$. $\triangle$

i	x_i	4-ary expansion	$\eta_2(x_i)$
1	239	$[3, 2, 3, 3]_4$	(1,0)
2	251	$[3, 3, 2, 3]_4$	(1,0)
3	90	$[1, 1, 2, 2]_4$	(1,1)
4	224	$[3, 2, 0, 0]_4$	(0,0)
5	226	$[3, 2, 0, 2]_4$	(1,1)
6	187	$[2, 3, 2, 3]_4$	(1,0)
7	229	$[3, 2, 1, 1]_4$	(0,1)
8	180	$[2, 3, 1, 0]_4$	(0,0)

Table 9.1: Values of $\eta_2(x_i)$ for some $x_i \in \{0, \ldots, 255\}$

Let $s \in \mathbb{Z}_2^{\gamma+2\delta}$ be a secret message, $\mathbf{s} = \Phi^{-1}(s) \in \mathbb{Z}_2^\gamma \times \mathbb{Z}_4^\delta$, and $x = (x_1, \ldots, x_N) \in \aleph^N$ the source of greyscale symbols. The proposed embedding process $e(x, s)$, shown in Figure 9.3, is based on the matrix encoding method. Vector $\epsilon h_j = \chi(\mathbf{s} - \eta(x)\mathcal{H}^T)$ indicates the change needed to embed s within x, where ϵ belongs to $\mathbb{Z}_4$, h_j is a column vector of $\chi(\mathcal{H})$, and $\eta(x)\mathcal{H}^T$ is computed by using the product defined in (3.9). We may have the following situations, depending on which column h_j needs to be multiplied by ϵ:

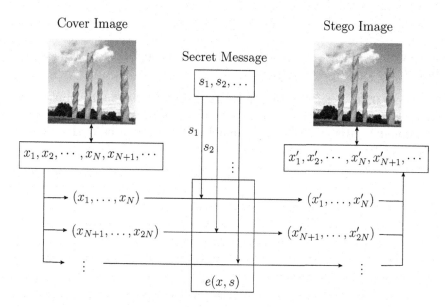

Figure 9.3: Steganographic scheme using a $\mathbb{Z}_2\mathbb{Z}_4$-additive perfect code $\mathcal{C}$ of type $(\alpha, \beta; \gamma, \delta)$, with $N = \frac{\alpha+1}{2} + \beta$ and $s_1, s_2, \ldots \in \mathbb{Z}_2^{\gamma+2\delta}$

i) If $h_j = h_1$, then the embedder has to change the least significant bit of x_1 by adding or subtracting one unit to x_1. It does not matter if we add or subtract, but when the value of the symbol x_1 is $2^B - 1$ we subtract 1, and when the value of the symbol x_1 is 0 we add 1.

ii) If h_j is among the first α column vectors of $\chi(\mathcal{H})$ and $2 \leq j \leq \alpha$, then ϵ can only be $\epsilon = 1$. Column h_j and its complementary column $\bar{h}_j$ are of the form h_{2i-2}, h_{2i-1}, for some index i with $1 < i \leq \frac{\alpha+1}{2}$. From (9.13), take the corresponding symbol x_i. If $\eta_2(x_i) = (b_1, b_2)$, then $\eta_2(x_i + 1) = (b_1, b_2 + 1)$ and $\eta_2(x_i - 1) = (b_1 + 1, b_2)$. Therefore, if $h_j = h_{2i-2}$, to change the value of h_j, we add one unit to x_i; otherwise, if $h_j = h_{2i-1}$, to change the value of h_j, we subtract one unit to x_i. Note that a problem may crop up at this point if we need to add 1 to a symbol x_i of value $2^B - 1$ or subtract 1 from a symbol of value 0.

iii) If h_j is one of the last β columns of $\chi(\mathcal{H})$, then this situation corresponds to add $\epsilon \in \{0, 1, 2, 3\}$. Note that because we use a $\mathbb{Z}_2\mathbb{Z}_4$-additive perfect code, ϵ will never be 2. Hence, the embedder should add ($\epsilon = 1$) or subtract ($\epsilon = 3$) one unit to the corresponding symbol x_i, where $i = j - \frac{\alpha-1}{2}$ from (9.13). Once again, a problem may arise from the boundary values.

Finally, if $y \in \aleph^N$ is a stego vector of greyscale symbols, the retrieval map is just $r(y) = \Phi(\eta(y)\mathcal{H}^T) \in \mathbb{Z}_2^M$, where $M = \gamma + 2\delta$.

Example 9.26. Let $x = (239, 251, 90, 224, 226, 187, 229, 180)$ be an N-length source of greyscale symbols, where $x_i \in \{0, \ldots, 255\}$ and $N = 8$, and let $\mathcal{H}$ be the matrix given in (9.12). By Example 9.25, we have that $\eta(x) = (0, 1, 0 \mid 2, 0, 2, 3, 1, 0)$.

i) Let $s = (0, 0, 1, 1)$ be the vector representing the secret message that we want to embed in x. Then, $\mathbf{s} = \Phi^{-1}(s) = (0, 2)$. We compute $\eta(x)\mathcal{H}^T = (2, 3)$ and find, using the matrix encoding method, that $\epsilon = 3$ and $\chi(\mathbf{s} - \eta(x)\mathcal{H}^T) = 3h_9$. According to the described method, we should subtract 1 from x_i where, following (9.13), $i = 9 - \frac{\alpha - 1}{2} = 8$. Thus, x_8 becomes 179 and the distorted sequence carrying on the secret message s is $y = e(x, s) = (239, 251, 90, 224, 226, 187, 229, 179)$.

Now, we check $r(y) = s$. Since the 4-ary expansion of 179 is $[2, 3, 0, 3]_4$, $\eta_2(179) = (1, 0)$. Then, $\eta(y) = \Phi^{-1}(0, 1, 0, 1, 1, 0, 0, 1, 1, 1, 0, 0, 1, 1, 0)$ $= (0, 1, 0 \mid 2, 0, 2, 3, 1, 3)$. Therefore, $r(y) = \Phi(\eta(y)\mathcal{H}^T) = \Phi(0, 2) = (0, 0, 1, 1) = s$, which is the secret message.

ii) Let $s = (0, 0, 1, 0)$. Then, $\mathbf{s} = \Phi^{-1}(s) = (0, 3)$. Since $\eta(x)\mathcal{H}^T = (2, 3)$, in this case, we find $\epsilon h_j = \chi(\mathbf{s} - \eta(x)\mathcal{H}^T) = (2, 0)$, so $\epsilon = 1$ and $h_j = h_3$. Following the method explained above, we should subtract one unit to x_i where, from (9.13), $3 = 2i - 1$ and so $i = 2$. Thus x_2 becomes 250 and the distorted sequence carrying on the secret message s is $y = e(x, s) = (239, 250, 90, 224, 226, 187, 229, 180)$.

Now, we check $r(y) = s$. Since the 4-ary expansion of 250 is $[3, 3, 2, 2]_4$, $\eta_2(250) = (1, 1)$. Then, $\eta(y) = \Phi^{-1}(0, 1, 1, 1, 1, 0, 0, 1, 1, 1, 0, 0, 1, 0, 0)$ $= (0, 1, 1 \mid 2, 0, 2, 3, 1, 0)$. Therefore, $r(y) = \Phi(\eta(y)\mathcal{H}^T) = \Phi(0, 3) = (0, 0, 1, 0) = s$, which is the hidden message.

$\triangle$

The method explained above uses a perfect code of length $2^M - 1$, which corresponds to $N = 2^{M-1}$ symbols. It can hide any secret vector $s \in \mathbb{Z}_2^{\gamma + 2\delta}$ into the given N symbols. Hence, the embedding rate is $M = \gamma + 2\delta$ bits per N symbols, $E = \dfrac{\gamma + 2\delta}{N} = \dfrac{M}{2^{M-1}}$. Concerning the average distortion D, there are $N - 1$ symbols x_i, for $2 \leq i \leq N$, with a probability $2/2^M$ of being subjected to a change; a symbol x_1 with a probability $1/2^M$ of being the one changed; and, finally, there is a probability of $1/2^M$ that neither of the symbols will need to be changed to embed the secret message s. Hence, $D = \dfrac{2N-1}{N2^M} = \dfrac{2^M-1}{2^{2M-1}}$. The described method has a CI-rate

$$(D, E) = \left(\frac{2^M - 1}{2^{2M-1}}, \frac{M}{2^{M-1}}\right),$$

where M is any integer $M \geq 2$.

Proposition 9.27. *[140] The proposed embedding method based on $\mathbb{Z}_2\mathbb{Z}_4$-additive perfect codes is optimal in the sense that it achieves the smallest possible distortion at a given embedding rate for a fixed block length.*

Note that we can only generate an embedding scheme for the natural values of $M \geq 2$. However, we can use the direct sum of codes [112] to obtain codes whose CI-rates are convex combinations of CI-rates of both codes. Thus, given any non-allowable parameter D for the average distortion, we can take two codes with CI-rates (D_1, E_1) and (D_2, E_2), respectively, where $D_1 < D < D_2$, and the direct sum of an appropriate number of copies of both codes generates a new code with a CI-rate (D, E), with $D = \lambda D_1 + (1-\lambda)D_2$, $E = \lambda E_1 + (1 - \lambda)E_2$, and $0 \leq \lambda \leq 1$. From a graphical point of view, this is equivalent to drawing a line between two contiguous points (D_1, E_1) and (D_2, E_2), as it is shown in Figure 9.4.

The use of perfect $\mathbb{Z}_2\mathbb{Z}_4$-linear codes in ±1-steganography was first proposed in [140]. This method performs better than those based on the direct sum of ternary Hamming codes, and also deals with the extreme greyscale values more efficiently. In [139] the use of product codes in steganography was first proposed. Subsequently, the use of product codes was extended and improved [168] and adapted to $\mathbb{Z}_2\mathbb{Z}_4$-linear codes [135] to obtain all advantages related to the performance and the processing of extreme greyscale values and performing better than the method in [140].

The problematic cases related to boundary values are also cited in [86, 160], but their authors assumed that the probability of grey value saturation was not too large. We argue that, although rare, grey saturation can still occur. It is possible to add some modifications to the method explained above to solve the extreme greyscale values problem [140] obtaining the same embedding rate E but a slightly worse average distortion D.

The problem may arise when the embedder is required to add one unit to a source symbol x_i containing the maximum allowed value, $2^B - 1$; or to subtract one unit from a symbol x_i containing the minimum allowed value, 0. To face this problem, we take advantage of the fact that we are working

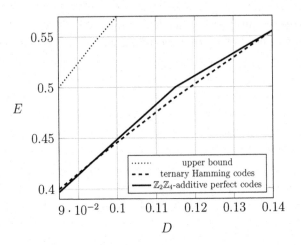

Figure 9.4: CI-rate (D, E), for $B = 8$, of steganographic methods based on ternary Hamming codes and on $\mathbb{Z}_2\mathbb{Z}_4$-additive perfect codes, compared with the upper bound $H(D) + D$, where $H(D)$ is the binary entropy function

with a 1-perfect code. We use a linked column vector h'_j for each one of the last β columns h_j of $\chi(\mathcal{H})$. For each one of these columns h_j, we consider $2h_j = h'_j$, where h'_j is one of the first α columns of $\chi(\mathcal{H})$.

The correspondence between each x_i and a column vector h_j of $\chi(\mathcal{H})$ is the same as in (9.13). However, given an N-length source of symbols $x = (x_1, \ldots, x_N)$, a secret message $s \in \mathbb{Z}_2^{\gamma+2\delta}$, $\mathbf{s} = \Phi^{-1}(s)$, and the vector ϵh_j, such that $\epsilon h_j = \chi(\mathbf{s} - \eta(x)\mathcal{H}^T)$, we can now make some variations on how the changes have to be done for the specific problematic cases:

i) If $h_j = h_1$, like in the general case shown before, it does not matter if we add or subtract one unit to x_i. However, we subtract 1 if $x_i = 2^B - 1$, and we add 1 if $x_i = 0$.

ii) If h_j is among the first α columns of $\chi(\mathcal{H})$, $2 \leq j \leq \alpha$ and, from (9.13), the embedder is required to add 1 to the corresponding symbol $x_i = 2^B - 1$ (or subtract 1 to the symbol $x_i = 0$), then the embedder should instead subtract 1 (or add 1, respectively) to x_i as well as perform the appropriate operation ($+1$ or -1) over x_1 to change $\eta(x_1)$. Indeed, the columns $h_1, h_j, \bar{h}_j$ are linearly dependents.

iii) If h_j is one of the last β columns of $\chi(\mathcal{H})$, and the embedder has to add 1 to the corresponding symbol $x_i = 2^B - 1$ according to (9.13) (or subtract 1 to the symbol $x_i = 0$), then the embedder should instead subtract 1 (or add 1, respectively) from the symbol associated to h_j and also go to the previous item to proceed with $h'_j = 2h_j$.

Example 9.28. Let $s = (0, 0, 1, 1)$, $\mathbf{s} = (0, 2)$ and $x = (239, 251, 90, 224, 226, 187, 229, 0)$ be as in Example 9.26, except for the value of x_8 which is now $x_8 = 0$. We have that $\eta(x) = (0, 1, 0 \mid 2, 0, 2, 3, 1, 0)$. As in Example 9.26, we compute $\eta(x)\mathcal{H}^T = (2, 3)$ and find that $\epsilon = 3$ and $\chi(\mathbf{s} - \eta(x)\mathcal{H}^T) = 3h_9$. According to the described method, we should subtract 1 from x_8. However, we are not able to compute $x_8 - 1$. Instead of this, taking into account that $h'_9 = 2h_9 = h_2$, we add one unit to x_8, and we also proceed with the column $h'_9 = h_2$. Then, according to (9.13), we should also add one unit to x_2. Therefore, we obtain the distorted sequence $y = e(x, s) = (239, 252, 90, 224, 226, 187, 229, 1)$.

Now, we check $r(y) = s$. Using Table 9.1 and since the 4-ary expansion of 252 and 1 are $[3, 3, 3, 0]_4$ and $[0, 0, 0, 1]_4$, respectively, we have $\eta_2(252) = (0, 0)$ and $\eta_2(1) = (0, 1)$. Then, $\eta(y) = \Phi^{-1}(0, 0, 0, 1, 1, 0, 0, 1, 1, 1, 0, 0, 1, 0, 1) = (0, 0, 0 \mid 2, 0, 2, 3, 1, 1)$. Therefore, $r(y) = \Phi(\eta(y)\mathcal{H}^T) = \Phi(0, 2) = (0, 0, 1, 1)$, which is the hidden message. $\hspace{1cm}$ $\triangle$

Bibliography

[1]   T. Abualrub, I. Siap, and N. Aydin. "$\mathbb{Z}_2\mathbb{Z}_4$-additive cyclic codes". In: *IEEE Transactions on Information Theory* 60.3 (2014), pp. 1508–1514.

[2]   T. Abualrub, I. Siap, and I. Aydogdu. "$\mathbb{Z}_2(\mathbb{Z}_2 + u\mathbb{Z}_2)$-linear cyclic codes". In: *Proc. of the International Multiconference of Engineers and Computing Scientists*. Vol. II. 2014.

[3]   N. Annamalai and C. Durairajan. "The structure of $\mathbb{Z}_2[u]\mathbb{Z}_2[u, v]$-additive codes". In: *2016 International Conference on Electrical, Electronics, and Optimization Techniques, ICEEOT*. 2016, pp. 11351–1356.

[4]   J. A. Armario, I. Bailera, and R. Egan. "Butson full propelinear codes". In: *arXiv:2010.06206* (2020).

[5]   E. F. Assmus and J. D. Key. *Designs and Their Codes*. 103. Cambridge University Press, 1994.

[6]   I. Aydogdu and T. Abualrub. "The structure of $\mathbb{Z}_2\mathbb{Z}_{2^s}$-additive cyclic codes". In: *Discrete Mathematics, Algorithms and Applications* 10.04 (2018), p. 1850048.

[7]   I. Aydogdu, T. Abualrub, and I. Siap. "On $\mathbb{Z}_2\mathbb{Z}_2[u]$-additive codes". In: *International Journal of Computer Mathematics* 92.9 (2015), pp. 1806–1814.

[8]   I. Aydogdu, T. Abualrub, and I. Siap. "$\mathbb{Z}_2\mathbb{Z}_2[u]$-cyclic and constacyclic codes". In: *IEEE Transactions on Information Theory* 63.8 (2016), pp. 4883–4893.

[9]   I. Aydogdu, T. Abualrub, and I. Siap. "$\mathbb{Z}_2\mathbb{Z}_2[u]$-cyclic and constacyclic codes". In: *IEEE Transactions on Information Theory* 63.8 (2017), pp. 4883–4893.

[10] I. Aydogdu and F. Gursoy. "$\mathbb{Z}_2\mathbb{Z}_4\mathbb{Z}_8$-cyclic codes". In: *Journal of Applied Mathematics and Computing* 60 (2017), pp. 327–341.

© Springer Nature Switzerland AG 2022
J. Borges et al., *Z₂Z₄-Linear Codes*, https://doi.org/10.1007/978-3-031-05441-9

[11]   I. Aydogdu and I. Siap. "Counting the generator matrices of $\mathbb{Z}_2\mathbb{Z}_8$-codes". In: *Mathematical Sciences And Applications E-Notes* 1.2 (2013), pp. 143–149.

[12]   I. Aydogdu and I. Siap. "The structure of $\mathbb{Z}_2\mathbb{Z}_{2^s}$-additive codes: bounds on the minimum distance". In: *Applied Mathematics and Information Sciences* 7.6 (2013), pp. 2271–2278.

[13]   I. Aydogdu and I. Siap. "On $\mathbb{Z}_{p^r}\mathbb{Z}_{p^s}$-additive codes". In: *Linear and Multilinear Algebra* 63.10 (2015), pp. 2089–2102.

[14]   I. Aydogdu, I. Siap, and R. Ten-Valls. "On the structure of $\mathbb{Z}_2\mathbb{Z}_2[u^3]$-linear and cyclic codes". In: *Finite Fields and Their Applications* 48 (2017), pp. 241–260.

[15]   I. Bailera, J. Borges, and J. Rifà. "On Hadamard full propelinear codes with associated group $C_{2t} \times C_2$". In: *Advances in Mathematics of Communications* 15.1 (2021), pp. 35–54.

[16]   M. Bajalan, A. F. Tabue, and E. Martínez-Moro. "Galois LCD codes over mixed alphabets". In: *arXiv:2202.028443v1* (2022).

[17]   E. Bannai, S. T. Dougherty, M. Harada, and M. Oura. "Type II codes, even unimodular lattices, and invariant rings". In: *IEEE Transactions on Information Theory* 45.4 (1999), pp. 1194–1205.

[18]   R. Barrolleta, J. Pernas, J. Pujol, and M. Villanueva. "Codes over $\mathbb{Z}_4$. A Magma package". In: *Edition* 2.1 (2017). URL: http://www.ccsg.uab.cat.

[19]   R. D. Barrolleta and M. Villanueva. "Partial permutation decoding for binary linear and $\mathbb{Z}_4$-linear Hadamard codes". In: *Designs, Codes and Cryptography* 86.3 (2018), pp. 569–586.

[20]   R. D. Barrolleta and M. Villanueva. "Partial permutation decoding for several families of linear and $\mathbb{Z}_4$-linear codes". In: *IEEE Transactions on Information Theory* 65.1 (2019), pp. 131–141.

[21]   H. Bauer, B. Ganter, and F. Hergert. "Algebraic techniques for nonlinear codes". In: *Combinatorica* 3.1 (1983), pp. 21–33.

[22]   J. J. Bernal, J. Borges, C. Fernández-Córdoba, and M. Villanueva. "Permutation decoding of $\mathbb{Z}_2\mathbb{Z}_4$-linear codes". In: *Designs, Codes and Cryptography* 76.2 (2015), pp. 269–277.

[23]   D. K. Bhunia, C. Fernández-Córdoba, C. Vela, and M. Villanueva. "Equivalences among $\mathbb{Z}_{p^s}$-linear Hadamard codes". In: *arXiv: 2203.15407* (2022).

[24] D. K. Bhunia, C. Fernández-Córdoba, and M. Villanueva. "Construction of $\mathbb{Z}_p\mathbb{Z}_{p^2}$-additive generalized Hadamard codes". In: *arXiv: 2203.15657* (2022).

[25] D. K. Bhunia, C. Fernández-Córdoba, and M. Villanueva. "On the linearity and classification of $\mathbb{Z}_{p^s}$-linear generalized Hadamard codes". In: *Designs, Codes and Cryptography* 90 (2022), pp. 1037–1058.

[26] J. Bierbrauer. *Introduction to Coding Theory*. Chapman & Hall/CRC, 2004.

[27] J. Bierbrauer and J. Fridrich. "Constructing good covering codes for applications in steganography". In: *Transactions on data hiding and multimedia security III*. Springer, 2008, pp. 1–22.

[28] M. Bilal, J. Borges, S. T. Dougherty, and C. Fernández-Córdoba. "Maximum distance separable codes over $\mathbb{Z}_4$ and $\mathbb{Z}_2 \times \mathbb{Z}_4$". In: *Designs, Codes and Cryptography* 61.1 (2011), pp. 31–40.

[29] J. Borges, C. Fernández, J. Pujol, J. Rifà, and M. Villanueva. "$\mathbb{Z}_2\mathbb{Z}_4$-additive codes. A Magma package". In: *Edition* 5.0 (2022). URL: http://www.ccsg.uab.cat.

[30] J. Borges and C. Fernández-Córdoba. "There is exactly one $\mathbb{Z}_2\mathbb{Z}_4$-cyclic 1-perfect code". In: *Designs, Codes and Cryptography* 85.3 (2017), pp. 557–566.

[31] J. Borges, C. Fernández-Córdoba, and R. Ten-Valls. "Binary images of $\mathbb{Z}_2\mathbb{Z}_4$-additive cyclic codes". In: *IEEE Transactions on Information Theory* 64.12 (2018), pp. 7551–7556.

[32] J. Borges, S. T. Dougherty, and C. Fernández-Córdoba. "Characterization and constructions of self-dual codes over $\mathbb{Z}_2 \times \mathbb{Z}_4$". In: *Advances in Mathematics of Communications* 6.3 (2012), pp. 287–303.

[33] J. Borges, S. T. Dougherty, C. Fernández-Córdoba, and R. Ten-Valls. "$\mathbb{Z}_2\mathbb{Z}_4$-additive cyclic codes: kernel and rank". In: *IEEE Transactions on Information Theory* 65.4 (2019), pp. 2119–2127.

[34] J. Borges, C. Fernandez, and K. T. Phelps. "Quaternary Reed–Muller codes". In: *IEEE Transactions on Information Theory* 51.7 (2005), pp. 2686–2691.

[35] J. Borges and C. Fernández-Córdoba. "A characterization of $\mathbb{Z}_2\mathbb{Z}_2[u]$-linear codes". In: *Designs, Codes and Cryptography* 86.7 (2018), pp. 1377–1389.

[36] J. Borges, C. Fernández-Córdoba, and K. T. Phelps. "ZRM codes". In: *IEEE Transactions on Information Theory* 54.1 (2008), pp. 380–386.

[37] J. Borges, C. Fernández-Córdoba, J. Pujol, J. Rifà, and M. Villanueva. "$\mathbb{Z}_2\mathbb{Z}_4$-linear codes: generator matrices and duality". In: *Designs, Codes and Cryptography* 54.2 (2010), pp. 167–179.

[38] J. Borges, C. Fernández-Córdoba, and J. Rifà. "Every $\mathbb{Z}_{2^k}$-code is a binary propelinear code". In: *Electronic Notes in Discrete Mathematics* 10 (2001), pp. 100–102.

[39] J. Borges, C. Fernández-Córdoba, and R. Ten-Valls. "$\mathbb{Z}_2\mathbb{Z}_4$-additive cyclic codes, generator polynomials, and dual codes". In: *IEEE Transactions on Information Theory* 62.11 (2016), pp. 6348–6354.

[40] J. Borges, C. Fernández-Córdoba, and R. Ten-Valls. "$\mathbb{Z}_2$-double cyclic codes". In: *Designs, Codes and Cryptography* 86.3 (2018), pp. 463–479.

[41] J. Borges, C. Fernández-Córdoba, and R. Ten-Valls. "Linear and cyclic codes over direct product of finite chain rings". In: *Mathematical Methods in the Applied Sciences* 41 (2018), pp. 6519–6529.

[42] J. Borges, C. Fernández-Córdoba, and R. Ten-Valls. "On $\mathbb{Z}_{p^r}\mathbb{Z}_{p^s}$-additive cyclic codes". In: *Advances in Mathematics of Communications* 12.1 (2018), pp. 169–179.

[43] J. Borges, K. T. Phelps, and J. Rifà. "The rank and kernel of extended 1-perfect $\mathbb{Z}_4$-linear and additive non-$\mathbb{Z}_4$-linear codes". In: *IEEE Transactions on Information Theory* 49.8 (2003), pp. 2028–2034.

[44] J. Borges, K. T. Phelps, J. Rifà, and V. A. Zinoviev. "On $\mathbb{Z}_4$-linear Preparata-like and Kerdock-like codes". In: *IEEE Transactions on Information Theory* 49.11 (2003), pp. 2834–2843.

[45] J. Borges and J. Rifà. "A characterization of 1-perfect additive codes". In: *IEEE Transactions on Information Theory* 45.5 (1999), pp. 1688–1697.

[46] W. Bosma, J. J. Cannon, C. Fieker, and A. Steel, eds. *Handbook of Magma functions*. Edition 2.26-4, 2021. 6347 pages. URL: http://magma.maths.usyd.edu.au/magma/.

[47] A. E. Brouwer and W. H. Haemers. "Distance-regular graphs". In: *Spectra of Graphs*. Springer, 2012, pp. 177–185.

[48] R. A. Brualdi and V. Pless. "Weight enumerators of self-dual codes". In: *IEEE Transactions on Information Theory* 37.4 (1991), pp. 1222–1225.

[49] A. R. Calderbank and N. J. Sloane. "Modular and p-adic cyclic codes". In: *Designs, Codes and Cryptography* 6.1 (1995), pp. 21–35.

[50] B. Calışkan and K. Balıkçı. "Counting $\mathbb{Z}_2\mathbb{Z}_4\mathbb{Z}_8$-additive codes". In: *European Journal of Pure and Applied Mathematics* 12.2 (2019), pp. 668–679.

[51] C. Carlet. "$\mathbb{Z}_{2^k}$-linear codes". In: *IEEE Transactions on Information Theory* 44.4 (1998), pp. 1543–1547.

[52] C. J. Colbourn and M. K. Gupta. "On quaternary MacDonald codes". In: *International Conference on Information Technology: Computers and Communications*. 2003.

[53] I. Constantinescu and W. Heise. "A metric for codes over residue class rings". In: *Problems of Information Transmission* 33.3 (1997), pp. 208–213.

[54] J. H. Conway and N. J. Sloane. "A new upper bound on the minimal distance of self-dual codes". In: *IEEE Transactions on Information Theory* 36.6 (1990), pp. 1319–1333.

[55] R. Crandall. "Some notes on steganography". In: *Posted on steganography mailing list* (1998), pp. 1–6. URL: http://os.inf.tu-dresden.de/westfeld/crandall.pdf.

[56] Á. del Rio and J. Rifà. "Families of Hadamard $\mathbb{Z}_2\mathbb{Z}_4\mathbb{Q}_8$-codes". In: *IEEE Transactions on Information Theory* 59.8 (2013), pp. 5140–5151.

[57] P. Delsarte. "An algebraic approach to the association schemes of coding theory". In: vol. 10. Philips Res. Rep. Suppl., 1973, 97 pages.

[58] P. Delsarte and V. I. Levenshtein. "Association schemes and coding theory". In: *IEEE Transactions on Information Theory* 44.6 (1998), pp. 2477–2504.

[59] T. Deng and J. Yang. "Double cyclic codes over $\mathbb{F}_q + v\mathbb{F}_q$". In: *Mathematics* 8.10 (2020).

[60] L. Diao, J. Gao, and J. Lu. "Some results on $\mathbb{Z}_p\mathbb{Z}_p[v]$-additive cyclic codes". In: *Advances in Mathematics of Communications* 14.4 (2020), pp. 555–572.

[61] H. Dinh and S. López-Permouth. "Cyclic and negacyclic codes over finite chain rings". In: *IEEE Transactions on Information Theory* 50.8 (2004), pp. 1728–1744.

[62] H. Q. Dinh, T. Bag, A. K. Upadhyay, R. Bandi, and W. Chinnakum. "On the structure of cyclic codes over $\mathbb{F}_q RS$ and applications in quantum and LCD codes constructions". In: *IEEE Access* 8 (2020), pp. 18902–18914.

[63] H. Q. Dinh, S. Pathak, A. K. Upadhyay, and W. Yamaka. "New DNA codes from cyclic codes over mixed alphabets". In: *Mathematics* 8.11 (2020).

[64] S. T. Dougherty and E. Saltürk. "Counting codes over rings". In: *Designs, Codes and Cryptography* 73.1 (2014), pp. 151–165.

[65] S. Dougherty, J.-L. Kim, and L. Hongwei. "Constructions of self-dual codes over finite commutative chain rings". In: *International Journal of Information and Coding Theory* 1.2 (2010), pp. 171–190.

[66] S. Dougherty, J.-L. Kim, H. Kulosman, and L. Hongwei. "Self-dual codes over commutative Frobenius rings". In: *Finite Fields and Their Applications* 16.1 (2010), pp. 14–16.

[67] S. T. Dougherty. "Shadow codes and their weight enumerators". In: *IEEE Transactions on Information Theory* 41.3 (1995), pp. 762–768.

[68] S. T. Dougherty. "Formally self-dual codes and Gray maps". In: *Proc. of the XIII International Workshop on Algebraic and Combinatorial Coding Theory, ACCT2012, Pomorie, Bulgaria*. 2012, pp. 136–141.

[69] S. T. Dougherty. *Algebraic Coding Theory Over Finite Commutative Rings*. Springer, 2017.

[70] S. T. Dougherty and C. Fernández-Córdoba. "Codes over $\mathbb{Z}_{2^k}$, Gray map and self-dual codes". In: *Advances in Mathematics of Communications* 5.4 (2011), pp. 571–588.

[71] S. T. Dougherty and C. Fernández-Córdoba. "$\mathbb{Z}_2\mathbb{Z}_4$-additive formally self-dual codes". In: *Designs, Codes and Cryptography* 72.2 (2014), pp. 435–453.

[72] S. T. Dougherty and C. Fernández-Córdoba. "Kernels and ranks of cyclic and negacyclic quaternary codes". In: *Designs, Codes and Cryptography* 81.2 (2016), pp. 347–364.

[73] S. T. Dougherty, H. Liu, and L. Yu. "One weight $\mathbb{Z}_2\mathbb{Z}_4$-additive codes". In: *Applicable Algebra in Engineering, Communications and Computing* 27.2 (2016), pp. 123–138.

[74] S. T. Dougherty and K. Shiromoto. "Maximum distance codes over rings of order 4". In: *IEEE Transactions on Information Theory* 47.1 (2001), pp. 400–404.

[75] S. T. Dougherty and P. Solé. "Shadows of codes and lattices". In: *Proc. of the Third Asian Mathematical Conference 2000*. World Scientific. 2002, pp. 139–152.

[76] C. Fernández-Córdoba, S. Pathak, and A. K. Upadhyay. "On $\mathbb{Z}_{p^r}\mathbb{Z}_{p^r}\mathbb{Z}_{p^s}$-additive cyclic codes". In: *arXiv:2202.11454* (2022).

[77] C. Fernández-Córdoba. "On Reed-Muller and Related Quaternary Codes". PhD thesis. Universitat Autònoma de Barcelona, 2005.

[78] C. Fernández-Córdoba, J. Pujol, and M. Villanueva. "On rank and kernel of $\mathbb{Z}_4$-linear codes". In: *Coding Theory and Applications*. Springer, 2008, pp. 46–55.

[79] C. Fernández-Córdoba, J. Pujol, and M. Villanueva. "$\mathbb{Z}_2\mathbb{Z}_4$-linear codes: rank and kernel". In: *Designs, Codes and Cryptography* 56.1 (2010), pp. 43–59.

[80] C. Fernández-Córdoba, C. Vela, and M. Villanueva. "On the kernel of $\mathbb{Z}_{2^s}$-linear Hadamard codes". In: *5th International Castle Meeting on Coding Theory and Applications, ICMCTA Vihula, Estonia*. Ed. by A. Barbero, V. Skachek, and Ø. Ytreus. LNCS, Springer Int. Publishing, 2017, pp. 107–117.

[81] C. Fernández-Córdoba, C. Vela, and M. Villanueva. "On $\mathbb{Z}_{2^s}$-linear Hadamard codes: kernel and partial classification". In: *Designs, Codes and Cryptography* 87.2-3 (2019), pp. 417–435.

[82] C. Fernández-Córdoba, C. Vela, and M. Villanueva. "Equivalences among $\mathbb{Z}_{2^s}$-linear Hadamard codes". In: *Discrete Mathematics* 343.3 (2020), p. 111721.

[83] C. Fernández-Córdoba, C. Vela, and M. Villanueva. "On $\mathbb{Z}_8$-linear Hadamard codes: rank and classification". In: *IEEE Transactions on Information Theory* 66.2 (2020), pp. 970–982.

[84] C. Fernández-Córdoba, C. Vela, and M. Villanueva. "Linearity and kernel of $\mathbb{Z}_{2^s}$-linear simplex and MacDonald codes". In: *submitted to IEEE Transactions on Information Theory, arXiv:1910.07911* (2021).

[85] W. Fish, J. D. Key, and E. Mwambene. "Partial permutation decoding for simplex codes". In: *Advances in Mathematics of Communications* 6.4 (2012), pp. 505–516.

[86] J. Fridrich and P. Lisonek. "Grid colorings in steganography". In: *IEEE Transactions on Information Theory* 53.4 (2007), pp. 1547–1549.

[87] J. Gao, M. Shi, T. Wu, and F.-W. Fu. "On double cyclic codes over $\mathbb{Z}_4$". In: *Finite Fields and Their Applications* 39 (2016), pp. 233–250.

[88] J. Geng, H. Wu, and P. Solé. "On one-Lee weight and two-Lee weight $\mathbb{Z}_2\mathbb{Z}_4[u]$-additive codes and their constructions". In: *Advances in Mathematics of Communications* (2021).

[89] A. Gleason. "Weight polynomials of self-dual codes and the MacWilliams identities". In: *Actes Congrès Internationel de Mathematique, Gauthier-Villars, Paris.* Vol. 3. 1970, pp. 211–215.

[90] M. Greferath and S. E. Schmidt. "Gray isometries for finite chain rings and a nonlinear ternary $(36, 3^{12}, 15)$ code". In: *IEEE Transactions on Information Theory* 45.7 (1999), pp. 2522–2524.

[91] M. K. Gupta, M. C. Bhandari, and A. K. Lal. "On some linear codes over $\mathbb{Z}_{2^s}$". In: *Designs, Codes and Cryptography* 36.3 (2005), pp. 227–244.

[92] A. R. Hammons, P. V. Kumar, A. R. Calderbank, N. J. Sloane, and P. Solé. "The $\mathbb{Z}_4$-linearity of Kerdock, Preparata, Goethals, and related codes". In: *IEEE Transactions on Information Theory* 40.2 (1994), pp. 301–319.

[93] O. Heden. "A new construction of group and nongroup perfect codes". In: *Information and Control* 34.4 (1977), pp. 314–323.

[94] X.-D. Hou, J. T. Lahtonen, and S. Koponen. "The Reed-Muller code $R(r, m)$ is not $\mathbb{Z}_4$-linear for $3 \leq r \leq m - 2$". In: *IEEE Transactions on Information Theory* 44.2 (1998), pp. 798–799.

[95] X.-D. Hou, X. Meng, and J. Gao. "On $\mathbb{Z}_2\mathbb{Z}_2[u^3]$-additive cyclic and complementary dual codes". In: *IEEE Access* 9 (2021), pp. 65914–65924.

[96] J. Howell. "Spans in the module $\mathbb{Z}_m^s$". In: *Linear and Multilinear Algebra* 19 (1986), pp. 67–77.

[97] W. C. Huffman and V. Pless. *Fundamentals of Error Correcting Codes.* Cambridge University Press, 2010.

[98] H. Islam, O. Prakash, and P. Solé. "$\mathbb{Z}_4\mathbb{Z}_4[u]$-additive cyclic and constacyclic codes". In: *Advances in Mathematics of Communications* 15.4 (2021), pp. 737–755.

[99] M. Kiermaier, A. Wassermann, and J. Zwanzger. "New upper bounds on binary linear codes and a $\mathbb{Z}_4$-code with a better-than-linear Gray image". In: *IEEE Transactions on Information Theory* 62.12 (2016), pp. 6768–6771.

[100] D. S. Krotov. "$\mathbb{Z}_4$-linear perfect codes". In: *Diskretnyj Analiz i Issledovanie Operatsij. Ser. 1* 7.4 (2000), pp. 78–90.

[101] D. S. Krotov. "Lower bounds for the number of m-quasigroups of order 4 and for the number of perfect binary codes". In: *Diskretnyj Analiz i Issledovanie Operatsij. Ser. 1* 7 (2000), pp. 47–53.

[102] D. S. Krotov. "$\mathbb{Z}_4$-linear Hadamard and extended perfect codes". In: *Electronic Notes in Discrete Mathematics* 6 (2001), pp. 107–112.

[103] D. S. Krotov. "On $\mathbb{Z}_{2^k}$-dual binary codes". In: *IEEE Transactions on Information Theory* 53.4 (2007), pp. 1532–1537.

[104] D. S. Krotov. "On the automorphism groups of the additive 1-perfect binary codes". In: *3rd International Castle Meeting on Coding Theory and Application*. Ed. by J. Borges and M. Villanueva. Universitat Autonoma de Barcelona. Servei de Publicacion, 2011, pp. 171–176.

[105] D. S. Krotov and M. Villanueva. "Classification of the $\mathbb{Z}_2\mathbb{Z}_4$-linear Hadamard codes and their automorphism groups". In: *IEEE Transactions on Information Theory* 61.2 (2015), pp. 887–894.

[106] D. S. Krotov and S. V. Avgustinovich. "On the number of 1-perfect binary codes: a lower bound". In: *IEEE Transactions on Information Theory* 54.4 (2008), pp. 1760–1765.

[107] M. LeVan. "Designs and Codes". PhD thesis. Auburn University, 1995.

[108] Z. Li. "$\mathbb{Z}_2(\mathbb{Z}_2 + u\mathbb{Z}_2)(\mathbb{Z}_2 + u\mathbb{Z}_2 + u^2\mathbb{Z}_2)$-additive cyclic codes". In: *Engineering Mathematics* 3.2 (2019), pp. 30–39.

[109] B. Lindström. "Group partitions and mixed perfect codes". In: *Canad. Math. Bull* 18.1 (1975), pp. 57–60.

[110] Z. Lu and S. Zhu. "$\mathbb{Z}_p\mathbb{Z}_p[u]$-additive codes". In: *arXiv:1510.08636* (2015).

[111] F. J. MacWilliams. "Permutation decoding of systematic codes". In: *The Bell System Technical Journal* 43 (1964), pp. 485–505.

[112] F. J. MacWilliams and N. J. A. Sloane. *The Theory of Error-Correcting Codes*. Elsevier, 1977.

[113] P. Montolio and J. Rifà. "Construction of Hadamard $\mathbb{Z}_2\mathbb{Z}_4\mathbb{Q}_8$-codes for each allowable value of the rank and dimension of the kernel". In: *IEEE Transactions on Information Theory* 61.4 (2015), pp. 1948–1958.

[114] H. Mostafanasab. "Triple cyclic codes over $\mathbb{Z}_2$". In: *Palestine journal of mathematics, Special Issue II* 6 (2017), pp. 123–134.

[115] P. Moulin and R. Koetter. "Data-hiding codes". In: *Proceedings of the IEEE* 93.12 (2005), pp. 2083–2126.

[116] H. Movahedi and L. Pourfaraj. "Self-dual double cyclic codes over $\mathbb{Z}_2$". In: *Journal of linear and topological algebra* 10.4 (2021), pp. 257–267.

[117]   A. A. Nechaev. "The Kerdock code in a cyclic form". In: *Discrete Mathematics and Applications* 1 (1991), pp. 365–384.

[118]   J. Pernas, J. Pujol, and M. Villanueva. "Kernel dimension for some families of quaternary Reed-Muller codes". In: *Mathematical Methods in Computer Science*. Springer, 2008, pp. 128–141.

[119]   J. Pernas, J. Pujol, and M. Villanueva. "Classification of some families of quaternary Reed–Muller codes". In: *IEEE Transactions on Information Theory* 57.9 (2011), pp. 6043–6051.

[120]   J. Pernas, J. Pujol, and M. Villanueva. "Characterization of the automorphism group of quaternary linear Hadamard codes". In: *Designs, Codes and Cryptography* 70.1-2 (2014), pp. 105–115.

[121]   K. T. Phelps and M. Levan. "Kernels of nonlinear Hamming codes". In: *Designs, Codes and Cryptography* 6.3 (1995), pp. 247–257.

[122]   K. T. Phelps, J. Rifà, and M. Villanueva. "On the additive ($\mathbb{Z}_4$-linear and non-$\mathbb{Z}_4$-linear) Hadamard codes: rank and kernel". In: *IEEE Transactions on Information Theory* 52.1 (2006), pp. 316–319.

[123]   K. T. Phelps and M. Villanueva. "On perfect codes: rank and kernel". In: *Designs, Codes and Cryptography* 27.3 (2002), pp. 183–194.

[124]   K. T. Phelps and J. Rifà. "On binary 1-perfect additive codes: some structural properties". In: *IEEE Transactions on Information Theory* 48.9 (2002), pp. 2587–2592.

[125]   V. Pless, R. A. Brualdi, and W. C. Huffman. *Handbook of Coding Theory*. Elsevier Science Inc., 1998.

[126]   V. S. Pless and Z. Qian. "Cyclic codes and quadratic residue codes over $\mathbb{Z}_4$". In: *IEEE Transactions on Information Theory* 42.5 (1996), pp. 1594–1600.

[127]   O. Prakash, S. Yadav, H. Islam, and P. Solé. "On $\mathbb{Z}_4\mathbb{Z}_4[u^3]$-additive constacyclic codes". In: *Advances in Mathematics of Communications* (2022). DOI: 10.3934/amc.2022017.

[128]   E. Prange. "The use of information sets in decoding cyclic codes". In: *IEEE Transactions on Information Theory* 8.5 (1962), pp. 55–59.

[129]   J. Pujol, J. Rifà, and L. Ronquillo. "Construction of additive Reed-Muller codes". In: *International Symposium on Applied Algebra, Algebraic Algorithms, and Error-Correcting Codes*. Vol. 5527. Lecture Notes Computer Science. Springer, 2009, pp. 223–226.

[130] J. Pujol, J. Rifà, and F. I. Solov'eva. "Construction of $\mathbb{Z}_4$-linear Reed–Muller codes". In: *IEEE Transactions on Information Theory* 55.1 (2009), pp. 99–104.

[131] L. Qian and X. Cao. "Bounds and optimal q-ary codes derived from the $\mathbb{Z}_q R$-cyclic codes". In: *IEEE Transactions on Information Theory* 66.2 (2020), pp. 923–935.

[132] E. M. Rains and N. J. Sloane. "Self-dual codes". In: *Handbook of Coding Theory I*. Ed. by V. Pless, R. A. Brualdi, and W. C. Huffman. Elsevier Science Inc., 1998, pp. 177–294.

[133] J. Rifà, J. M. Basart, and L. Huguet. "On completely regular propelinear codes". In: *International Conference on Applied Algebra, Algebraic Algorithms, and Error-Correcting Codes*. Vol. 357. Lecture Notes Computer Science. Springer, 1988, pp. 341–355.

[134] J. Rifà and J. Pujol. "Translation-invariant propelinear codes". In: *IEEE Transactions on Information Theory* 43.2 (1997), pp. 590–598.

[135] J. Rifà and L. Ronquillo. "Product perfect $\mathbb{Z}_2\mathbb{Z}_4$-linear codes in steganography". In: *2010 International Symposium On Information Theory & Its Applications*. IEEE. 2010, pp. 696–701.

[136] J. Rifà, F. I. Solov'eva, and M. Villanueva. "On the intersection of $\mathbb{Z}_2\mathbb{Z}_4$-additive Hadamard codes". In: *IEEE Transactions on Information Theory* 55.4 (2009), pp. 1766–1774.

[137] J. Rifà, F. I. Solov'eva, and M. Villanueva. "On the intersection of $\mathbb{Z}_2\mathbb{Z}_4$-additive perfect codes". In: *IEEE Transactions on Information Theory* 54.3 (2008), pp. 1346–1356.

[138] J. Rifà and E. Suárez-Canedo. "Hadamard full propelinear codes of type Q; rank and kernel". In: *Designs, Codes and Cryptography* 86.9 (2018), pp. 1905–1921.

[139] H. Rifà-Pous and J. Rifà. "Product perfect codes and steganography". In: *Digital Signal Processing* 19.4 (2009), pp. 764–769.

[140] H. Rifà-Pous, J. Rifà, and L. Ronquillo. "$\mathbb{Z}_2\mathbb{Z}_4$-additive perfect codes in steganography". In: *Advances in Mathematics of Communications* 5.3 (2011), pp. 425–433.

[141] M. Shi, S. Li, and P. Solé. "$\mathbb{Z}_2\mathbb{Z}_4$-additive quasi-cyclic codes". In: *IEEE Transactions on Information Theory* 67.11 (2021), pp. 7232–7239.

[142] M. Shi, R. Wu, and D. S. Krotov. "On $\mathbb{Z}_p\mathbb{Z}_{p^k}$-additive codes and their duality". In: *IEEE Transactions on Information Theory* 65.6 (2019), pp. 3841–3847.

[143] M. J. Shi, A. Alahmadi, and P. Solè. *Codes and Rings: Theory and Practice*. Academic Press, 2017.

[144] M. Shi, Z. Sepasdar, A. Alahmadi, and P. Solé. "On two-weight $\mathbb{Z}_{2^k}$-codes". In: *Designs, Codes and Cryptography* 86.6 (2018), pp. 1201–1209.

[145] M. Shi, C. Wang, R. Wu, Y. Hu, and Y. Chang. "One-weight and two-weight $\mathbb{Z}_2\mathbb{Z}_2[u,v]$-additive codes". In: *Cryptography and Communications* 12.3 (2020), pp. 443–454.

[146] M. Shi, R. Wu, and P. Solé. "Asymptotically good additive cyclic codes exist". In: *IEEE Communications Letters* 22.10 (2018), pp. 1980–1983.

[147] M. Shi et al. "The geometry of two-weight codes over $\mathbb{Z}_{p^m}$". In: *IEEE Transactions on Information Theory* 67.12 (2021), pp. 7769–7781.

[148] I. Siap and N. Kulhan. "The structure of generalized quasi-cyclic codes". In: *Applied Mathematics E-Notes* 5 (2005), pp. 24–30.

[149] R. Singleton. "Maximum distance q-ary codes". In: *IEEE Transactions on Information Theory* 10.2 (1964), pp. 116–118.

[150] F. I. Solov'eva. "On $\mathbb{Z}_4$-linear codes with the parameters of Reed-Muller codes". In: *Problems of Information Transmission* 43.1 (2007), pp. 26–32.

[151] B. Srinivasulu and M. Bhaintwal. "$\mathbb{Z}_2$-Triple cyclic codes and their duals". In: *European Journal of Pure and Applied Mathematics* 10.2 (2017), pp. 392–409.

[152] B. Srinivasulu and B. Maheshanand. "$\mathbb{Z}_2(\mathbb{Z}_2 + u\mathbb{Z}_2)$-additive cyclic codes and their duals". In: *Discrete Mathematics, Algorithms and Applications* 8.2 (2016), 19 pages.

[153] A. Storjohann and T. Mulders. "Fast algorithms for linear algebra modulo N". In: *European Symposium on Algorithms*. Lecture Notes Computer Science. Springer, 1998, pp. 139–150.

[154] J. J. Sylvester. "LX. Thoughts on inverse orthogonal matrices, simultaneous sign successions, and tessellated pavements in two or more colours, with applications to Newton's rule, ornamental tile-work, and the theory of numbers". In: *The London, Edinburgh, and Dublin Philosophical Magazine and Journal of Science* 34.232 (1867), pp. 461–475.

[155] H. Tapia-Recillas and G. Vega. "On $\mathbb{Z}_{2^k}$-linear and quaternary codes". In: *SIAM Journal on Discrete Mathematics* 17.1 (2003), pp. 103–113.

[156] A. Torres-Martín and M. Villanueva. "Systematic encoding and permutation decoding for $\mathbb{Z}_{p^s}$-linear codes". In: *IEEE Transactions on Information Theory* (2022). DOI: 10.1109/TIT.2022.3157192.

[157] Z. Wan. *Quaternary Codes*. Vol. 8. World Scientific, 1997.

[158] H. N. Ward. "A restriction on the weight enumerator of a self-dual code". In: *Journal of Combinatorial Theory, Series A* 21.2 (1976), pp. 253–255.

[159] A. Westfeld. "F5–A steganographic algorithm". In: *Information Hiding: 4th International Workshop, IH 2001, Pittsburgh, PA, USA, April 25-27, 2001. Proceedings*. Vol. 2137. Springer. 2001, p. 289.

[160] F. M. Willems and M. Van Dijk. "Capacity and codes for embedding information in gray-scale signals". In: *IEEE Transactions on Information Theory* 51.3 (2005), pp. 1209–1214.

[161] J. Wolfmann. "Binary images of cyclic codes over $\mathbb{Z}_4$". In: *IEEE Transactions on Information Theory* 47.5 (2001), pp. 1773–1779.

[162] J. A. Wood. "Duality for modules over finite chain rings and applications to coding theory". In: *American Journal of Mathematics* 121.3 (1999), pp. 555–575.

[163] T. Wu, J. Gao, and F.-W. Fu. "Some results on triple cyclic codes over $\mathbb{Z}_4$". In: *IEICE Transactions on Fundamentals* 99.5 (2016), pp. 998–1004.

[164] T. Wu, J. Gao, Y. Gao, and F.-W. Fu. "$\mathbb{Z}_2\mathbb{Z}_2\mathbb{Z}_4$-additive cyclic codes". In: *Advances in Mathematics of Communications* 12.4 (2018), pp. 641–657.

[165] T. Yao, M. Shi, and P. Solé. "Double cyclic codes over $\mathbb{F}_q + u\mathbb{F}_q + u^2\mathbb{F}_q$". In: *International Journal of Information and Coding Theory* 3.2 (2015), pp. 145–157.

[166] T. Yao, S. Zhu, and B. Pang. "Triple cyclic codes over $\mathbb{F}_q + u\mathbb{F}_q$". In: *International Journal of Foundations of Computer Science* 32.2 (2021), pp. 115–135.

[167] B. Yildiz and Z. Ö. Özger. "Generalization of the Lee weight to $\mathbb{Z}_{p^k}$". In: *TWMS Journal of Applied and Engineering Mathematics* 2.2 (2012), pp. 145–153.

[168] Z.-L. Zhao and F. Gao. "An improved steganographic method of product perfect codes". In: *2011 IEEE International Conference on Signal Processing, Communications and Computing (ICSPCC)*. IEEE. 2011, pp. 1–5.

Printed in the United States
by Baker & Taylor Publisher Services